High Accuracy Algorithm for the Differential Equations Governing Anomalous Diffusion

Algorithm and Models for Anomalous Diffusion

High Accuracy Algorithm for the Differential Equations Governing Anomalous Diffusion

Algorithm and Models for Anomalous Diffusion

Weihua Deng • Zhijiang Zhang
Lanzhou University, China

World Scientific

NEW JERSEY • LONDON • SINGAPORE • BEIJING • SHANGHAI • HONG KONG • TAIPEI • CHENNAI • TOKYO

Published by

World Scientific Publishing Co. Pte. Ltd.
5 Toh Tuck Link, Singapore 596224
USA office: 27 Warren Street, Suite 401-402, Hackensack, NJ 07601
UK office: 57 Shelton Street, Covent Garden, London WC2H 9HE

Library of Congress Cataloging-in-Publication Data
Names: Deng, Weihua, author. | Zhang, Zhijiang (Mathematician), author.
Title: High accuracy algorithm for the differential equations governing anomalous diffusion :
algorithm and models for anomalous diffusion / by Weihua Deng (Lanzhou University, China),
Zhijiang Zhang (Lanzhou University, China).
Description: New Jersey : World Scientific, 2018. | Includes bibliographical references and index.
Identifiers: LCCN 2018043039 | ISBN 9789813142206 (hardcover : alk. paper)
Subjects: LCSH: Diffusion. | Nonlinear mechanics. | Fractional differential equations. |
Differential equations, Partial. | Differential equations.
Classification: LCC QC185 .D46 2018 | DDC 519.2/33--dc23
LC record available at https://lccn.loc.gov/2018043039

British Library Cataloguing-in-Publication Data
A catalogue record for this book is available from the British Library.

For any available supplementary material, please visit
https://www.worldscientific.com/worldscibooks/10.1142/10095#t=suppl

Printed in Singapore

Preface

The aim of this book is to physically introduce the anomalous diffusion processes and the models governing the physical phenomena, and provide efficient and reliable algorithms for numerically solving the models.

The anomalous diffusions concerned in this book are the stochastic processes $\mathbf{X}(t)$ with their second order moment being the power of the time t, i.e., $\langle|\mathbf{X}(t)|^2\rangle \sim t^\gamma$ with $\gamma \in [0,1) \cup (1,2]$. More concretely, we are considering two types of stochastic processes: the Gaussian processes with long-range dependent covariance functions; and the Lévy processes with the probability measure having divergent second moment, including inverse subordinated Lévy processes and tempered Lévy processes. In the part of numerical methods, almost all the efforts of the book are put on the models describing the distribution of the positions of the particles or the functional distribution of the trajectories of particles undergoing Lévy type motion.

The objects of this study are stochastic processes. So, we start this journey from the microscopic (stochastic) models, introducing fractional Brownian motion, tempered fractional Brownian motion, Lévy processes, tempered Lévy processes, subordinated Lévy processes, inverse subordinated Lévy processes, and continuous time random walks. Some statistical properties are mentioned. By generating random variables, the stochastic processes can be simulated and then the statistical quantities can be computed. However, the biggest challenge of Monte Carlo method is its computational cost. It is natural to turn to derive the deterministic equations governing the statistical quantities. After introducing the microscopic models, we derive the macroscopic (deterministic) integro-differential equations, characterizing the statistical quantities covering the probability density function (PDF) of the positions of the particles, the PDF of the functional of the particles' trajectories, mean exit time, and escape probability.

The derived models are partial differential equations (PDEs) with integral-differential operators, commonly referred to as fractional operators. First, we focus on the time fractional differential equations. The numerical discretizations include predictor-corrector algorithm, discontinuous Galerkin (DG) method, numerical inversion of Laplace transform. The detailed numerical analyses are provided and numerical experiments are made to verify theoretical results.

For the numerical approximations of space fractional PDEs, we use two chapters to discuss them. The first one is about the finite difference methods with detailed theoretical analysis and numerical experiments, including first order scheme, second order scheme, higher order schemes, and compact schemes. The second one concentrates on the variational methods, i.e., finite element methods, spectral methods, and discontinuous Galerkin methods.

Acknowledgments

The material of this book is mainly to summarize the works of the authors' group in the past ten years. We would like to thank all the group members and their collaborators. In fact, even after the book was completed, they made great efforts for improving the presentation, including Can Li, Shuqin Wang, Yanyan Yu, Lijing Zhao, Yajing Li, Wanli Wang, Xudong Wang, Yaoqiang Yan.

The research works of the authors' group are built on or closely related to the research papers of hundreds of talented authors. It seems inevitable that some important contributors are left out, even though we did attempt to recognize the contributions of all the related authors. While expressing our thanks, we beg their forgiveness.

Contents

Chapter 1

Introduction

This chapter is to introduce the normal and anomalous stochastic processes in physics. We first briefly discuss normal Gaussian diffusion, that is, Brownian motion, which has independent stationary increments and the increment is of Gaussian distribution. Then we turn to the anomalous Gaussian processes, i.e., fractional Brownian motion and tempered fractional Brownian motion, having long-range dependent covariance function. Another direction of generalization of Brownian motion is to keep its independence and stationarity of increments, namely, Lévy process; the process mainly concerned in this book is the one with Lévy measure, having divergent second moment. Then we introduce the continuous time random walk (CTRW) models, which characterize the stochastic processes, and the scaling limits of the CTRW models are the corresponding stochastic processes.

1.1 Brownian motion

The continuous probability distribution function, Gaussian distribution, is so common in probability theory that it is also called normal distribution and a random variable with a Gaussian distribution is said to be normally distributed. The probability density function (PDF) of the Gaussian distribution is

$$u(x) = \frac{1}{\sqrt{2\pi}\sigma} e^{-\frac{(x-\mu)^2}{2\sigma^2}}, \tag{1.1}$$

where μ is mean or expectation and σ^2 is variance.

Definition 1.1. Brownian motion, termed as Wiener process in mathematics, is a stochastic process $\mathbf{X}(t)$ characterized by four axioms:

1. $\mathbf{X}(0) = 0$;
2. $\mathbf{X}(t)$ is almost surely continuous;

3. $\mathbf{X}(t)$ has stationary increments, i.e., for $0 \leq t_1 \leq t_2$, $\mathbf{X}(t_2) - \mathbf{X}(t_1)$ is a Gaussian variable with mean 0 and variance $t_2 - t_1$;

4. $\mathbf{X}(t)$ has independent increments, namely, if $t_1 < t_2 \leq t_3 < t_4$, then the random variables $\mathbf{X}(t_2) - \mathbf{X}(t_1)$ and $\mathbf{X}(t_4) - \mathbf{X}(t_3)$ are independent.

The Brownian motion is named after the botanist Robert Brown. Einstein [Einstein (1905)] derives the diffusion equation, governing the PDF of Brownian particles,

$$\frac{\partial p}{\partial t} = D\frac{\partial^2 p}{\partial x^2}, \tag{1.2}$$

where the diffusion coefficient D is related to the mean squared displacement of a Brownian particle, i.e.,

$$\langle x^2(t) \rangle = 2Dt. \tag{1.3}$$

Assuming N particles starting from original at time $t = 0$, the solution of (1.2) is

$$p(x,t) = \frac{N}{\sqrt{4\pi Dt}} e^{-\frac{x^2}{4Dt}}.$$

In fact, Einstein's theory also relates the diffusion coefficient to measurable physical quantities, which can determine the size of atoms and the number of atoms in a mole, etc.

1.2 Fractional and tempered fractional Brownian motion

For fractional Brownian motion (fBm) $\mathbf{X}_H(t)$, the generalization of Brownian motion, its increments need not to be independent, having the covariance function

$$\mathbf{E}\left[\mathbf{X}_H(t)\mathbf{X}_H(s)\right] = D_H(|t|^{2H} + |s|^{2H} - |t-s|^{2H}), \tag{1.4}$$

where H is a real number in $(0, 1)$ and $D_H = \Gamma(1 - 2H)\cos(H\pi)/(2H\pi)$; but it keeps the properties of continuity and has stationary increments. The fBm is defined as [Mandelbrot and Van-Ness (1968)]

$$\mathbf{X}_H(t) = \mathbf{X}_H(0) \tag{1.5}$$

$$+ \frac{1}{\Gamma(H+\frac{1}{2})} \int_{-\infty}^{0} \left[(t-s)^{H-\frac{1}{2}} - (-s)^{H-\frac{1}{2}}\right] d\mathbf{X}(s) \tag{1.6}$$

$$+ \frac{1}{\Gamma(H+\frac{1}{2})} \int_{0}^{t} (t-s)^{H-\frac{1}{2}} d\mathbf{X}(s), \tag{1.7}$$

with $t > 0$ (similarly for $t < 0$); it is self-similar in the sense that $\mathbf{X}_H(at)$ and $a^H\mathbf{X}_H(t)$ have the same probability distribution, because of the homogeneousity (of order $2H$) of the covariance function; in fact, fBm is the only self-similar Gaussian process. The increment process, $\xi(t) = \mathbf{X}_H(t+1) - \mathbf{X}_H(t)$, is defined as fractional Gaussian noise. The mean $\langle \xi(t) \rangle = 0$ and its covariance is

$$\langle \xi(t_1)\xi(t_2) \rangle = 2D_H H(2H-1)|t_1 - t_2|^{2H-2}, \quad t_1, t_2 > 0. \tag{1.8}$$

The fBm is ergodic with slow convergence, and $H = \frac{3}{4}$ marks the critical point of the speed of convergence [Deng and Barkai (2009)]. In the ballistic limit $H \to 1$, ergodicity is broken.

Tempered fractional Brownian motion (tfBm) [Meerschaert and Sabzikar (2013)] is a modified fBm, adding an exponential tempering to the power law kernel in the moving average representation. The tfBm is defined as

$$\mathbf{X}_{H,\lambda}(t) = \int_{-\infty}^{+\infty} \left[e^{-\lambda(t-s)_+}(t-s)_+^{H-\frac{1}{2}} - e^{-\lambda(-s)_+}(-s)_+^{H-\frac{1}{2}} \right] d\mathbf{X}_H(s), \tag{1.9}$$

where $(s)_+ = sI\,(s > 0)$, $H > 0$, and $t \in \mathbb{R}$. The covariance function of $\mathbf{X}_{H,\lambda}(t)$ is

$$\mathbf{E}\left[\mathbf{X}_{H,\lambda}(t)\mathbf{X}_{H,\lambda}(s)\right] = \frac{1}{2}\left[C_t^2|t|^{2H} + C_s^2|s|^{2H} - C_{t-s}^2|t-s|^{2H}\right], \tag{1.10}$$

where $s, t \in \mathbb{R}$, and $C_t^2 = \frac{2\Gamma(2H)}{(2\lambda|t|)^{2H}} - \frac{2\Gamma(H+1/2)}{\sqrt{\pi}} \frac{1}{(2\lambda|t|)^H} K_H(\lambda|t|)$ for $t \neq 0$, and $K_\nu(z)$ is the modified Bessel function of the second kind and $C_0^2 = 0$.

The tempered fractional Brownian noise, the increment process, is $\xi_\lambda(t) = \mathbf{X}_{H,\lambda}(t+1) - \mathbf{X}_{H,\lambda}(t)$, which is a stationary Gaussian time series with mean zero and covariance function

$$\langle \xi_\lambda(t_1)\xi_\lambda(t_2) \rangle = \frac{1}{2}\Big[|t_1 - t_2 + 1|^{2H} C_{t_1-t_2+1}^2 - 2|t_1 - t_2|^{2H} C_{t_1-t_2}^2 \tag{1.11}$$

$$+ |t_1 - t_2 - 1|^{2H} C_{t_1-t_2-1}^2 \Big]. \tag{1.12}$$

From above, one can see that tfBm is still a Gaussian process with stationary increments, but it is not self-similar or it is a generalized self-similar process in the sense that $\mathbf{X}_{H,\lambda}(ct)$ and $C^H\mathbf{X}_{H,c\lambda}(t)$ have the same probability distribution.

1.3 Lévy process

Lévy process can be considered as a generalization of Brownian motion in another direction, compared with (tempered) fBm. While keeping the other properties of Brownian motion, Lévy process is not longer a Gaussian process and the continuity of its path is replaced by continuity in probability. More concretely, for $t \geq 0$, $\mathbf{X}(t)$ is said to be a Lévy process if it satisfies [Applebaum (2009)]:

1. $\mathbf{X}(0) = 0$;

2. $\mathbf{X}(t)$ has stationary increments, i.e., for any $s < t$, $\mathbf{X}(t) - \mathbf{X}(s)$ is equal in distribution to $\mathbf{X}(t-s)$;

3. $\mathbf{X}(t)$ has independent increments;

4. $\mathbf{X}(t)$ is continuous in probability, i.e., for any $\varepsilon > 0$ and $t \geq 0$ it holds that $\lim_{h \to 0} P(|\mathbf{X}(t+h) - \mathbf{X}(t)| > \varepsilon) = 0$.

Remark 1.1. 1. For a Lévy process, there is a corresponding version that is almost surely right continuous with left limits; both Brownian and Poisson processes are specific Lévy processes; for Poisson process, $\mathbf{X}(t) - \mathbf{X}(s)$ is a Poisson distribution with expectation $\lambda(t-s)$, $\lambda > 0$, being the intensity or rate of the process. 2. Because of the stationary and independence of the increments, the distribution of a Lévy process has the property of infinite divisibility; and each infinite divisible distribution naturally corresponds to Lévy process.

The distribution of a Lévy process can be explicitly characterized by its characteristic function, given by the Lévy-Khintchine formula. That is,

$$\begin{aligned}\mathbf{E}\left[e^{i\mathbf{k}\cdot\mathbf{X(t)}}\right] = &\exp\Big(t\big(i(\mathbf{b},\mathbf{k}) - \frac{\mathbf{1}}{\mathbf{2}}(\mathbf{k},\mathbf{A}\mathbf{k}) \\ &+ \int_{\mathbb{R}^n\setminus\{0\}} \left(e^{i(\mathbf{k},\mathbf{y})} - 1 - i(\mathbf{k},\mathbf{y})\chi_{\{|\mathbf{y}|<1\}}\right)\nu(d\mathbf{y})\big)\Big), \end{aligned} \tag{1.13}$$

where $\mathbf{b} \in \mathbb{R}^n$ and $\mathbf{A}$ is a positive symmetric $n \times n$ matrix and ν is sigma-finite Lévy measure on $\mathbb{R}^n \setminus \{0\}$, satisfying

$$\int_{\mathbb{R}^n\setminus\{0\}} (1 \wedge \mathbf{y}^T\mathbf{y})\nu(d\mathbf{y}) < \infty.$$

Any Lévy process may be decomposed into the sum of Brownian motion, a linear drift, and a pure jump process which captures all jumps of the original Lévy process.

1.4 Continuous time random walk

British statistician Karl Pearson in 1905 asked the question [Pearson (1905)]: A man starts from a point O and walks l yards in a straight line; he then turns through any angle whatever and walks another l yards in a second straight line; he repeats the process n times; I require the probability that after these n stretches he is at a distance between r and $r+dr$ from the starting point, O.

After Pearson coined the name 'random walk', the random walk is further generalized to CTRW [Montroll and Weiss (1965)], where the waiting (holding) time between two renewals is random. A simple formulation of a CTRW is to consider the stochastic process $\mathbf{X}(t)$ defined by [Klafter and Sokolov (2011)]

$$\mathbf{X}(t) = \mathbf{X}(0) + S_{N_t}, \tag{1.14}$$

where $S_{N_t} = \sum_{i=1}^{N_t} \xi_i$, $N_t = \max\{n : T_n \leq t\}$, $T_n = \sum_{j=1}^{n} \eta_j$, and ξ_i and η_j are, respectively, the ith displacement and jth waiting time. In the following discussion, if not specifically pointed out, ξ_i and η_j are assumed to be independent identically distributed (i.i.d.) random variables; and their PDFs are respectively denoted as $\psi(x)$ and $\phi(t)$. Following the convention of physical and mathematical communities, we use $\psi(k)$ and $\phi(s)$ to the Fourier and Laplace transforms of $\psi(x)$ and $\phi(t)$ in Chapters 1 and 2, but in the other chapters, we use the notations $\hat{\psi}(\omega)$ (or $\mathscr{F}[\psi(x)](\omega)$) and $\mathscr{L}[\phi(t)](s)$ to denote the Fourier and Laplace transforms of the corresponding functions (see Appendix A for the detailed definitions of the Fourier and Laplace transforms).

The Montroll-Weiss formula [Montroll and Weiss (1965)] well answers the Pearson's question, which shows the Fourier-Laplace transform of the PDF $p(\mathbf{X}, t)$ taking the value $\mathbf{X}$ at time t satisfies

$$p(\mathbf{k}, s) = \frac{1 - \phi(s)}{s} \cdot \frac{1}{1 - \psi(\mathbf{k})\phi(s)}. \tag{1.15}$$

If $\phi(t)$ is exponential and $\psi(x)$ is normally distributed, the CTRW leads to Brownian motion. If both the means of $|\xi_i|^2$ and η_j are finite, the scaling limit of the CTRW is Brownian motion.

1.5 Anomalous diffusion

Diffusion is the movement of the particles from a region of high concentration to a region of low concentration. The types of diffusions are usually classified according to the relation of the mean squared displacement (MSD) of a particle to the time t [Metzler and Klafter (2000)], i.e., $\langle[\mathbf{X}(t) - \langle\mathbf{X}(t)\rangle]^2\rangle \sim t^{\alpha}$; it is normal diffusion if $\alpha = 1$, otherwise it is anomalous diffusion, being called subdiffusion if $\alpha \in (0,1)$ and superdiffusion if $\alpha \in (1,2)$; in particular, it is called localization diffusion if $\alpha = 0$ and ballistic diffusion if $\alpha = 2$.

From (1.4), it can be noted that fBm exhibits subdiffusion for $H \in (0, \frac{1}{2})$ and superdiffusion for $H \in (\frac{1}{2}, 1)$. The tfBm performs localization diffusion. In (1.13), for $\nu(d\mathbf{y})$, if $\int_{\mathbb{R}^n \setminus \{0\}} (\mathbf{y}^T \mathbf{y}) \nu(d\mathbf{y})$ diverges, the Lévy process $\mathbf{X}(t)$ displays superdiffusion. Generally, the subordinated Lévy process can be used to model subdiffusion. For the CTRW model (1.14), if the second moment of ξ_i diverges and the first moment of η_j is finite, it describes superdiffusion; it characterizes subdiffusion if the second moment of ξ_i is bounded and the first moment of η_j diverges; if both the second moment of ξ_i and the first moment of η_j are infinite, the CTRW model may describe subdiffusion or superdiffusion or even normal diffusion, depending on that the dominant role played by ξ_i or η_j or balanced between ξ_i and η_j.

Chapter 2

Mathematical models

In the last chapter, we introduce the anomalous diffusion processes. This book mainly focuses on the Lévy processes and subordinated Lévy processes with the Lévy measure, having divergent second moment; these processes are also the scaling limits of the CTRW models with their jump length having divergent second moment, and the scaling limits of the ones that both their first moment of waiting time and second moment of jump length are unbounded. This chapter introduces the macroscopic models and their derivations.

2.1 Fractional Fokker-Planck equations and their derivations

The first step to study anomalous diffusions is to discuss the PDF of the position of the particles, which is governed by the fractional Fokker-Planck equations (or fractional diffusion equations), including the one with time fractional derivative

$$\frac{\partial p(x,t)}{\partial t} = {}_0D_t^{1-\alpha} K_\alpha \frac{\partial^2}{\partial x^2} p(x,t), \tag{2.1}$$

the one with space fractional derivative

$$\frac{\partial p(x,t)}{\partial t} = K_\beta \nabla_x^\beta p(x,t), \tag{2.2}$$

and the one with both the time and space fractional derivatives

$$\frac{\partial p(x,t)}{\partial t} = {}_0D_t^{1-\alpha} K_{\alpha\beta} \nabla_x^\beta p(x,t), \tag{2.3}$$

where $p(x,t)$ is the PDF of being in position x at time t, $0 < \alpha < 1$, and $1 < \beta < 2$; ${}_0D_t^{1-\alpha}$ is the Riemann-Liouville derivative, defined by [Samko *et al.* (1993)]

$$ {}_0D_t^{1-\alpha}p(x,t) = \frac{1}{\Gamma(\alpha)}\frac{\partial}{\partial t}\int_0^t \frac{p(x,\tau)}{(t-\tau)^{1-\alpha}}d\tau; \tag{2.4} $$

and ∇_x^β is the Riesz space fractional derivative, given as [Yang *et al.* (2010)]

$$ \nabla_x^\beta f(x) = -\frac{1}{2\cos\frac{\beta\pi}{2}}\left[{}_{-\infty}D_x^\beta f(x) + {}_xD_{+\infty}^\beta f(x)\right], \tag{2.5} $$

with $(n-1<\beta<n)$

$$ {}_{-\infty}D_x^\beta f(x) := \frac{1}{\Gamma(n-\beta)}\frac{d^n}{dx^n}\int_{-\infty}^x \frac{f(\xi)}{(x-\xi)^{\beta+1-n}}d\xi, \tag{2.6} $$

$$ {}_xD_{+\infty}^\beta f(x) := \frac{(-1)^n}{\Gamma(n-\beta)}\frac{d^n}{dx^n}\int_x^{+\infty} \frac{f(\xi)}{(\xi-x)^{\beta+1-n}}d\xi. \tag{2.7} $$

Besides, K_α, K_β and $K_{\alpha\beta}$ are the generalized diffusion constants.

The above equations can be derived from the Montroll-Weiss formula [Montroll and Weiss (1965)], i.e., in Fourier-Laplace space ($x \to k$, $t \to s$), the PDF $p(x,t)$ obeys (see (1.15) for high dimensional case)

$$ p(k,s) = \frac{1-\phi(s)}{s}\cdot\frac{p_0(k)}{1-\varphi(k,s)}, \tag{2.8} $$

where $p_0(k)$ is the Fourier transform of the initial distribution $p_0(x)$, $\varphi(k,s)$ the Fourier transform of the joint PDF $\varphi(x,t)$ of jump length and waiting time, and $\phi(t) = \int_{-\infty}^{+\infty}\varphi(x,t)dx$. Obviously, if the jump length and waiting time are independent variables, then $\varphi(x,t) = \psi(x)\phi(t)$, where $\psi(x) = \int_0^\infty \varphi(x,t)dt$. Then (2.8) can be written as

$$ p(k,s) = \frac{1-\phi(s)}{s}\cdot\frac{p_0(k)}{1-\psi(k)\phi(s)}. \tag{2.9} $$

Here we show the procedure of deriving the above macroscopic equations. For more detailed derivations, one can refer to [Metzler and Klafter (2000)]. A CTRW process can be described by

$$ \eta(x,t) = \int_{-\infty}^{+\infty}dx'\int_0^\infty dt'\eta(x',t')\varphi(x-x',t-t') + p_0(x)\delta(t), \tag{2.10} $$

where $\eta(x,t)$ is the PDF of just arriving at position x at time t, $\delta(t)$ is the delta function, and the second summand denotes the initial distribution of the random walk. We usually choose $p_0(x)$ as $\delta(x)$, i.e., the process starts at $x=0$. Then there exists

$$ p(x,t) = \int_0^t \eta(x,t')\Psi(t-t')dt', \tag{2.11} $$

where $\Psi(t) = 1 - \int_0^t \phi(t')dt'$ is the survival probability, i.e., the probability of no jump events during the time interval $(0, t)$. Further using the fact that $\Psi(s) = \frac{1-\phi(s)}{s}$, and combining (2.10) and (2.11), it leads to (2.8).

Now, we consider the more concrete situations. The types of anomalous diffusions can be categorized by the characteristic waiting time $T = \int_0^\infty t\phi(t)dt$ and the jump length variance $\Sigma^2 = \int_{-\infty}^{+\infty} x^2\psi(x)dx$ being finite or diverging, respectively. We first consider the situation that Σ^2 is finite, but T diverges. Taking Gaussian jump length PDF $\psi(x) = (4\pi\sigma^2)^{-\frac{1}{2}}\exp(-x^2/4\sigma^2)$ leading to $\Sigma^2 = 2\sigma^2$, and the long-tailed waiting time PDF with the asymptotic behavior $\phi(t) \sim \frac{\alpha\tau^\alpha}{\Gamma(1-\alpha)}t^{-1-\alpha}$ for $0 < \alpha < 1$, their Fourier and Laplace transforms are, respectively,

$$\psi(k) \sim 1 - \sigma^2 k^2, \quad \phi(s) \sim 1 - (\tau s)^\alpha, \tag{2.12}$$

for small k and s. Substituting (2.12) into (2.9) leads to

$$p(k, s) = \frac{p_0(k)/s}{1 + K_\alpha s^{-\alpha}k^2}. \tag{2.13}$$

Employing the integration formula of fractional integrals [Samko *et al.* (1993)], $\mathscr{L}[{}_0D_t^{-\alpha}p(x,t)] = s^{-\alpha}p(x,s)$ with $\alpha \geq 0$, and making the inverse Fourier-Laplace transforms of (2.13), one finally arrives at the fractional diffusion equation (2.1), i.e.,

$$\frac{\partial p(x,t)}{\partial t} = {}_0D_t^{1-\alpha}K_\alpha \frac{\partial^2}{\partial x^2}p(x,t), \tag{2.14}$$

where $K_\alpha = \frac{\sigma^2}{\tau^\alpha}$; this model describes subdiffusion. If $\alpha \to 1$, it recovers the classical diffusion equation for Brownian motion

$$\frac{\partial p(x,t)}{\partial t} = K_1 \frac{\partial^2}{\partial x^2}p(x,t) \tag{2.15}$$

with $K_1 = \frac{\sigma^2}{\tau}$.

If T is finite but Σ^2 diverges, one can choose Poisson waiting time distribution $\phi(t) = \tau^{-1}\exp(-t/\tau)$ with $T = \tau$ and its Laplace transform is $\phi(s) \sim 1 - \tau s$; and the Lévy distribution can be used as the distribution of jump length, i.e., $\psi(k) = \exp(-\sigma^\beta|k|^\beta) \sim 1 - \sigma^\beta|k|^\beta$ for small k, with $0 < \beta < 2$, corresponding to the asymptotic behaviour $\psi(x) \sim A_\beta\sigma^{-\beta}|x|^{-1-\beta}$ for $|x| \gg \sigma$. Then we obtain that

$$p(k, s) = \frac{p_0(k)}{s + K_\beta|k|^\beta}. \tag{2.16}$$

Performing the inverse Fourier and Laplace transforms on (2.16), we get

$$\frac{\partial p(x,t)}{\partial t} = K_\beta \nabla_x^\beta p(x,t), \tag{2.17}$$

where $K_\beta = \frac{\sigma^\beta}{\tau}$ and $\mathscr{F}[\nabla_x^\beta p(x,t)](k) = -|k|^\beta p(k,t)$. This model describes superdiffusion. If $\beta \to 2$, (2.15) is recovered again.

If both T and Σ^2 diverge, following the above procedure, one can get (2.3). For the high dimensional cases, (2.1), (2.2) and (2.3), respectively, become

$$\frac{\partial p(\mathbf{X},t)}{\partial t} = {}_0D_t^{1-\alpha} K_\alpha \Delta p(\mathbf{X},t), \tag{2.18}$$

$$\frac{\partial p(\mathbf{X},t)}{\partial t} = K_\beta \Delta^{\frac{\beta}{2}} p(\mathbf{X},t), \tag{2.19}$$

and

$$\frac{\partial p(\mathbf{X},t)}{\partial t} = {}_0D_t^{1-\alpha} K_\beta \Delta^{\frac{\beta}{2}} p(\mathbf{X},t), \tag{2.20}$$

where Δ is Laplace operator, $\Delta^{\frac{\beta}{2}}$ is the fractional Laplace operator with the Fourier transform [Silvestre (2007)] $\mathscr{F}[\Delta^{\frac{\beta}{2}} p(\mathbf{X},t)](\mathbf{k}) = -|\mathbf{k}|^\beta \mathbf{p}(\mathbf{k},\mathbf{t})$; and the power-law jump length PDF $\psi(\mathbf{X}) \sim C_{n,\beta}|\mathbf{X}|^{-n-\beta}$.

2.2 Tempered fractional Feynman-Kac equations and their derivations

The tempered anomalous diffusion describes the very slow transition from anomalous to normal diffusion. In this section, we consider the tempered anomalous diffusion which is described by the CTRW model with truncated power-law waiting time and/or jump length distribution(s). Sometimes it is more reasonable because of the finite lifespan of the particles and the bounded physical space. Furthermore, if one wants to dig out more information of the corresponding stochastic processes, analyzing the distribution of the functional defined by $A = \int_0^t U[x(\tau)]d\tau$ is one of the choices, where $U(x)$ is a prescribed function and $x(t)$ is a trajectory of the particle. In the following, we review the forward and backward tempered fractional Feynman-Kac equations briefly. For the detailed derivations, one can refer to [Wu *et al.* (2016)].

Consider the CTRW on an infinite one-dimensional lattice with spacing a, and the waiting time PDF $\phi(t)$ is the exponentially truncated power-law

$$\phi(t) \sim \alpha B_\alpha e^{-\lambda t} t^{-\alpha-1}/\Gamma(1-\alpha), \ 0 < \alpha < 1 \tag{2.21}$$

with its Laplace transform

$$\phi(s) \sim -B_\alpha(\lambda+s)^\alpha + B_\alpha \lambda^\alpha + 1. \tag{2.22}$$

The process starts at $x = 0$ and the particle waits at 0 for time t drawn from $\phi(t)$ and then jumps with probability $\frac{1}{2}$ to either a or $-a$, after which the process is renewed. Let $\eta(x, A, t)dt$ be the probability of the particle to jump into (x, A) in the time interval $[t, t+dt]$, and $\Psi(t) = 1 - \int_0^t \phi(t')dt'$ is the survival probability. Then we have

$$\begin{aligned}\eta(x, A, t) =& \delta(x)\delta(A)\delta(t) \\ &+ \int_0^t \phi(\tau)\left\{\frac{1}{2}\eta[x + a, A - \tau U(x + a), t - \tau]\right. \\ &\left.+\frac{1}{2}\eta[x - a, A - \tau U(x - a), t - \tau]\right\} d\tau \end{aligned} \quad (2.23)$$

with the initial condition $\delta(x)\delta(A)\delta(t)$. And letting $G(x, A, t)$ be the joint PDF of finding the particle at position x and time t with the functional value A, then there exists

$$G(x, A, t) = \int_0^t \Psi(\tau)\eta[x, A - \tau U(x), t - \tau]d\tau. \quad (2.24)$$

Taking Fourier transform ($x \to k$, $A \to p$) and Laplace transform ($t \to s$), we get

$$G(k, p, s) = \frac{1 - \phi[s - ipU(-i\frac{\partial}{\partial k})]}{s - ipU(-i\frac{\partial}{\partial k})} \cdot \frac{1}{1 - \cos(ka)\phi[s - ipU(-i\frac{\partial}{\partial k})]}, \quad (2.25)$$

where we use $\mathscr{F}[g_1(x)g_2(x)](k) = g_1(-i\frac{\partial}{\partial k})\mathscr{F}[g_2(x)](k)$. Substituting (2.22) and $\cos(ka) \sim 1 - \frac{a^2k^2}{2}$ for the limit $k \to 0$ into (2.25), and making some rearrangements, then performing inverse Fourier and Laplace transforms, we finally get the forward tempered fractional Feynman-Kac equation:

$$\begin{aligned}&-\frac{a^2}{2B_\alpha}\frac{\partial^2}{\partial x^2}D_t^{1-\alpha,\lambda}G(x, p, t) + \left[\lambda - ipU(x) + \frac{\partial}{\partial t}\right]G(x, p, t) \\ &\quad = \lambda^\alpha D_t^{1-\alpha,\lambda}[G(x, p, t) - e^{ipU(x)t}\delta(x)] + \lambda e^{ipU(x)t}\delta(x), \end{aligned} \quad (2.26)$$

with the initial condition $G(x, A, t = 0) = \delta(x)\delta(A)$ or $G(x, p, t = 0) = \delta(x)$; $D_t^{1-\alpha,\lambda}$ is the tempered fractional substantial derivative, defined by

$$\begin{aligned}&D_t^{1-\alpha,\lambda}G(x, p, t) \\ &= \frac{1}{\Gamma(\alpha)}\left[\lambda - ipU(x) + \frac{\partial}{\partial t}\right]\int_0^t \frac{e^{-(t-\tau)\cdot(\lambda - ipU(x))}}{(t-\tau)^{1-\alpha}}G(x, p, \tau)d\tau;\end{aligned} \quad (2.27)$$

in Laplace space, $D_t^{1-\alpha,\lambda} \to [\lambda + s - ipU(x)]^{1-\alpha}$.

If $\lambda = 0$, (2.26) reduces to the imaginary time fractional Schrödinger equation [Turgeman *et al.* (2009)]

$$-\frac{a^2}{2B_\alpha}\frac{\partial^2}{\partial x^2}D_t^{1-\alpha}G(x, p, t) + \left[\frac{\partial}{\partial t} - ipU(x)\right]G(x, p, t) = 0. \quad (2.28)$$

If $\lambda = p = 0$, (2.26) turns to the fractional diffusion equation

$$\frac{\partial}{\partial t}G(x, p = 0, t) = \frac{a^2}{2B_\alpha}\frac{\partial^2}{\partial x^2}D_t^{1-\alpha}G(x, p = 0, t). \tag{2.29}$$

In some cases may we just be interested in the distribution of A, so it would be convenient to obtain an equation for $G_{x_0}(A, t)$, which is the PDF of the functional A at time t for a process starting at x_0 and obeys

$$\begin{aligned} G_{x_0}(A, t) = &\int_0^t d\tau\phi(\tau)\frac{1}{2}\{G_{x_0+a}[A - \tau U(x_0), t - \tau] \\ &+G_{x_0-a}[A - \tau U(x_0), t - \tau]\} \\ &+ \Psi(t)\delta[A - tU(x_0)], \end{aligned} \tag{2.30}$$

where $\tau U(x_0)$ is the contribution to A from the pausing time on x_0 in the time interval $(0, \tau)$; the last term on the right hand side of (2.30) shows motionless particles, for which $A(t) = tU(x_0)$. After performing Fourier transform ($x_0 \to k$, $A \to p$) and Laplace transform ($t \to s$), we have

$$\begin{aligned} G_k(p, s) = &\phi\left[-ipU\left(-i\frac{\partial}{\partial k}\right) + s\right]\cos(ka)G_k(p, s) \\ &+ \Psi\left[-ipU\left(-i\frac{\partial}{\partial k}\right) + s\right]\delta(k). \end{aligned} \tag{2.31}$$

Being similar to the above process, the backward tempered fractional Feynman-Kac equation can be obtained as

$$\begin{aligned} &-\frac{a^2}{2B_\alpha}D_t^{1-\alpha,\lambda}\frac{\partial^2}{\partial x_0^2}G_{x_0}(p, t) + \left[\lambda - ipU(x_0) + \frac{\partial}{\partial t}\right]G_{x_0}(p, t) \\ &\qquad = \lambda e^{ipU(x_0)t} + \lambda^\alpha D_t^{1-\alpha,\lambda}\left[G_{x_0}(p, t) - e^{ipU(x_0)t}\right] \end{aligned} \tag{2.32}$$

with the initial condition $G_{x_0}(A, t = 0) = \delta(A)$ or $G_{x_0}(p, t = 0) = 1$. When $\lambda = 0$, (2.32) turns to the backward fractional Feynman-Kac equation [Turgeman *et al.* (2009)],

$$-\frac{a^2}{2B_\alpha}D_t^{1-\alpha}\frac{\partial^2}{\partial x_0^2}G_{x_0}(p, t) + \left[-ipU(x_0) + \frac{\partial}{\partial t}\right]G_{x_0}(p, t) = 0. \tag{2.33}$$

Instead of discussing the tempered CTRW on a lattice, we further analyze the tempered CTRW with a tempered power-law jump length distribution, $\psi(x) \sim \frac{A_\beta}{|\Gamma(-\beta)|}e^{-\gamma|x|}|x|^{-\beta-1}$, $\gamma > 0$, $0 < \beta < 2$; and its Fourier transform is

$$\psi(k) \sim 1 - A_\beta^\theta(\gamma^2 + k^2)^{\beta/2} + 2A_\beta\gamma^\beta, \tag{2.34}$$

where the coefficients are given by $\theta = \arg(\gamma + ik)$ and $A_\beta^\theta = 2A_\beta\cos(\beta\theta)$.

Substituting (2.22) and (2.34) into

$$G(k,p,s)=\frac{1-\phi[s-ipU(-i\frac{\partial}{\partial k})]}{s-ipU(-i\frac{\partial}{\partial k})}\frac{1}{1-\psi(k)\phi[s-ipU(-i\frac{\partial}{\partial k})]}, \tag{2.35}$$

making some rearrangements and inverse transforms, we obtain the forward tempered fractional Feynman-Kac equation with the tempered power-law jump length distribution,

$$\begin{aligned}-K_{\alpha,\beta}[\nabla_x^{\beta,\gamma}+\gamma^\beta]D_t^{1-\alpha,\lambda}G(x,p,t)+\left[\lambda-ipU(x)+\frac{\partial}{\partial t}\right]G(x,p,t)\\ =\lambda e^{ipU(x)t}\delta(x)+\lambda^\alpha D_t^{1-\alpha,\lambda}\left[G(x,p,t)-e^{ipU(x)t}\delta(x)\right],\end{aligned} \tag{2.36}$$

where $K_{\alpha,\beta}=\frac{2A_\beta}{B_\alpha}$. The tempered fractional Riesz derivative $\nabla_x^{\beta,\gamma}$ is defined in Fourier $x\to k$ space as $\nabla_x^{\beta,\gamma}\to-(\gamma^2+k^2)^{\beta/2}$; and in x space, the operator is defined as

$$\nabla_x^{\beta,\gamma}f(x)=-\frac{1}{2\cos(\frac{\pi\beta}{2})}[{}_{-\infty}\mathbb{D}_x^{\beta,\gamma}f(x)+{}_x\mathbb{D}_{+\infty}^{\beta,\gamma}f(x)], \tag{2.37}$$

where ${}_{-\infty}\mathbb{D}_x^{\beta,\gamma}$ and ${}_x\mathbb{D}_{+\infty}^{\beta,\gamma}$ are the Riemann-Liouville tempered fractional derivatives (see Appendix B).

Next, we substitute (2.22) and (2.34) into

$$\begin{aligned}G_k(p,s)=&\phi\left[-ipU\left(-i\frac{\partial}{\partial k}\right)+s\right]\psi(k)G_k(p,s)\\ &+\Psi\left[-ipU\left(-i\frac{\partial}{\partial k}+s\right)\right]\delta(k),\end{aligned} \tag{2.38}$$

and make inverse Fourier and Laplace transforms after some rearrangements, getting the backward tempered fractional Feynman-Kac equation with tempered power-law jump length distribution,

$$\begin{aligned}-K_{\alpha,\beta}D_t^{1-\alpha,\lambda}[\nabla_{x_0}^{\beta,\gamma}+\gamma^\beta]G_{x_0}(p,t)+\left[\lambda-ipU(x_0)+\frac{\partial}{\partial t}\right]G_{x_0}(p,t)\\ =\lambda e^{ipU(x_0)t}+\lambda^\alpha D_t^{1-\alpha,\lambda}\left[G_{x_0}(p,t)-e^{ipU(x_0)t}\right].\end{aligned} \tag{2.39}$$

For high dimensions, (2.26), (2.32), (2.36) and (2.39) become [Deng *et al.* (2018)]

$$\begin{aligned}&-\frac{a^2}{2B_\alpha}\Delta D_t^{1-\alpha,\lambda}G(\mathbf{X},p,t)+\left[\lambda-ipU(\mathbf{X})+\frac{\partial}{\partial t}\right]G(\mathbf{X},p,t)\\ &=\lambda^\alpha D_t^{1-\alpha,\lambda}\left[G(\mathbf{X},p,t)-e^{ipU(\mathbf{X})t}\delta(\mathbf{X})\right]+\lambda e^{ipU(\mathbf{X})t}\delta(\mathbf{X}),\end{aligned} \tag{2.40}$$

$$-\frac{a^2}{2B_\alpha} D_t^{1-\alpha,\lambda} \Delta G_{\mathbf{X_0}}(p,t) + \left[\lambda - ipU(\mathbf{X_0}) + \frac{\partial}{\partial t}\right] G_{\mathbf{X_0}}(p,t)$$
$$= \lambda e^{ipU(\mathbf{X_0})t} + \lambda^\alpha D_t^{1-\alpha,\lambda} \left[G_{\mathbf{X_0}}(p,t) - e^{ipU(\mathbf{X_0})t}\right], \quad (2.41)$$

$$-K_{\alpha,\beta}(\Delta+\gamma)^{\beta/2} D_t^{1-\alpha,\lambda} G(\mathbf{X},p,t) + \left[\lambda - ipU(\mathbf{X}) + \frac{\partial}{\partial t}\right] G(\mathbf{X},p,t)$$
$$= \lambda e^{ipU(\mathbf{X})t}\delta(\mathbf{X}) + \lambda^\alpha D_t^{1-\alpha,\lambda} \left[G(\mathbf{X},p,t) - e^{ipU(\mathbf{X})t}\delta(\mathbf{X})\right], \quad (2.42)$$

and

$$-K_{\alpha,\beta} D_t^{1-\alpha,\lambda}(\Delta+\gamma)^{\beta/2} G_{\mathbf{X_0}}(p,t) + \left[\lambda - ipU(\mathbf{X_0}) + \frac{\partial}{\partial t}\right] G_{\mathbf{X_0}}(p,t)$$
$$= \lambda e^{ipU(\mathbf{X_0})t} + \lambda^\alpha D_t^{1-\alpha,\lambda} \left[G_{\mathbf{X_0}}(p,t) - e^{ipU(\mathbf{X_0})t}\right], \quad (2.43)$$

where Δ is the Laplacian operator, and $(\Delta+\gamma)^{\beta/2}$ is the tempered fractional Laplacian operator, with the Fourier transform [Deng *et al.* (2018)] $\mathscr{F}[(\Delta+\gamma)^{\beta/2} f(\mathbf{X})] = (\gamma^\beta - (\gamma^2 + |\mathbf{k}|^2)^{\beta/2} + O(|\mathbf{k}|^2))f(\mathbf{k})$ for $\beta \in (0,1) \cup (1,2)$. The derivation process is similar to the above one, except that the asymptotic behaviour of the tempered power-law jump length PDF $\psi(\mathbf{X}) \sim D_{n,\beta,\lambda} e^{-\gamma|\mathbf{X}|}|\mathbf{X}|^{-\beta-n}$.

Besides, there are some applications for the distribution of the functionals of the paths of particles performing tempered anomalous dynamics, for example, the occupation time in half-space, the first passage time, the maximal displacement, and the fluctuations of the occupation fraction. One can view the explicit derivations from [Wu *et al.* (2016)].

2.3 Governing equations for mean exit time and escape probability

The mean first exit (passage) time characterizes the average time of a stochastic process never leaving a fixed region in the state space, while the escape probability describes the likelihood of a transition from one region to another for a stochastic system driven by discontinuous (with jumps) Lévy motion. In the past decades, most of the research works on the mean first exit time or escape probability are for the uncoupled Langevin type dynamical system [Friedman (1975)]

$$dX_t = F(X_t)dt + \varepsilon\sigma(X_t)dW_t, \quad (2.44)$$

where W_t indicates the Gaussian or non-Gaussian β-stable type Lévy process, and ε is a parameter that measures the strength of the noise. The first exit time from the spatial domain Ω is defined as follows:

$$\tau(\mathbf{x}) := \inf\{t \geq 0, X_t(\mathbf{x}, \mathbf{y}) \notin \Omega\}, \tag{2.45}$$

and the mean first exit time $u(\mathbf{x}) := \langle \tau(\mathbf{x}) \rangle$.

If W_t is a (tempered) non-Gaussian β-stable type Lévy process, we could also define another concept to quantify the exit phenomenon: escape probability, because of the discontinuity of the stochastic paths; the probability of a particle starting at a point $\mathbf{x}$, first escaping a domain Ω and landing in a subset E of Ω^c (the complement of Ω), is called escape probability and denoted as $P_E(\mathbf{x})$.

Let us firstly review the system with the Gaussian white noise $\xi(t)$. The stochastic trajectory of a n-dimensional CTRW $Y(t)$ is expressed in terms of the coupled Langevin equation [Fogedby (1994)]

$$\begin{aligned} \dot{X}(\zeta) &= F(X(\zeta)) + \sqrt{2\varepsilon}\sigma(X(\zeta))\xi(\zeta), \\ \dot{T}(\zeta) &= \eta(\zeta), \end{aligned} \tag{2.46}$$

where $F(\mathbf{x})$ is a smooth vector field in a bounded domain Ω in $\mathbb{R}^n$; $\sigma(\mathbf{x})$ is a $n \times k$ matrix of smooth noise coefficient, with ε a parameter that measures the strength of the noise; $\xi(\zeta)$ is k-dimensional white Gaussian noise with $\langle \xi(\zeta) \rangle = 0$ and $\langle \xi^T(\zeta_1)\xi(\zeta_2) \rangle = \delta(\zeta_2 - \zeta_1)$ $(k \leq n)$; $\eta(\zeta)$ models the waiting times of the tempered anomalous diffusion process, being assumed to be independent from the X process, and we take $\eta(\zeta)$ as a tempered one-sided Lévy-stable noise with tempering index μ and stability index $0 < \alpha < 1$, obtained by the characteristic function of T: $\langle e^{-sT(\zeta)} \rangle = e^{-\zeta((s+\mu)^\alpha - \mu^\alpha)}$.

The above system (2.46) are determined by its transition PDF

$$p(\mathbf{x}, \mathbf{y}, t)\mathbf{dy} \equiv Pr\{\mathbf{x(t)} \in \mathbf{y} + \mathbf{dy} \,|\, \mathbf{x(0)} = \mathbf{x}\}, \tag{2.47}$$

which satisfies the tempered fractional backward Kolmogorov equation [Risken and Frank (1996)]

$$\frac{\partial}{\partial t} p(\mathbf{x}, \mathbf{y}, t) = \frac{\partial}{\partial t} \int_0^t K(t - t', \mu) L^*_{\mathbf{x}} p(\mathbf{x}, \mathbf{y}, t') dt', \tag{2.48}$$

with the absorbing boundary condition

$$p(\mathbf{x}, \mathbf{y}, t)|_{\mathbf{x}\in\Omega,\, \mathbf{y}\in\partial\Omega} = 0 \tag{2.49}$$

and the initial condition

$$p(\mathbf{x}, \mathbf{y}, 0) = \delta(\mathbf{y} - \mathbf{x}), \tag{2.50}$$

where the Laplace transform of the memory kernel is given by $K(s,\mu) = \frac{1}{(s+\mu)^\alpha - \mu^\alpha}$, and the diffusion matrix $a(\mathbf{y})$ is defined by $a(\mathbf{y}) = \frac{1}{2}\sigma(\mathbf{y})\sigma(\mathbf{y})^T$; the Laplacian operator is

$$L_{\mathbf{x}}^* = \sum_{i=1}^{n} F^i(\mathbf{x})\frac{\partial}{\partial x^i} + \varepsilon \sum_{i,j=1}^{n} a^{ij}(\mathbf{x})\frac{\partial^2}{\partial x^i \partial x^j}. \tag{2.51}$$

Evidently, τ, the first exit time to the boundary, is independent of the boundary behavior of the process. The exit time distribution is given by

$$P_r\{\tau > t \,|\, \mathbf{x}(0) = \mathbf{x}\} = \int_\Omega p(\mathbf{x},\mathbf{y},t)d\mathbf{y}, \tag{2.52}$$

where $p(\mathbf{x},\mathbf{y},t)$ is the solution of (2.48) with (2.49) and (2.50) as the boundary and initial conditions. Now the mean first exit time of trajectories that start at $\mathbf{x} \in \Omega$, is given by

$$\begin{aligned} u(\mathbf{x}) &\equiv E(\tau \,|\, \mathbf{x}(0) = \mathbf{x}) \\ &= \int_0^\infty t d_t [P_r(\tau < t \,|\, \mathbf{x}(0) = \mathbf{x}) - 1] \\ &= \int_0^\infty P_r(\tau > t \,|\, \mathbf{x}(0) = \mathbf{x}) dt \\ &= \int_0^\infty \int_\Omega p(\mathbf{x},\mathbf{y},t) d\mathbf{y} dt. \end{aligned} \tag{2.53}$$

Defining

$$P(\mathbf{x},\mathbf{y},t) = \int_0^t p(\mathbf{x},\mathbf{y},t')dt', \tag{2.54}$$

we have

$$\begin{aligned} P(\mathbf{x},\mathbf{y}) :&= P(\mathbf{x},\mathbf{y},t=\infty) = \lim_{s\to 0} s \cdot P(\mathbf{x},\mathbf{y},s) \\ &= \lim_{s\to 0} p(\mathbf{x},\mathbf{y},s) \end{aligned} \tag{2.55}$$

by using the final value theorem of Laplace transform $(\lim_{t\to\infty} f(t) = \lim_{s\to 0} sf(s))$. Taking Laplace transform for (2.48) and then letting $s \to 0$ results in

$$L_{\mathbf{x}}^* P(\mathbf{x},\mathbf{y}) = -\alpha\mu^{\alpha-1}\delta(\mathbf{y}-\mathbf{x}). \tag{2.56}$$

Using

$$u(\mathbf{x}) = \int_\Omega \int_0^\infty p(\mathbf{x},\mathbf{y},t)dtd\mathbf{y} = \int_\Omega P(\mathbf{x},\mathbf{y})d\mathbf{y}, \tag{2.57}$$

we finally have the governing equation for the mean first exit time

$$L_{\mathbf{x}}^{*}u(\mathbf{x}) = -\alpha\mu^{\alpha-1} \quad \text{for} \quad \mathbf{x} \in \Omega \tag{2.58}$$

with the boundary condition

$$u(\mathbf{x}) = 0 \quad \text{for} \quad \mathbf{x} \in \partial\Omega. \tag{2.59}$$

Next we consider the stochastic dynamics driven by non-Gaussian Lévy noises:

$$\begin{aligned} \dot{X}(\zeta) &= F(X(\zeta)) + \varepsilon \dot{L}(\zeta), \\ \dot{T}(\zeta) &= \eta(\zeta), \end{aligned} \tag{2.60}$$

where F is a vector field (or drift); $L = (L(t), t \geq 0)$ is a Lévy process which can be explicitly characterized by its characteristic function (1.13); ε and $\eta(\zeta)$ are the same as the ones used in (2.46); $L(\zeta)$ and $\eta(\zeta)$ are statistically independent. And the pseudo-differential generator A is defined as [Applebaum (2009)]

$$\begin{aligned} (Af)(\mathbf{x}) =& F^i(\partial_i f)(\mathbf{x}) + a^i(\partial_i f)(\mathbf{x}) + \frac{1}{2}b^{ij}(\partial_i\partial_j f)(\mathbf{x}) \\ &+ \int_{\mathbb{R}^n\setminus\{0\}} \left[f(\mathbf{x}+\mathbf{y}) - f(\mathbf{x}) - \mathbf{y}^i(\partial_i f)(\mathbf{x})\chi_{\{|\mathbf{y}|<1\}}\right] \nu(d\mathbf{y}), \end{aligned} \tag{2.61}$$

with $\nu(d\mathbf{y}) = \frac{\beta\Gamma(\frac{n+\beta}{2})}{2^{1-\beta}\pi^{n/2}\Gamma(1-\beta/2)}|\mathbf{y}|^{-\beta-n}d\mathbf{y}$ and $f \in C_0^\infty(\mathbb{R}^n)$.

The transition PDF $p(\mathbf{x}, \mathbf{y}, t)$ for system (2.60) satisfies the tempered fractional Kolmogorov equation [Applebaum (2009)]

$$\frac{\partial}{\partial t}p(\mathbf{x}, \mathbf{y}, t) = \frac{\partial}{\partial t}\int_0^t K(t-t', \mu)(Ap)(\mathbf{x}, \mathbf{y}, t')dt', \tag{2.62}$$

with the initial condition (2.50) and absorbing boundary condition

$$p(\mathbf{x}, \mathbf{y}, t)|_{\mathbf{x}\in\Omega,\, \mathbf{y}\in\Omega^c} = 0. \tag{2.63}$$

Similar to the discussion above, we have

$$AP(\mathbf{x}, \mathbf{y}) = -\alpha\mu^{\alpha-1}\delta(\mathbf{y}-\mathbf{x}); \tag{2.64}$$

and the governing equation of the mean exit time is

$$Au(\mathbf{x}) = -\alpha\mu^{\alpha-1} \quad \text{for} \quad \mathbf{x} \in \Omega \tag{2.65}$$

and

$$u(\mathbf{x}) = 0 \quad \text{for} \quad \mathbf{x} \in \Omega^c. \tag{2.66}$$

We then consider the escape probability of a particle whose motion is described by the coupled Langevin equation (2.60). Basing on (2.64), (2.50) and (2.63), we have

$$AP_E(\mathbf{x}) = A\int_E P(\mathbf{x},\mathbf{y})d\mathbf{y} = \int_E -\alpha\mu^{\alpha-1}\delta(\mathbf{y}-\mathbf{x})d\mathbf{y} = 0, \tag{2.67}$$

for all $\mathbf{x} \in \Omega$. Hence, the escape probability $P_E(\mathbf{x})$ solves

$$\begin{aligned} &AP_E(\mathbf{x}) = 0, && \mathbf{x} \in \Omega, \\ &P_E(\mathbf{x})|_{\mathbf{x}\in E} = 1, && P_E(\mathbf{x})|_{\mathbf{x}\in\Omega^c\setminus E} = 0. \end{aligned} \tag{2.68}$$

For the solutions of (2.58), (2.65), and (2.68) and the other details, one can refer to [Deng *et al.* (2017)].

Chapter 3

Numerical methods for the time fractional differential equations

Since the obtained analytical solutions of fractional differential equations are usually in the form of transcendental functions or infinite series, and in much more cases, the analytical solutions are not available, these naturally motivate the developments of numerical methods for fractional differential equations. Comparing with the integer-order differential equations, one of the bigger challenges we have to face is the expensiveness of its computation cost besides its complexity, since fractional operators are pseudodifferential operators which are non-local; for time fractional differential equations, the computation cost increases much faster than classical ones with the increase of the time. One can refer to [Diethelm (2010); Guo *et al.* (2015); Weilbee (2005)] for a deep understanding.

In this chapter, we present some new numerical schemes for solving the fractional ordinary differential equations (ODEs) or the time fractional PDEs, which effectively reduce the computation cost or increase the convergence speed. In Sec. 3.1, we provide the Jacobian preditor-corrector method for solving the linear or nonlinear fractional ODEs, compared with the existed numerical schemes (with the cost $\mathcal{O}(N_E^2)$ or $\mathcal{O}(N_E \log N_E)$), which only has the computation cost $\mathcal{O}(N_E)$. Here N_E denotes the steps in time. In Sec. 3.2, we develop the local discontinuous Galerkin (LDG) method for solving the fractional ODEs; the superconvergence of the proposed numerical schemes is also considered. In the last section, we present the numerical schemes for solving time fractional PDEs based on the numerical inverse Laplace transform, which have been well developed in solving the integer-order partial differential equation, but much less for fractional ones.

3.1 Jacobi preditor-corrector method

In this section, we develop a new type of predictor-corrector approach for the initial value problem

$$\begin{cases} {}_0^C D_t^\alpha x(t) := f\big(t, x(t)\big), \\ x^{(k)}(0) = x_0^{(k)}, \quad k = 0, 1, \cdots, \lceil \alpha \rceil - 1, \end{cases} \tag{3.1}$$

where $\alpha > 0$, $\lceil \alpha \rceil = \{z \in \mathbb{N} : 0 \le z - \alpha < 1\}$, and ${}_0^C D_t^\alpha x(t)$ is the Caputo fractional derivative defined by

$${}_0^C D_t^\alpha x(t) = \begin{cases} \frac{1}{\Gamma(\lceil \alpha \rceil - \alpha)} \int_0^t (t-s)^{\lceil \alpha \rceil - \alpha - 1} \frac{d^{\lceil \alpha \rceil} x(s)}{ds^{\lceil \alpha \rceil}} ds, & \alpha \notin \mathbb{N}^+, \\ \frac{d^{\lceil \alpha \rceil} x(s)}{ds^{\lceil \alpha \rceil}}, & \alpha \in \mathbb{N}^+. \end{cases}$$

It is well known that problem (3.1) is equivalent to the Volterra integral equation [Daftardar-Gejji and Babakhani (2004); Diethelm *et al.* (2002, 2004)]

$$x(t) = \sum_{k=0}^{\lceil \alpha \rceil - 1} \frac{t^k}{k!} x_0^{(k)} + \frac{1}{\Gamma(\alpha)} \int_0^t (t-\tau)^{\alpha-1} f\big(\tau, x(\tau)\big) d\tau, \quad t \in [0, T] \tag{3.2}$$

in the sense that a continuous function solves (3.2) if and only if it solves (3.1). The existence and uniqueness of the solution of (3.2) have been established in [Diethelm and Ford (2002)] under the conditions: (a) the continuity of f with respect to both its arguments; (b) f is Lipschitz continue with respect to the second argument.

Many numerical approaches have been proposed to solve (3.1) or (3.2), such as [Ford and Simpson (2001); Diethelm *et al.* (2002, 2004); Deng (2007a,b)]. Diethelm, Ford and their coauthors successfully present the numerical approximation of (3.2) using Adams-type predictor-corrector approach and give the corresponding detailed error analysis in [Diethelm *et al.* (2002)] and [Diethelm *et al.* (2004)], respectively. The convergence order of their approach is proved to be $\min\{2, 1+\alpha\}$, and the arithmetic complexity of their algorithm with steps N_E is $O(N_E^2)$, whereas a comparable algorithm for a classical initial value problem only gives rise to $O(N_E)$. The challenge of the computation cost is essential due to the non-local characteristic of fractional derivatives. Diethelm's method has been modified in [Deng (2007a)], where the convergence order is improved to $\min\{2, 1+2\alpha\}$ and almost half of the computation cost is reduced, but the cost is still $O(N_E^2)$. There are already two typical ways have been suggested to overcome this challenge. One is the fixed memory principle of Podlubny [Podlubny (1999)]. However, it is shown that the fixed memory principle is not suitable for Caputo derivative, because we can not reduce the computation cost significantly for preserving the convergence order [Diethelm *et al.* (2002); Ford and Simpson (2001)]. The other more promising idea is the nested memory concept of Ford and Simpson [Ford and Simpson (2001); Deng (2007b)] which can lead to $O(N_E \log N_E)$ complexity, but still retain the order of convergence. However, the convergence order there can not exceed 2. It should be noted that for the effectiveness of the short memory principle, in [Ford and Simpson (2001)], α has to belong to the interval $(0,1)$; and in [Deng (2007b)], α can be within $(0,2)$.

Here, the Jacobi-Gauss-Lobatto quadrature, rather than the composite rectangular formula in [Diethelm *et al.* (2002, 2004)] or the composite trapezoidal formula in [Deng (2007a)], is chosen to approximate the Riemann-Liouville integral in (3.2). In other words, we apprehend the Riemann-Liouville integral from the viewpoint of a normal integral with a special weight function. Thus it can be treated basing on the theories of the classical numerical integration and of polynomial interpolation [Quarteroni *et al.* (2007)]. The result turns out that a good numerical approximation to (3.2) can be obtained with a convergence order N_I, which is the number of interpolating points been used and N_I does not depend on the fractional order α. Moreover, the computation complexity is reduced to $O(N_E)$, the same as classical initial value problem, which is one of the main advantages of the algorithm. We call the whole procedure as Jacobi predictor-corrector approach [Zhao and Deng (2014)].

3.1.1 *Algorithms*

In this subsection, we describe the basic process of the Jacobi predictor-corrector approach for solving (3.2). Rewrite (3.2) as

$$x(t) = \sum_{k=0}^{\lceil\alpha\rceil-1} \frac{t^k}{k!} x_0^{(k)} + \frac{1}{\Gamma(\alpha)} \left(\frac{t}{2}\right)^{\alpha} \int_{-1}^{1} (1-s)^{\alpha-1} \tilde{f}\big(s, \tilde{x}(s)\big)\, ds, \tag{3.3}$$

where

$$\tilde{f}\big(s,\tilde{x}(s)\big) = f\Big(\frac{t}{2}(1+s), x\Big(\frac{t}{2}(1+s)\Big)\Big), \quad -1 \le s \le 1;$$

$$\tilde{x}(s) = x\Big(\frac{t}{2}(1+s)\Big), \quad -1 \le s \le 1.$$

By the theory of the classical numerical integration [Quarteroni *et al.* (2007)], we can approximate the integral in (3.3) by using the Jacobi-Gauss-Lobatto quadrature (see Appendix A) w.r.t. the weight function

$\omega(s) = (1-s)^{\alpha-1}(1+s)^0$. That is,

$$x(t) \approx \sum_{k=0}^{\lceil\alpha\rceil-1} \frac{t^k}{k!} x_0^{(k)} + \frac{1}{\Gamma(\alpha)}\Big(\frac{t}{2}\Big)^{\alpha} \sum_{j=0}^{N_J} \omega_j \tilde{f}(s_j, \tilde{x}(s_j)), \tag{3.4}$$

where $N_J + 1$, $\{\omega_j\}_{j=0}^{N_J}$, $\{s_j\}_{j=0}^{N_J}$ in (3.4) denote the number, the weights, and the value of the Jacobi-Gauss-Lobatto nodes in the interval $[-1, 1]$, respectively.

Let us define a grid in the interval $[0, T]$ with $N_E + 1$ equi-spaced nodes t_i, given by

$$t_i = ih, \quad i = 0, \cdots, N_E, \tag{3.5}$$

where $h = T/N_E$ is the stepsize. Supposing that the numerical values of $x(t)$ at $t_0, t_1, \cdots, t_n$ have been obtained, which are denoted as $x_0, x_1, \cdots, x_n$, separately ($x_0 = x_0^{(0)}$), we are going to compute the value of $x(t)$ at t_{n+1}, i.e. x_{n+1}. By (3.4), we have

$$x(t_{n+1}) \approx \sum_{k=0}^{\lceil\alpha\rceil-1} \frac{t_{n+1}^k}{k!} x_0^{(k)} + \frac{1}{\Gamma(\alpha)}\Big(\frac{t_{n+1}}{2}\Big)^{\alpha} \sum_{j=0}^{N_J} \omega_j \tilde{f}_{n+1}(s_j, \tilde{x}_{n+1}(s_j)), \tag{3.6}$$

where

$$\tilde{f}_{n+1}(s, \tilde{x}_{n+1}(s)) = f\Big(\frac{t_{n+1}}{2}(1+s), x\Big(\frac{t_{n+1}}{2}(1+s)\Big)\Big), \quad -1 \le s \le 1;$$

$$\tilde{x}_{n+1}(s) = x\Big(\frac{t_{n+1}}{2}(1+s)\Big), \quad -1 \le s \le 1.$$

Note that $\tilde{f}_{n+1}(s_{N_J}, \tilde{x}_{n+1}(s_{N_J})) = f(t_{n+1}, x(t_{n+1}))$. In order to avoid solving the possible non-linear algebraic equations (when f is non-linear w.r.t. x), we develop the Jacobi predictor-corrector scheme of (3.3) as

$$x_{n+1} = \sum_{k=0}^{\lceil\alpha\rceil-1} \frac{t_{n+1}^k}{k!} x_0^{(k)} + \frac{1}{\Gamma(\alpha)}\Big(\frac{t_{n+1}}{2}\Big)^{\alpha}\Big(\sum_{j=0}^{N_J-1} \omega_j \tilde{f}_{n+1,j} + \omega_{N_J} f(t_{n+1}, x_{n+1}^P)\Big), \tag{3.7}$$

and

$$x_{n+1}^P = \sum_{k=0}^{\lceil\alpha\rceil-1} \frac{t_{n+1}^k}{k!} x_0^{(k)} + \frac{1}{\Gamma(\alpha)}\Big(\frac{t_{n+1}}{2}\Big)^{\alpha} \sum_{j=0}^{N_J} \omega_j \tilde{f}_{n+1,j}^P, \tag{3.8}$$

where $\{\tilde{f}_{n+1,j}^P\}_{j=0}^{N_J}$ in (3.8) means that all of the values $\tilde{f}_{n+1}$ at the Jacobi-Gauss-Lobatto nodes are got by using the interpolations based on the values of $\{f(t_i, x_i)\}_{i=0}^{n}$; whereas $\{\tilde{f}_{n+1,j}\}_{j=0}^{N_J-1}$ in (3.7) are obtained by using the interpolations based on the values of $\{f(t_i, x_i)\}_{i=0}^{n}$ and $f(t_{n+1}, x_{n+1}^P)$. The algorithm for realizing (3.7) and (3.8) is detailedly described in Algorithm 1, where the equi-spaced nodes located in the neighborhood of s_j (should be $(1 + s_j)t_{n+1}/2$ as to variable t) have been used to obtain $\tilde{f}_{n+1,j}^P$ and $\tilde{f}_{n+1,j}$.

Algorithm 1 The approach for realizing (3.7) and (3.8)

Step 0. Some notations:

N_I : the number of equi-spaced nodes used for the interpolation;

$n_l = \lceil N_I/2 \rceil$: to evaluate $\tilde{f}_{n+1}(s_j, \tilde{x}_{n+1}(s_j))$, the expected number of the interpolating equi-spaced nodes on the left hand side of s_j;

$n_r = \lfloor N_I/2 \rfloor$: to evaluate $\tilde{f}_{n+1}(s_j, \tilde{x}_{n+1}(s_j))$, the expected number of the interpolating equi-spaced nodes on the right hand side of s_j. Here $\lfloor N_I/2 \rfloor := \{z \in \mathbb{N} : 0 < \alpha - z \leq 1\}$;

n_e : the number of the interpolating equi-spaced nodes that can be used on the left hand side of s_j;

$P_l(t) := L\big[(t_0, f(t_0, x_0)), (t_1, f(t_1, x_1)), \cdots, (t_{N_I-1}, f(t_{N_I-1}, x_{N_I-1}))\big](t)$;

$$P_{r,n+1}(t) := L\big[(t_{n-N_I+1}, f(t_{n-N_I+1}, x_{n-N_I+1}))\\ (t_{n-N_I+2}, f(t_{n-N_I+2}, x_{n-N_I+2})), \cdots, (t_n, f(t_n, x_n))\big](t);$$

$$C_{r,n+1}(t) := L\big[(t_{n-N_I+2}, f(t_{n-N_I+2}, x_{n-N_I+2})),\\ (t_{n-N_I+3}, f(t_{n-N_I+3}, x_{n-N_I+3})), \cdots, (t_{n+1}, f(t_{n+1}, x_{n+1}^P))\big](t);$$

$$P_{n+1}(t) := L\big[(t_{n_e-n_l}, f(t_{n_e-n_l}, x_{n_e-n_l})), (t_{n_e-n_l+1}, f(t_{n_e-n_l+1}, x_{n_e-n_l+1})),\\ \cdots, (t_{n_e+n_r-1}, f(t_{n_e+n_r-1}, x_{n_e+n_r-1}))\big](t),$$

where $L\big[(t_0, f(t_0, x_0)), \cdots, (t_{N_I-1}, f(t_{N_I-1}, x_{N_I-1}))\big](t)$ denotes the N_I order Lagrange interpolation polynomial passing through the points $(t_0, f(t_0, x_0)), \cdots, (t_{N_I-1}, f(t_{N_I-1}, x_{N_I-1}))$, etc.

Step 1. To start the procedure:

Compute $x_1, x_2, \cdots, x_{N_I-1}$ by a single step method (e.g., the Improved-Adams' methods in [Deng (2007a)]) with a sufficiently small step-length h_0 such that x_i, $i = 1, 2, \cdots, N_I - 1$, are accurate enough for not deteriorating the accuracy of the method we are discussing.

Step 2. To predict:

$sum = 0$

do $j = 0, \cdots, N_J$

if $n_e \leq n_l$ (the number of the equi-spaced nodes located on the left hand side of s_j (should be $(1+s_j)t_{n+1}/2$ as to variable t) is equal to / less than what we expect)

$sum = sum + \omega_j P_l\big((1+s_j)t_{n+1}/2\big)$

else if $n_e + n_r \geq n+1$ (the number of the equi-spaced nodes located on the right hand side of s_j (should be $(1+s_j)t_{n+1}/2$ as to variable t) is equal to / less than what we expect)

$sum = sum + \omega_j P_{r,n+1}\big((1+s_j)t_{n+1}/2\big)$

else

$sum = sum + \omega_j P_{n+1}\big((1+s_j)t_{n+1}/2\big)$

enddo

$x_{n+1}^P = \sum_{k=0}^{\lceil\alpha\rceil-1} \frac{t_{n+1}^k}{k!} x_0^{(k)} + \frac{1}{\Gamma(\alpha)}\left(\frac{t_{n+1}}{2}\right)^\alpha \cdot sum;$

Step 3. To correct:

$sum = 0$

do $j = 0, \cdots, N_J - 1$

if $n_e \leq n_l$

$sum = sum + \omega_j P_l\big((1+s_j)t_{n+1}/2\big)$

else if $n_e + n_r \geq n+2$

$sum = sum + \omega_j C_{r,n+1}\big((1+s_j)t_{n+1}/2\big)$

else

$sum = sum + \omega_j P_{n+1}\big((1+s_j)t_{n+1}/2\big)$

enddo

$x_{n+1} = \sum_{k=0}^{\lceil\alpha\rceil-1} \frac{t_{n+1}^k}{k!} x_0^{(k)} + \frac{1}{\Gamma(\alpha)}\left(\frac{t_{n+1}}{2}\right)^\alpha \cdot \big(sum + \omega_{N_J} f(t_{n+1}, x_{n+1}^P)\big)$

Although the description of this algorithm seems tedious, its operation is simple and mechanical. It can be observed that for the computation of x_{n+1}, only changing $2(N_J+1)$ values is needed, each of which can be obtained by interpolating N_I nearby values; whereas in [Diethelm *et al.* (2002); Deng (2007a)], it should take $O(n+1)$ multiplications and divisions. In other words, the computational cost here has no relationship with the variable $n+1$, just depends on the number of the interpolation nodes N_I and the number of Jacobi-Gauss-Lobatto nodes N_J+1. So, for fixed N_J and N_I, to approximate $x(T)$, the total computational cost is $O(N_E)$, comparing with $O(N_E^2)$ in [Diethelm *et al.* (2002); Deng (2007a)] and $O(N_E \log N_E)$ in [Ford and Simpson (2001); Deng (2007b)], which is one of the most significant advantages of this algorithm. In addition, during the estimation for x_{n+1}, actually the values of $\{f(t_i, x_i)\}_{i=0}^n$, $\{\tilde{f}_{n+1,j}^P\}_{j=0}^{N_J}$, $f(t_{n+1}, x_{n+1}^P)$

and $\{\tilde{f}_{n+1,j}\}_{j=0}^{N_J-1}$ should be stored in sequence, so, for a fixed N_J, the memory requirements are $O(N_E)$, just as those in [Diethelm *et al.* (2002, 2004); Deng (2007a)].

3.1.2 *Error analysis*

First, we introduce four notations. The piecewise interpolating polynomial based on the N_I nodes of $\{(t_i, f(t_i, x_i))\}_{i=0}^{n}$ is denoted by $F_{A,N_I}^{P}\tilde{f}_{n+1}(s)$, where $-1 \leq s \leq 1$; the one based on the N_I nodes of $\{(t_i, f(t_i, x_i))\}_{i=0}^{n}$ and $(t_{n+1}, f(t_{n+1}, x_{n+1}^{P}))$ is written as $F_{A,N_I}\tilde{f}_{n+1}(s)$, where $-1 \leq s \leq 1$; the one based on the N_I nodes of $\{(t_i, f(t_i, x(t_i)))\}_{i=0}^{n}$ is signified by $F_{N_I}^{P}\tilde{f}_{n+1}(s)$, where $-1 \leq s \leq 1$; the one based on the N_I nodes of $\{(t_i, f(t_i, x(t_i)))\}_{i=0}^{n+1}$ is denoted as $F_{N_I}\tilde{f}_{n+1}(s)$, where $-1 \leq s \leq 1$. Note that x_i is the numerical solution and $x(t_i)$ is the exact solution.

The error analysis below is inspired by [Diethelm *et al.* (2004)].

3.1.2.1 *Some preliminaries and a useful lemma*

We know, the set of Jacobi polynomials $\{J_n^{\alpha,\beta}(x)\}_{n=0}^{\infty}$ (see Appendix A) with weight $\omega^{\alpha,\beta}(x) = (1-x)^{\alpha}(1+x)^{\beta}, \alpha, \beta > -1$ forms a complete $L^2_{\omega^{\alpha,\beta}}(-1,1)$-orthogonal system, where $L^2_{\omega^{\alpha,\beta}}(-1,1)$ is a weighted space defined by

$$L^2_{\omega^{\alpha,\beta}}(-1,1) = \{v : v \text{ is measurable and } \| v \|_{\omega^{\alpha,\beta}} < \infty\}, \tag{3.9}$$

equipped with the norm

$$\| v \|_{\omega^{\alpha,\beta}} = \left(\int_{-1}^{1} |v(x)|^2 \omega^{\alpha,\beta}(x) dx \right)^{\frac{1}{2}}, \tag{3.10}$$

and the inner product

$$(u, v)_{\omega^{\alpha,\beta}} = \int_{-1}^{1} u(x)v(x)\omega^{\alpha,\beta}(x)dx. \tag{3.11}$$

For bounding the approximation error of Jacobi polynomials, we need the following non-uniformly-weighted Sobolev spaces as in [Shen *et al.* (2011)]:

$$B^m_{\alpha,\beta}(-1,1) := \{v : \partial_x^k v \in L^2_{\omega^{\alpha+k,\beta+k}}(-1,1),\ 0 \leq k \leq m\}, \tag{3.12}$$

equipped with the inner product and the norm as

$$(u, v)_{B^m_{\alpha,\beta}} = \sum_{k=0}^{m} (\partial_x^k u, \partial_x^k v)_{\omega^{\alpha+k,\beta+k}}, \tag{3.13}$$

and

$$\| v \|_{B^m_{\alpha,\beta}} = \sqrt{(v, v)_{B^m_{\alpha,\beta}}}. \tag{3.14}$$

For any continuous functions u and v on $[-1, 1]$, we define a discrete inner product as

$$(u, v)_N = \sum_{j=0}^{N} u(x_j)v(x_j)\omega_j, \tag{3.15}$$

where $\{\omega_j\}_{j=0}^N$ is a set of Jacobi weights. The following result follows from Theorem 3.4.3 and Remark 3.7 in [Shen *et al.* (2011)].

Lemma 3.1. *For $\alpha, \beta > -1$ and any $u \in B^m_{\alpha,\beta}(-1, 1)$, $m \leq N + 1$, if $m = o(N)$ (in particular for fixed m), then we have*

$$\| u - I_N^{\alpha,\beta} u \|_{\omega^{\alpha,\beta}} \leq CN^{-m} \| \partial_x^m u \|_{\omega^{\alpha+m,\beta+m}}, \tag{3.16}$$

where C is a positive constant independent of m, N and u, $I_N^{\alpha,\beta} u$ is the N-th Lagrange interpolation polynomial of u on the set of the $(N + 1)$ Gauss-Lobatto points w.r.t the weight function $\omega^{\alpha,\beta}(x)$.

From the above Lemma, we can get a useful corollary:

Corollary 3.1. *If $u \in B^m_{\alpha,\beta}(-1, 1)$ for some $m \geq 0$ and $\phi \in \mathcal{P}_{N-1}$, then for the Jacobi-Gauss-Lobatto integration, we have*

$$| (u, \phi)_{\omega^{\alpha,\beta}} - (u, \phi)_N | \leq CN^{-m} \| \partial_x^m v \|_{\omega^{\alpha+m,\beta+m}} \| \phi \|_{\omega^{\alpha,\beta}} . \tag{3.17}$$

Proof. By (3.15), the theory of Jacobi-Gauss-Lobatto quadrature [Quarteroni *et al.* (2007)] and Lemma 3.1, it yields that

$$\begin{aligned}
&| (u, \phi)_{\omega^{\alpha,\beta}} - (u, \phi)_N | \\
&= | (u, \phi)_{\omega^{\alpha,\beta}} - \sum_{j=0}^{N} u(x_j)\phi(x_j)\omega_j | \\
&= | (u, \phi)_{\omega^{\alpha,\beta}} - \sum_{j=0}^{N} I_N^{\alpha,\beta} u(x_j)\phi(x_j)\omega_j | \\
&= | (u, \phi)_{\omega^{\alpha,\beta}} - (I_N^{\alpha,\beta} u, \phi)_N | \\
&= | (u, \phi)_{\omega^{\alpha,\beta}} - (I_N^{\alpha,\beta} u, \phi)_{\omega^{\alpha,\beta}} | \\
&\leq \| u - I_N^{\alpha,\beta} u \|_{\omega^{\alpha,\beta}} \| \phi \|_{\omega^{\alpha,\beta}} \\
&\leq CN^{-m} \| \partial_x^m u \|_{\omega^{\alpha+m,\beta+m}} \| \phi \|_{\omega^{\alpha,\beta}} .
\end{aligned} \tag{3.18}$$

□

3.1.2.2 *Auxiliary results*

By the definitions of the inner product (3.11), the discrete inner product (3.15), and the notations given at the beginning of this subsection, we can rewrite (3.3) at $t = t_{n+1}$, (3.7), and (3.8), respectively, as

$$x(t_{n+1}) = \sum_{k=0}^{\lceil\alpha\rceil-1} \frac{t_{n+1}^k}{k!} x_0^{(k)} + \frac{1}{\Gamma(\alpha)}\left(\frac{t_{n+1}}{2}\right)^{\alpha} \left(\tilde{f}_{n+1}(\cdot, \tilde{x}_{n+1}(\cdot)), 1\right)_{\omega^{\alpha-1,0}}, \tag{3.19}$$

$$x_{n+1} = \sum_{k=0}^{\lceil\alpha\rceil-1} \frac{t_{n+1}^k}{k!} x_0^{(k)} + \frac{1}{\Gamma(\alpha)}\left(\frac{t_{n+1}}{2}\right)^{\alpha} \left(F_{A,N_I}\tilde{f}_{n+1}(\cdot), 1\right)_{N_J}, \tag{3.20}$$

and

$$x_{n+1}^P = \sum_{k=0}^{\lceil\alpha\rceil-1} \frac{t_{n+1}^k}{k!} x_0^{(k)} + \frac{1}{\Gamma(\alpha)}\left(\frac{t_{n+1}}{2}\right)^{\alpha} \left(F_{A,N_I}^P\tilde{f}_{n+1}(\cdot), 1\right)_{N_J}. \tag{3.21}$$

On the other hand, since each $\{F_{A,N_I}^P\tilde{f}_{n+1}(s_j)\}$ or $\{F_{A,N_I}\tilde{f}_{n+1}(s_j)\}$ is essentially a linear combination of parts of $\{f(t_i, x_i)\}_{i=0}^n$ or $\{f(t_i, x_i)\}_{i=0}^n$ and $f(t_{n+1}, x_{n+1}^P)$, we can also formally rewrite (3.7) and (3.8) as convolution quadratures

$$x_{n+1} = \sum_{k=0}^{\lceil\alpha\rceil-1} \frac{t_{n+1}^k}{k!} x_0^{(k)} + C_\alpha\left[\sum_{i=0}^{n} a_{i,n+1} f(t_i, x_i) + a_{n+1,n+1} f(t_{n+1}, x_{n+1}^P)\right], \tag{3.22}$$

and

$$x_{n+1}^P = \sum_{k=0}^{\lceil\alpha\rceil-1} \frac{t_{n+1}^k}{k!} x_0^{(k)} + C_\alpha \sum_{i=0}^{n} b_{i,n+1} f(t_i, x_i), \tag{3.23}$$

where $C_\alpha = \frac{1}{\Gamma(\alpha)}\left(\frac{t_{n+1}}{2}\right)^{\alpha}$, and $\{a_{i,n+1}\}_{i=0}^{n+1}$ and $\{b_{i,n+1}\}_{i=0}^{n}$ are sets of real numbers depending on the number of the interpolating nodes N_I and the positions of those Jacobi nodes in the interval $[0, t_{n+1}]$. The formulae (3.22) and (3.23) can help us to understand the error analysis that we will be performing.

Theorem 3.1. *Assume that* $f(t, x(t)) = \sum_{l=1}^m t^{\gamma_l} g_l(t) + C_0$ *with* g_l *smooth,* $\gamma_m > \cdots > \gamma_1 \geq N_I \geq 1$, C_0 *be a constant. Then there is a constant* C_1, *independent of* n, h *and* N_J, *such that*

$$\frac{1}{\Gamma(\alpha)}\left|\int_0^{t_{n+1}} (t_{n+1}-\tau)^{\alpha-1} f(\tau, x(\tau))\, d\tau - \left(\frac{t_{n+1}}{2}\right)^{\alpha} \sum_{i=0}^{n} b_{i,n+1} f(t_i, x(t_i))\right|$$
$$\leq C_1 t_{n+1}^{\alpha}\left(t_{n+1}^{\gamma_1} N_J^{-\lceil 2\gamma_1\rceil} + h^{N_I}\right). \tag{3.24}$$

Proof. By the definitions of $\{F_{N_I}^P \tilde{f}_{n+1}(s_j)\}$ and (3.15), we have

$$\frac{1}{\Gamma(\alpha)}\int_0^{t_{n+1}}(t_{n+1}-\tau)^{\alpha-1}f\big(\tau,x(\tau)\big)d\tau - C_\alpha\sum_{i=0}^{n}b_{i,n+1}f(t_i,x_i)$$
$$= C_\alpha\left[\left(\tilde{f}_{n+1}\big(\cdot,\tilde{x}_{n+1}(\cdot)\big),1\right)_{\omega^{\alpha-1,0}} - \left(F_{N_I}^P\tilde{f}_{n+1}(\cdot),1\right)_{N_J}\right]$$
$$= C_\alpha\left[\left(\tilde{f}_{n+1}\big(\cdot,\tilde{x}_{n+1}(\cdot)\big),1\right)_{\omega^{\alpha-1,0}} - \left(\tilde{f}_{n+1}\big(\cdot,\tilde{x}_{n+1}(\cdot)\big),1\right)_{N_J}\right.$$
$$\left. + \left(\tilde{f}_{n+1}\big(\cdot,\tilde{x}_{n+1}(\cdot)\big),1\right)_{N_J} - \left(F_{N_I}^P\tilde{f}_{n+1}(\cdot),1\right)_{N_J}\right]$$
$$= I_{n+1,1} + I_{n+1,2}, \tag{3.25}$$

where

$$I_{n+1,1} = C_\alpha\left[\left(\tilde{f}_{n+1}\big(\cdot,\tilde{x}_{n+1}(\cdot)\big),1\right)_{\omega^{\alpha-1,0}} - \left(\tilde{f}_{n+1}\big(\cdot,\tilde{x}_{n+1}(\cdot)\big),1\right)_{N_J}\right],$$
$$I_{n+1,2} = C_\alpha\left[\left(\tilde{f}_{n+1}\big(\cdot,\tilde{x}_{n+1}(\cdot)\big),1\right)_{N_J} - \left(F_{N_I}^P\tilde{f}_{n+1}(\cdot),1\right)_{N_J}\right]. \tag{3.26}$$

Since

$$f\big(t,x(t)\big) = \sum_{l=1}^{m} t^{\gamma_l}g_l(t) + C_0$$

with g_l smooth, and

$$\tilde{f}_{n+1}\big(s,\tilde{x}_{n+1}(s)\big) = \sum_{l=1}^{m}(1+s)^{\gamma_l}t_{n+1}^{\gamma_l}\tilde{g}_{n+1,l}(s) + C_0$$

with $\tilde{g}_{n+1,l}$ smooth. So, by the definition of the beta function in [Podlubny (1999)] and non-uniformly-weighted Sobolev spaces in (3.12)–(3.14), $\tilde{f}_{n+1}\big(\cdot,\tilde{x}_{n+1}(\cdot)\big) \in B_{\alpha-1,0}^{\lceil 2\gamma_1\rceil}(-1,1)$. Because of

$$\|\ \partial_s^{\lceil 2\gamma_1\rceil}\tilde{f}_{n+1}\ \|^2_{\omega^{\alpha-1+\lceil 2\gamma_1\rceil,\lceil 2\gamma_1\rceil}}$$
$$\le Ct_{n+1}^{\gamma_l}\ \|\ (1+s)^{\gamma_1-\lceil 2\gamma_1\rceil}\ \|^2_{\omega^{\alpha-1+\lceil 2\gamma_1\rceil,\lceil 2\gamma_1\rceil}}$$
$$= Ct_{n+1}^{\gamma_l}\cdot\frac{\Gamma(\alpha+\lceil 2\gamma_1\rceil)\Gamma(2\gamma_1-\lceil 2\gamma_1\rceil+1)}{\Gamma(\alpha+2\gamma_1+1)}, \tag{3.27}$$

$$\|\ 1\ \|^2_{\omega^{\alpha-1,0}} = \frac{\Gamma(\alpha)}{\Gamma(\alpha+1)} = \frac{1}{\alpha}, \tag{3.28}$$

i.e., both $\|\ \partial_s^{\lceil 2\gamma_1\rceil}\tilde{f}_{n+1}\ \|_{\omega^{\alpha-1+\lceil 2\gamma_1\rceil,\lceil 2\gamma_1\rceil}}$ and $\|\ 1\ \|_{\omega^{\alpha-1,0}}$ can be controlled by a constant. From Lemma 3.1, we can obtain that

$$|\ I_{n+1,1}\ | \le Ct_{n+1}^{\alpha+\gamma_1}N_J^{-\lceil 2\gamma_1\rceil}. \tag{3.29}$$

Using the theories of the Lagrange interpolation and of Gaussian quadrature [Quarteroni *et al.* (2007)], it gets

$$
\begin{aligned}
| I_{n+1,2} |= & C_\alpha \left| \sum_{j=0}^{N_J} \omega_j \left(\tilde{f}_{n+1}(s_j, \tilde{x}_{n+1}(s_j)) - F_{N_I}^P \tilde{f}_{n+1}(s_j) \right) \right| \\
\leq & C_\alpha \max_{0 \leq j \leq N_J} \left| f\Big(\frac{t_{n+1}}{2}(1+s_j), x\Big(\frac{t_{n+1}}{2}(1+s_j)\Big)\Big) - F_{N_I}^P \tilde{f}_{n+1}(s_j) \right| \sum_{j=0}^{N_J} \omega_j \\
\leq & \frac{1}{\Gamma(\alpha+1)} t_{n+1}^\alpha \cdot C(N_I) \parallel f^{(N_I)} \parallel_\infty h^{N_I} \\
\leq & (f, \alpha, N_I) t_{n+1}^\alpha h^{N_I},
\end{aligned}
\tag{3.30}
$$

where the positiveness of the Gaussian quadrature and $\sum_{j=0}^{N_J} \omega_j = \int_{-1}^{1} (1-s)^{\alpha-1} ds = 2^\alpha/\alpha$ have been used. The upper limit $C(N_I)$ does not depend on N_J, because it is independent of the position of $\frac{t_{n+1}}{2}(1+s_j)$, or it is uniform about N_J.

Thus, the proof is completed by combining (3.25), (3.29) and (3.30) together. □

Remark 3.1. Firstly, it is known that the two-parameter function of the Mittag-Leffler type plays a very important role in the fractional calculus, which is defined as

$$
E_{\alpha,\beta}(z) = \sum_{k=0}^{\infty} \frac{z^k}{\Gamma(\alpha k + \beta)}, \quad (\alpha > 0, \ \beta > 0). \tag{3.31}
$$

By the fundamental formula (1.82) in [Podlubny (1999)], we have

$$
{}_0D_t^{-\gamma}(t^{\beta-1} E_{\alpha,\beta}(\lambda t^\alpha)) = t^{\beta+\gamma-1} E_{\alpha,\beta+\gamma}(\lambda t^\alpha), \tag{3.32}
$$

where

$$
{}_0D_t^{-\gamma}(t^{\beta-1} E_{\alpha,\beta}(\lambda t^\alpha)) = \frac{1}{\Gamma(\gamma)} \int_0^t (t-\xi)^{\gamma-1} (\xi^{\beta-1} E_{\alpha,\beta}(\lambda \xi^\alpha)) d\xi.
$$

It can be seen that an equivalent expression of $t^{\beta-1} E_{\alpha,\beta}(\lambda t^\alpha)$ is $\sum_{l=1}^{\infty} t_l^{\gamma_l} g_l(t) + C_0$, $(m = \infty)$. (Actually, for $t^{\beta-1} E_{\alpha,\beta}(\lambda t^\alpha)$, the proof of Theorem 3.1 is also true.) In fact, in papers such as [Lubich (1986)], the discussed function $f(x)$ takes the similar assumption (more simpler than in the draft) $f(x) = x^{\beta-1} g(x)$, $\beta \neq 0, -1, -2, \cdots$; g is sufficiently differentiable.

Also, by Theorem 2.8 in [Deng (2010)], the knowledge of the smoothness properties of $f(t, x(t))$ is indispensable for the construction of good numerical schemes. Thus, for a numerical method with an convergence order N_I, it is not harsh to demand $f(t,x(t)) = \sum_{l=1}^{m} t_l^{\gamma_l} g_l(t) + C_0$, where the smallest index $\gamma_1 \geq N_I \geq 1$.

Remark 3.2. That the inequality (3.30) holds also needs that the order of the piecewise interpolating polynomials $N_I - 1$ is not very big, considering the Runge's phenomenon. In fact, we can see later that the choice of N_I is limited by the regularity of the function $f\big(t,x(t)\big)$, besides, it is not necessary to take N_I as a very big number.

Next we come to a result corresponding to the corrector formula. Since the proof of Theorem 3.2 is very similar to Theorem 3.1, we omit it.

Theorem 3.2. *Assume that* $f\big(t,x(t)\big) = \sum_{l=1}^{m} t^{\gamma_l} g_l(t) + C_0$ *with* g_l *smooth,* $\gamma_m > \cdots > \gamma_1 \geq N_I \geq 1$, C_0 *be a constant. Then there is a constant* C_2, *independent of* n, h *and* N_J, *such that*

$$\left| \frac{1}{\Gamma(\alpha)} \int_0^{t_{n+1}} (t_{n+1} - \tau)^{\alpha-1} f\big(\tau, x(\tau)\big) d\tau - C_\alpha \sum_{i=0}^{n+1} a_{i,n+1} f\big(t_i, x(t_i)\big) \right|$$
$$\leq C_2 t_{n+1}^{\alpha} \big(t_{n+1}^{\gamma_1} N_J^{-\lceil 2\gamma_1 \rceil} + h^{N_I}\big). \quad (3.33)$$

3.1.2.3 *Error analysis for the Jacobi predictor-corrector approach*

In the following, we give the error estimate of the Jacobi predictor-corrector approach. We can observe another advantage of the presented method via the following result, the convergence order, which has no relation with the fractional order α, is just the same as the number of the interpolation nodes N_I. In other words, one can get the desired convergence order just by choosing as a big number of interpolation nodes as one can.

Theorem 3.3. *Assume that* $f(t,x)$ *is Lipschitz continuous w.r.t. its second parameter* x. *If* $f\big(t,x(t)\big) = \sum_{l=1}^{m} t^{\gamma_l} g_l(t) + C_0$, *with* g_l *smooth,* $\gamma_m > \cdots > \gamma_1 \geq N_I \geq 1$, C_0 *be a constant, and* $h \leq 1$. *Then for the Jacobi predictor-corrector approach (3.7) and (3.8) and for some suitably chosen* T, *there is a constant* C *(depending on* α, N_I *and* f*), independent of* n, h *and* N_J, *such that*

$$\max_{1 \leq n+1 \leq N_E} \mid x(t_{n+1}) - x_{n+1} \mid \leq C h^{N_I}, \quad (3.34)$$

where $N_E = T/h$.

Proof. We use the mathematical induction to prove this theorem.

a) First we prove that (3.34) holds when $n+1 = N_I$: Denoting L as the Lipschitz constant of f w.r.t. its second parameter x, then by (3.3), (3.23) and Theorem 3.1, there exists

$$
\begin{aligned}
&\mid x(t_{n+1}) - x_{n+1}^P \mid \\
&= \frac{1}{\Gamma(\alpha)} \left| \int_0^{t_{n+1}} (t_{n+1} - \tau)^{\alpha-1} f(\tau, x(\tau))\, d\tau - C_\alpha \sum_{i=0}^{n} b_{i,n+1} f(t_i, x_i) \right| \\
&\leq \frac{1}{\Gamma(\alpha)} \left| \int_0^{t_{n+1}} (t_{n+1} - \tau)^{\alpha-1} f(\tau, x(\tau))\, d\tau - C_\alpha \sum_{i=0}^{n} b_{i,n+1} f(t_i, x(t_i)) \right| \\
&+ C_\alpha \left| \sum_{j=0}^{N_J} \omega_j \left(F_{N_I}^P \tilde{f}_{n+1}(s_j) - F_{A,N_I}^P \tilde{f}_{n+1}(s_j) \right) \right| \\
&\leq C_1 t_{n+1}^\alpha (t_{n+1}^{\gamma_1} N_J^{-\lceil 2\gamma_1 \rceil} + h^{N_I}) \\
&\quad + C_\alpha \max_{0 \leq j \leq N_J} \mid F_{N_I}^P \tilde{f}_{n+1}(s_j) - F_{A,N_I}^P \tilde{f}_{n+1}(s_j) \mid \cdot \sum_{j=0}^{N_J} \omega_j \\
&\leq C_1 t_{n+1}^\alpha (t_{n+1}^{\gamma_1} N_J^{-\lceil 2\gamma_1 \rceil} + h^{N_I}) \\
&\quad + \frac{t_{n+1}^\alpha}{\Gamma(\alpha+1)} C(N_I) \cdot \max_{0 \leq i \leq n} \mid f(t_i, x(t_i)) - f(t_i, x_i) \mid \\
&\leq C_1 t_{n+1}^\alpha (t_{n+1}^{\gamma_1} N_J^{-\lceil 2\gamma_1 \rceil} + h^{N_I}) + \frac{L \cdot t_{n+1}^\alpha}{\Gamma(\alpha+1)} C(N_I) \cdot \max_{0 \leq i \leq n} \mid x(t_i) - x_i \mid \\
&= C_1 t_{n+1}^\alpha (t_{n+1}^{\gamma_1} N_J^{-\lceil 2\gamma_1 \rceil} + h^{N_I}) + C_3 L t_{n+1}^\alpha \cdot \max_{0 \leq i \leq n} \mid x(t_i) - x_i \mid . \qquad (3.35)
\end{aligned}
$$

We have assumed that the starting error $\max_{0 \leq i \leq n = N_I - 1} \mid x(t_i) - x_i \mid$ is very small (not deteriorating the accuracy of the present algorithm), so the first term in the right hand side of the last formula in (3.35) plays the leading role. Thus

$$
\mid x(t_{n+1}) - x_{n+1}^P \mid \leq C_4 t_{n+1}^\alpha (t_{n+1}^{\gamma_1} N_J^{-\lceil 2\gamma_1 \rceil} + h^{N_I}). \qquad (3.36)
$$

Combining the above estimate with (3.3), (3.22) and Theorem 3.2,

$$
\begin{aligned}
&| x(t_{n+1}) - x_{n+1} | \\
&= \frac{1}{\Gamma(\alpha)} \Bigg| \int_0^{t_{n+1}} (t_{n+1}-\tau)^{\alpha-1} f(\tau, x(\tau))\, d\tau \\
&\quad -\Big(\frac{t_{n+1}}{2}\Big)^{\alpha}\Big[\sum_{i=0}^{n} a_{i,n+1} f(t_i, x_i) + a_{n+1,n+1} f(t_{n+1}, x_{n+1}^P)\Big]\Bigg| \\
&\le \Bigg| \frac{1}{\Gamma(\alpha)} \int_0^{t_{n+1}} (t_{n+1}-\tau)^{\alpha-1} f(\tau, x(\tau))\, d\tau - C_\alpha \sum_{i=0}^{n+1} a_{i,n+1} f(t_i, x(t_i)) \Bigg| \\
&\quad + C_\alpha \Bigg| \sum_{j=0}^{N_J} \omega_j \Big(F_{N_I} \tilde{f}_{n+1}(s_j) - F_{A,N_I} \tilde{f}_{n+1}(s_j) \Big) \Bigg| \\
&\le C_2 t_{n+1}^{\alpha} \big(t_{n+1}^{\gamma_1} N_J^{-\lceil 2\gamma_1 \rceil} + h^{N_I} \big) \\
&\quad + C_\alpha \max_{0 \le j \le N_J} | F_{N_I} \tilde{f}_{n+1}(s_j) - F_{A,N_I} \tilde{f}_{n+1}(s_j) | \cdot \sum_{j=0}^{N_J} \omega_j \\
&\le C_2 t_{n+1}^{\alpha} \big(t_{n+1}^{\gamma_1} N_J^{-\lceil 2\gamma_1 \rceil} + h^{N_I} \big) + \frac{t_{n+1}^{\alpha}}{\Gamma(\alpha+1)} C(N_I) \cdot \\
&\max\Big\{ \max_{0 \le i \le n} | f(t_i, x(t_i)) - f(t_i, x_i) |, | f(t_{n+1}, x(t_{n+1})) - f(t_{n+1}, x_{n+1}^P) | \Big\} \\
&\le C_2 t_{n+1}^{\alpha} \big(t_{n+1}^{\gamma_1} N_J^{-\lceil 2\gamma_1 \rceil} + h^{N_I} \big) \\
&\quad + \frac{L \cdot t_{n+1}^{\alpha}}{\Gamma(\alpha+1)} C(N_I) \cdot \max\Big\{ \max_{0 \le i \le n} | x(t_i) - x_i |, | x(t_{n+1}) - x_{n+1}^P | \Big\} \\
&\le C_2 t_{n+1}^{\alpha} \big(t_{n+1}^{\gamma_1} N_J^{-\lceil 2\gamma_1 \rceil} + h^{N_I} \big) \\
&\quad + C_5 L t_{n+1}^{\alpha} \cdot \max\Big\{ \max_{0 \le i \le n} | x(t_i) - x_i |, C_4 t_{n+1}^{\alpha} \big(t_{n+1}^{\gamma_1} N_J^{-\lceil 2\gamma_1 \rceil} + h^{N_I} \big) \Big\} \\
&\le C_6 (1 + L t_{n+1}^{\alpha}) t_{n+1}^{\alpha} \big(t_{n+1}^{\gamma_1} N_J^{-\lceil 2\gamma_1 \rceil} + h^{N_I} \big) \\
&= C_6 (1 + L t_{N_I}^{\alpha}) t_{N_I}^{\alpha} \big(t_{N_I}^{\gamma_1} N_J^{-\lceil 2\gamma_1 \rceil} + h^{N_I} \big) \\
&= C_6 (1 + L h^{\alpha} N_I^{\alpha}) h^{\alpha} N_I^{\alpha} \big(h^{\gamma_1} N_I^{\gamma_1} N_J^{-\lceil 2\gamma_1 \rceil} + h^{N_I} \big) \\
&\le C_6 (1 + L N_I^{\alpha}) N_I^{\alpha} h^{\alpha+N_I} \cdot \big(h^{\gamma_1 - N_I} N_I^{\gamma_1} N_J^{-\lceil 2\gamma_1 \rceil} + 1 \big) \\
&\le 2 C_6 (1 + L N_I^{\alpha}) N_I^{\alpha} h^{\alpha+N_I} \\
&= C_7 h^{\alpha+N_I} \\
&\le C_7 h^{N_I} := C h^{N_I},
\end{aligned}
\tag{3.37}
$$

where the last inequality holds since $h \le 1,\ \gamma_1 \ge N_I \ge 1,\ \alpha > 0$, besides, it can always be taken that $N_I/N_J^2 < 1$.

b) We further prove that (3.34) holds for any $n+1 > N_I$: Assume that $\max_{0\le i\le n+1} \mid x(t_{n+1}) - x_{n+1} \mid\le Ch^{N_I}$. Then we are going to prove that $\mid x(t_{n+2}) - x_{n+2} \mid\le Ch^{N_I}$. Since the discussions are similar to the ones given in **a)**, we briefly present them,

$$\begin{aligned}
&\mid x(t_{n+2}) - x^P_{n+2} \mid \\
&\le C_1 t^\alpha_{n+2}(t^{\gamma_1}_{n+2} N_J^{-\lceil 2\gamma_1\rceil} + h^{N_I}) + C_3 L t^\alpha_{n+2} \cdot \max_{0\le i\le n+1} \mid x(t_i) - x_i \mid \\
&\le C_1 T^\alpha (T^{\gamma_1} N_J^{-\lceil 2\gamma_1\rceil} + h^{N_I}) + C_3 L T^\alpha \cdot \max_{0\le i\le n+1} \mid x(t_i) - x_i \mid \\
&\le C_1 T^\alpha (T^{\gamma_1} N_J^{-\lceil 2\gamma_1\rceil} + h^{N_I}) + C_8 L T^\alpha h^{N_I} \\
&\le C_9 T^\alpha (T^{\gamma_1} N_J^{-\lceil 2\gamma_1\rceil} + (1+L) h^{N_I}) \\
&\le C_{10} T^\alpha (T^{\gamma_1} N_J^{\lceil 2\gamma_1\rceil} + h^{N_I}),
\end{aligned} \tag{3.38}$$

and

$$\begin{aligned}
\mid x(t_{n+2}) - x_{n+2} \mid &\le C_2 t^\alpha_{n+2}(t^{\gamma_1}_{n+2} N_J^{-\lceil 2\gamma_1\rceil} + h^{N_I}) \\
&\quad + C_5 L t^\alpha_{n+2} \cdot \max\{ \max_{0\le i\le n+1} \mid x(t_i) - x_i \mid, \mid x(t_{n+2}) - x^P_{n+2} \mid\} \\
&\le C_2 T^\alpha (T^{\gamma_1} N_J^{-\lceil 2\gamma_1\rceil} + h^{N_I}) \\
&\quad + C_5 L T^\alpha \cdot \max\{ C h^{N_I}, C_{10} T^\alpha (T^{\gamma_1} N_J^{-\lceil 2\gamma_1\rceil} + h^{N_I})\}.
\end{aligned} \tag{3.39}$$

By choosing T sufficiently small, we can make sure that $C_{10}T^\alpha$, $C_5 C L T^\alpha$ as well as $2C_2 T^\alpha$ are all bounded by $C/2$. Having fixed the value of T, by choosing a suitable N_J, we can make $T^{\gamma_1} N_J^{-\lceil 2\gamma_1\rceil} < h^{N_I}$ hold. Thus we have

$$\begin{aligned}
&\mid x(t_{n+2}) - x_{n+2} \mid \\
&\le C_2 T^\alpha (h^{\gamma_1} + h^{N_I}) + C_5 L T^\alpha \cdot \max\{ C h^{N_I}, C_{10} T^\alpha (h^{\gamma_1} + h^{N_I})\} \\
&\le 2C_2 T^\alpha h^{N_I} + C_5 L T^\alpha \cdot \max\{ C h^{N_I}, 2C_{10} T^\alpha h^{N_I}\} \\
&\le C h^{N_I}.
\end{aligned} \tag{3.40}$$

Now we complete the proof of Theorem 3.3. □

Theorem 3.3 tells that the error of the Jacobi predictor-corrector depends on the smoothness of $f(t, x(t))$, and it is insensitive to the regularity of the solution $x(t)$. It also implies that the order of the convergence is N_I.

In the following, we show a direct corollary of Theorem 3.3:

Corollary 3.2. *Assume that $f(t,x)$ is Lipschitz continuous w.r.t. its second parameter x and take $N_I = 2$. If $f(t, x(t)) \in \mathbb{C}^2[0,T]$ for some suitably* *chosen T, and $h \leq 1$, then there is a constant C (depending on α, N_I and f), independent of n, h and N_J, such that*

$$\max_{1\leq n+1\leq N_E} \mid x(t_{n+1}) - x_{n+1} \mid\leq Ch^2, \tag{3.41}$$

where $N_E = T/h$.

Under the same conditions as in Corollary 3.2, the convergence orders of Adams-type predictor-corrector approach in [Diethelm *et al.* (2002, 2004)] and the Improved-Adams' methods in [Deng (2007a)] are respectively $\min\{2, 1+\alpha\}$ and $\min\{2, 1+2\alpha\}$, which depend on the fractional order α and are not able to exceed 2. While Corollary 3.2 implies that the error of the Jacobi predictor-corrector method can have no relation with α, and its convergence order can at least reach the optimal one of those in [Diethelm *et al.* (2002, 2004)] and [Deng (2007a)]. We say that this is the third contribution of the method discussed here.

Remark 3.3. We have referred in the previous analysis that N_I can not be very large due to the Runge's phenomenon. In reality, by Theorem 3.3, the choice of N_I is limited by the regularity of $f\big(t, x(t)\big)$. While considering the two main contributions of the method, as we mentioned above, the computational cost being $O(N_E)$, as well as the freedom of the error estimate about α, we don't think it is very necessary to improve the order of the convergence to a very high level. So, even though $f\big(t, x(t)\big)$ is very smooth, there is no need to let N_I be a very large value.

Remark 3.4. In practical computations, the Improved-Adams' methods proposed in [Deng (2007a)] can be used to start the algorithm. We can take the step-length h_0 discussed in the algorithm given in Subsec. 3.1.1 as $h \cdot 10^{-k}$, where $k \geq 1$ is a given integer. Then by the result in [Deng (2007a)], there exists

$$\begin{aligned}
&\max_{0\leq i\leq N_I-1} \mid x(t_i) - x_i \mid \\
&= O\left(h_0^{\min\{1+2\alpha,2\}}\right) \\
&= \begin{cases} 10^{-k(1+2\alpha)} \cdot O(h^{1+2\alpha}), & \text{if } \ 0 < \alpha \leq 0.5; \\ 10^{-2k} \cdot O(h^2), & \text{if } \ \alpha > 0.5. \end{cases}
\end{aligned} \tag{3.42}$$

If taking $h = 10^{-m}$, where m is a given positive integer, by a simple computation, we obtain that $\mid x(t_{n+1}) - x_{n+1} \mid= O(h^{N_I})$, as long as the integers k and m satisfy

$$N_I < \begin{cases} (1+2\alpha)(1+\frac{k}{m}), & \text{if } 0 < \alpha \leq 0.5; \\ 2 + \frac{2k}{m}, & \text{if } \ \alpha > 0.5. \end{cases} \tag{3.43}$$

3.1.3 *Modifications of the algorithm*

Since the error estimate depends on the property of $f(t,(x(t)))$, we shall simply discuss in the following case that the smoothness of f at $t = 0$ is weaker than other places. More information about the smoothness of f and its relationship with the property of the solution can be found in [Lubich (1983, 1985); Deng (2010)]. Another issue we will also mention is how to use this algorithm when α is very close to zero.

3.1.3.1 *The function f is not very smooth at the starting point*

When the smoothness of f is weaker at the initial time zero than other time, by Theorem 3.3, the convergence order N_I can not be improved directly. A sensible way is to divide the interval $[0, T]$ into two parts $[0, T_0]$ and $[T_0, T]$, where T_0 is a small positive real number, thus f would be smooth on the interval $[T_0, T]$, and Theorem 3.3 can be employed. For the small interval $[0, T_0]$, we use the Jacobi-Gauss-Lobatto quadrature with the weight function $\omega(s) = 1$. For the remaining part $[T_0, T]$, the algorithm provided above is employed. That is,

$$\begin{aligned}
x(t_{n+1}) &= \sum_{k=0}^{\lceil\alpha\rceil-1} \frac{t_{n+1}^k}{k!} x_0^{(k)} + \frac{1}{\Gamma(\alpha)} \int_0^{T_0} (t_{n+1}-\tau)^{\alpha-1} f(\tau, x(\tau)) d\tau \\
&\quad + \frac{1}{\Gamma(\alpha)} \int_{T_0}^{t_{n+1}} (t_{n+1}-\tau)^{\alpha-1} f(\tau, x(\tau)) d\tau \\
&= \sum_{k=0}^{\lceil\alpha\rceil-1} \frac{t_{n+1}^k}{k!} x_0^{(k)} + \frac{1}{\Gamma(\alpha)} \int_0^{T_0} (t_{n+1}-\tau)^{\alpha-1} f(\tau, x(\tau)) d\tau \\
&\quad + \frac{1}{\Gamma(\alpha)} \left(\frac{t_{n+1}-T_0}{2}\right)^{\alpha} \int_{-1}^{1} (1-s)^{\alpha-1} \tilde{f}_{n+1}(s, \tilde{x}_{n+1}(s)) ds \\
&\approx \sum_{k=0}^{\lceil\alpha\rceil-1} \frac{t_{n+1}^k}{k!} x_0^{(k)} + \frac{1}{\Gamma(\alpha)} \sum_{j=0}^{\tilde{N}_J} \tilde{\omega}_j (t_{n+1}-\tau_j)^{\alpha-1} f(\tau_j, x(\tau_j)) \\
&\quad + \frac{1}{\Gamma(\alpha)} \left(\frac{t_{n+1}-T_0}{2}\right)^{\alpha} \sum_{j=0}^{N_J} \omega_j \tilde{f}_{n+1}(s_j, \tilde{x}_{n+1}(s_j)),
\end{aligned} \tag{3.44}$$

where $\tilde{N}_J$, $\{\tilde{\omega}_j\}_{j=0}^{\tilde{N}_J}$ and $\{\tau_j\}_{j=0}^{\tilde{N}_J}$ correspond to the number, the weights, and the values of the Gauss-Lobatto nodes with the weight $\omega(s) = 1$ in the interval $[0, T_0]$, respectively. The values of $\{f(\tau_j, x(\tau_j))\}_{j=0}^{\tilde{N}_J}$ can be computed as in the starting procedure. Since f and x are continuous in the interval $[0, T_0]$, by the theory of Gaussian quadrature [Quarteroni *et al.* (2007)] and the analysis above, we can see that if $\tilde{N}_J$ is a big number, then the accuracy of the total error still can be remained.

3.1.3.2 *The value of α is very small*

In our opinion, the purpose of the correction procedure, including that in [Diethelm *et al.* (2002)], [Deng (2007a)] as well as in this work, is to try to impair the role of $f(t_{n+1}, x_{n+1}^P)$ played in the numerical integration, since $f(t_{n+1}, x_{n+1}^P)$ is also predicted using the numerical integration, and might be unstable due to the nonlinearity of f. During the Jacobi predictor-corrector procedure, when α becomes bigger, the weight of the Jacobi-Gauss-Lobatto quadrature at the endpoint of the right hand side becomes smaller, the provided algorithm becomes more robust. Whereas α is very small (or $\alpha - 1$ is very close to -1), say, α smaller than 0.1, the weight at the endpoint of the right hand side of the interval will become much bigger than at other places, which may impact the robustness of the algorithm. There are two choices to deal with this problem: one is to try to avoid using the high order interpolation in the algorithm; another is to divide the interval $[0, T]$ into two subintervals $[0, T_1]$ and $[T_1, T]$, and let

$$
\begin{aligned}
x(t) &= \sum_{k=0}^{\lceil\alpha\rceil-1} \frac{t^k}{k!} x_0^{(k)} + \frac{1}{\Gamma(\alpha)} \int_0^{T_1} (t-\tau)^{\alpha-1} f(\tau, x(\tau))\, d\tau \\
&\quad + \frac{1}{\Gamma(\alpha)} \int_{T_1}^{t} (t-\tau)^{\alpha-1} f(\tau, x(\tau))\, d\tau \\
&= \sum_{k=0}^{\lceil\alpha\rceil-1} \frac{t^k}{k!} x_0^{(k)} + \frac{1}{\Gamma(\alpha)} \int_0^{T_1} (t-\tau)^{\alpha-1} f(\tau, x(\tau))\, d\tau \\
&\quad + \frac{1}{\Gamma(\alpha)} \left(\frac{t-T_1}{2}\right)^{\alpha} \int_{-1}^{1} (1-s)^{\alpha-1} \tilde{f}(s, \tilde{x}(s))\, ds \\
&\approx \sum_{k=0}^{\lceil\alpha\rceil-1} \frac{t^k}{k!} x_0^{(k)} + \frac{1}{\Gamma(\alpha)} \sum_{j=0}^{\tilde{N}_J} \tilde{\omega}_j (t-\tau_j)^{\alpha-1} f(\tau_j, x(\tau_j)) \\
&\quad + \frac{1}{\Gamma(\alpha)} \left(\frac{t-T_1}{2}\right)^{\alpha} \sum_{j=0}^{N_J} \omega_j \tilde{f}(s_j, \tilde{x}(s_j)),
\end{aligned}
\tag{3.45}
$$

where $\left\{f(\tau_j, x(\tau_j))\right\}_{j=0}^{\tilde{N}_J}$ in the second term of the right hand side of the last formula can be computed by interpolation.

3.1.4 *Numerical results*

We only consider the examples with $0 < \alpha < 2$, since the algorithm will be more robust for $\alpha \geq 2$. All numerical computations are done in Matlab 7.5.0 on a normal laptop with 1GB of memory.

Example 3.1. The first example is to verify the high order convergence of the Jacobi predictor-corrector method by

$$ {}_0^C D_t^\alpha x(t) = \frac{\Gamma(9)}{\Gamma(9-\alpha)} t^{8-\alpha} + \frac{9}{4}\Gamma(\alpha+1) + t^8 + \frac{9}{4} t^\alpha - x(t) \tag{3.46} $$

with the initial condition(s) $x(0) = 0$ (and $x'(0) = 0$ if $1 < \alpha \leq 2$).

The exact solution of this initial value problem is

$$ x(t) = t^8 + \frac{9}{4} t^\alpha, \tag{3.47} $$

so, the right function $f\big(t, x(t)\big) = \frac{\Gamma(9)}{\Gamma(9-\alpha)} t^{8-\alpha} + \frac{9}{4}\Gamma(\alpha+1)$. Since $8-\alpha > 6$ if $0 < \alpha < 2$, by Theorem 3.3, we expect the convergence order can reach to as high as 6.

We start the procedure with the Improved-Adams' methods in [Deng (2007a)] as discussed in Remark 3.4, i.e., the values of $x(t)$ at $t_0, t_1, \cdots, t_{N_I-1}$ are computed by the Improved-Adams' methods. The convergence orders are verified at $T = 1$, and the number of the Jacobi nodes is taken as $N_J + 1 = 20 + 1 = 21$. The number of the interpolating nodes N_I is respectively taken as 2, 3 and 4 expecting that the corresponding convergence order is also 2, 3 and 4. The numerical results of the maximum errors for the Jacobi predictor-corrector approach are showed in the following tables, where 'rate' means the convergence order. From the results in Table 3.1 to Table 3.3, we can see that the datum confirm the theoretical results.

The results in Table 3.3 for $\alpha = 0.1$ show that we must be more careful to use the provided algorithm when α is very small (letting N_I be small or dividing the original interval into subintervals). However, we still confirm the convergence order by taking small T ($T = 0.1$) for $N_I = 4$ in Table 3.4.

Table 3.5 shows the CPU time and the steps N_E needed to solve (3.46) when $\alpha = 0.5$ and the maximum error is 1×10^{-3}, for the fractional Adams' methods in [Diethelm *et al.* (2002, 2004)], the Improved-Adams methods in [Deng (2007a)] and the Jacobi predictor-corrector approach here when N_I = 2, 3, 4, 5, 6. The consumed CPU time presented in Table 3.5

Table 3.1: The maximum errors for (3.46) when $t \in [0, 1]$ and $N_I = 2$.

h	$\alpha = 0.1$	rate	$\alpha = 0.3$	rate	$\alpha = 0.5$	rate	$\alpha = 0.9$	rate
1/40	2.59 1e-2	-	4.81 1e-3	-	2.32 1e-3	-	1.91 1e-3	-
1/80	5.62 1e-3	2.20	9.96 1e-4	2.27	6.58 1e-4	1.82	6.03 1e-4	1.66
1/160	1.12 1e-3	2.33	1.81 1e-4	2.46	1.20 1e-4	2.45	1.66 1e-4	1.86
h	$\alpha = 1.2$	rate	$\alpha = 1.5$	rate	$\alpha = 1.7$	rate	$\alpha = 1.9$	rate
1/40	3.09 1e-3	-	3.62 1e-3	-	3.42 1e-3	-	3.07 1e-3	-
1/80	8.07 1e-4	1.94	7.33 1e-4	2.30	6.20 1e-4	2.46	8.95 1e-4	1.78
1/160	1.67 1e-4	2.28	2.00 1e-4	1.87	1.70 1e-4	1.87	2.01 1e-4	2.16

Table 3.2: The maximum errors for (3.46) when $t \in [0, 1]$ and $N_I = 3$.

h	$\alpha = 0.1$	rate	$\alpha = 0.3$	rate	$\alpha = 0.5$	rate	$\alpha = 0.9$	rate
1/40	3.32 1e-3	-	5.08 1e-4	-	1.96 1e-4	-	1.46 1e-4	-
1/80	3.63 1e-4	3.19	5.04 1e-5	3.33	2.64 1e-5	2.89	2.24 1e-5	2.70
1/160	3.60 1e-5	3.34	4.35 1e-6	3.54	2.35 1e-6	3.49	3.22 1e-6	2.80
h	$\alpha = 1.2$	rate	$\alpha = 1.5$	rate	$\alpha = 1.7$	rate	$\alpha = 1.9$	rate
1/40	2.33 1e-4	-	2.57 1e-4	-	2.36 1e-4	-	2.05 1e-4	-
1/80	3.08 1e-5	2.92	2.57 1e-5	3.33	2.30 1e-5	3.36	3.32 1e-5	2.62
1/160	3.06 1e-6	3.33	3.58 1e-6	2.84	3.37 1e-6	2.77	3.61 1e-6	3.20

Table 3.3: The maximum errors for (3.46) when $t \in [0, 1]$ and $N_I = 4$.

h	$\alpha = 0.1$	rate	$\alpha = 0.3$	rate	$\alpha = 0.5$	rate	$\alpha = 0.9$	rate
1/20	6.33 1e-3	-	9.83 1e-4	-	2.56 1e-4	-	2.82 1e-5	-
1/40	3.77 1e-4	4.07	4.57 1e-5	4.43	7.21 1e-6	5.15	4.71 1e-6	2.58
1/80	2.89 1e-5	3.70	1.97 1e-6	4.54	2.71 1e-7	4.74	4.19 1e-7	3.49
1/160	3.89 1e-4	-3.75	6.35 1e-8	4.95	1.80 1e-8	3.91	3.54 1e-8	3.57
1/320	3.69 1e-1	-9.89	2.51 1e-9	4.66	1.34 1e-9	3.75	2.33 1e-9	3.92

shows that the fractional Adams methods generate the numerical solution with the same accuracy as the other two methods, but use much less CPU time. This advantage is more obvious as the terminate time goes long. It demonstrates the efficiency of the Jacobi predictor-corrector method.

Table 3.4: The maximum errors for (3.46) when $\alpha = 0.1$, $N_I = 4$.

h	$T = 0.1, N_I = 4$	rate
T/20	3.90 1e-11	-
T/40	2.51 1e-12	3.96
T/80	2.47 1e-13	3.35
T/160	1.15 1e-14	4.42

Table 3.5: The CPU time (sec) and the steps N_E needed to solve (3.46) when $\alpha = 0.5$ and the maximum error is 1×10^{-3}, for the fractional Adams methods in [Diethelm *et al.* (2002)], the Improved-Adams' methods in [Deng (2007a)] and the Jacobi predictor-corrector approach here when $N_I = 2, \cdots, 6$.

methods		terminal time			
		$T = 1.5$		$T = 2.0$	
		N_E	CPU time (sec)	N_E	CPU time (sec)
fractional Adams		1711	8.38 1e+1	8200	1.93 1e+3
Improved Adams		591	4.92 1e+0	1926	5.39 1e+1
Jacobi-	$N_I = 2$	286	1.00 1e+0	955	3.56 1e+0
predictor-	$N_I = 3$	72	4.53 1e -1	132	5.00 1e -1
	$N_I = 4$	30	9.38 1e -2	46	1.88 1e -1
corrector	$N_I = 5$	21	6.25 1e -2	31	1.09 1e -1
methods	$N_I = 6$	15	6.25 1e -2	22	1.09 1e -1

Example 3.2. We study the following equation as an example to show the robustness of the algorithm,

$$ {}_0^C D_t^\alpha x(t) = -x(t), \qquad x(0) = 1, \quad x'(0) = 0 \;\; (\text{if } 1 < \alpha \le 2). \tag{3.48}$$

It is well known that the exact solution of (3.48) is

$$x(t) = E_\alpha(-t^\alpha), \tag{3.49}$$

where

$$E_\alpha(z) = \sum_{k=0}^{\infty} \frac{z^k}{\Gamma(\alpha k + 1)}, \quad Re(\alpha) > 0, z \in \mathbb{C} \tag{3.50}$$

is the Mittag-Leffler function of order α. It is obvious that neither $x(t)$ nor $D_*^\alpha x(t)$ has a bounded first (second) derivative at $t = 0$ when $0 < \alpha \le 1$ $(1 < \alpha \le 2)$, so we deal with (3.48) as we discussed in Subsec.3.1.3. Here we take $T_0 = 0.1$, $N_J = 26$, $\tilde{N}_J = 2N_J$, where T_0, N_J and $\tilde{N}_J$ are defined as in Subsec.3.1.3, and the exact solutions are calculated using the function "$mlf.m$" [Podlubny (2012)]. The convergence order is also simply verified in Tables 3.6 and 3.7.

Table 3.6: The maximum errors for (3.48) when $t \in [0, 1.1]$ and $N_I = 2$.

h	$\alpha = 0.2$	rate	$\alpha = 0.5$	rate	$\alpha = 1.2$	rate	$\alpha = 1.8$	rate
1/10	4.84 1e-3	-	2.30 1e-3	-	1.20 1e-4	-	3.72 1e-4	-
1/20	1.49 1e-3	1.70	5.10 1e-4	2.17	3.47 1e-5	1.79	1.06 1e-4	1.81
1/40	4.04 1e-4	1.88	1.02 1e-4	2.33	7.83 1e-6	2.15	2.62 1e-5	2.02
1/80	9.96 1e-5	2.02	1.89 1e-5	2.43	1.87 1e-6	2.07	6.35 1e-6	2.05

Table 3.7: The maximum errors for (3.48) when $t \in [0, 1.1]$ and $N_I = 3$.

h	$\alpha = 0.2$	rate	$\alpha = 0.5$	rate	$\alpha = 1.2$	rate	$\alpha = 1.8$	rate
1/10	2.77 1e-3	-	7.30 1e-4	-	1.11 1e-5	-	1.64 1e-5	-
1/20	6.04 1e-4	2.20	1.14 1e-4	2.68	3.23 1e-6	1.78	3.00 1e-6	2.45
1/40	1.06 1e-4	2.51	1.43 1e-5	2.99	5.48 1e-7	2.56	4.64 1e-7	2.69
1/80	1.29 1e-5	3.04	1.40 1e-6	3.36	8.88 1e-8	2.63	5.91 1e-8	2.97

Further we compute (3.48) with a large time interval, $T = 50$; the parameters N_I, N_J, $\tilde{N}_J$ are taken the same as the above ones and h is taken as $1/10$. The exact solutions and relative errors are shown in Fig. 3.1 with $\alpha = 0.5$. It can be seen that the relative errors in the interval are less than $O(10^{-4})$ when time is long, which suggests that the method discussed is suitable for the long-time computation.

3.2 Local discontinuous Galerkin methods for fractional ordinary differential equations

In this section, we develop a LDG methods for the one-term and multi-term initial value problems for fractional ODEs. For ease of presentation, we shall focus the discussion on the following two types of fractional ODEs

$$ {}_0^C D_t^\alpha x(t) = f(x, t), \tag{3.51} $$

and

$$ {}_0^C D_t^\alpha x(t) + d(t)\frac{d^m x(t)}{dt^m} = f(x, t), \tag{3.52} $$

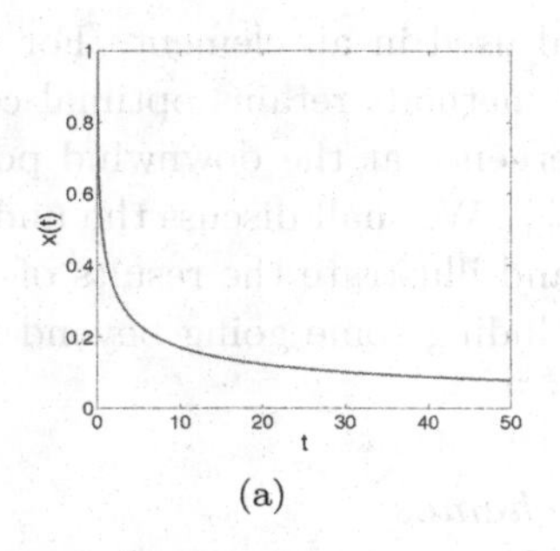

(a)

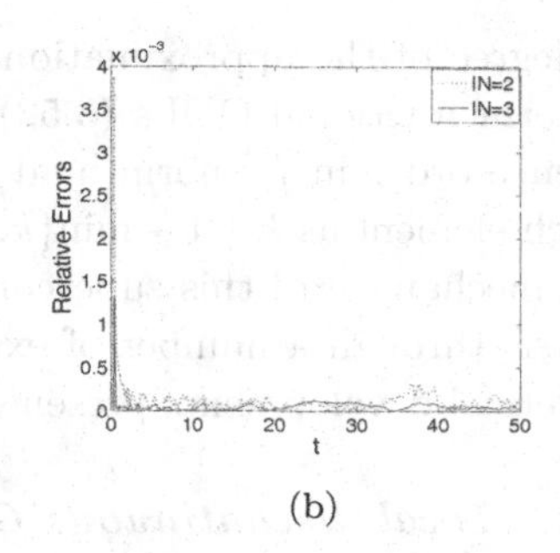

(b)

Fig. 3.1: (a) Exact solution of (3.48) for $\alpha = 0.5$; (b) Relative errors with $\alpha = 0.5$, $h = 1/10$, $N_I = 2$ or 3.

where $\alpha \in (0, 1]$, and m is a positive integer and also include examples of $\alpha \in [1, 2]$ in the interest of generalization. For (3.51) and (3.52), the initial conditions can be specified exactly as for the classical ODEs, i.e., the values of $x^{(j)}(a)$ must be given, where $j = 0, 1, 2, \cdots, \lfloor\alpha\rfloor$ for (3.51) and $j = 0, 1, 2, \cdots, \max\{\lfloor\alpha\rfloor, m-1\}$ for (3.52). Here $\lfloor\alpha\rfloor = \{z \in \mathbb{N} : 0 < \alpha - z \leq 1\}$.

The LDG methods have been well developed to solve classical differential equations [Hesthaven and Warburton (2008)], initiated for the classical ODEs [Delfour *et al.* (1981)] with substantial later work, mostly related to DG methods for the related Volterra integro-differential equation, including a priori analysis [Schötzau and Schwab (2000)], *hp*-adaptive methods [Brunner and Schötzau (2006); Mustapha *et al.* (2011)] and recent work on superconvergence in the h-version [Mustapha (2013)]. This has been extended to approximate the fractional spatial derivatives [Deng and Hesthaven (2013)] to solve fractional diffusion equation by using the idea of LDG methods [Bassi and Rebay (1997); Cockburn and Shu (1998)]. We will discuss LDG methods to allow for the approximate solution of general fractional ODEs. All the advantages/characteristics of the spatial DG methods carries over to this case with a central one being the ability to solve the equation interval by interval when the upwind flux, taking the value of $x(t)$ at a discontinuity point t_j as $x(t_j^-)$, is used. However, this is a natural choice, since for the initial value problems for the fractional (or classical) ODEs, the information travels forward in time. This implies that we just invert a local low order matrix rather than a global *full* matrix. The LDG methods for the first order fractional ODEs (3.51) have optimal order of convergence $k+1$ in the L^2 norm and we observe superconvergence of order $k + 1 + \min\{k, \alpha\}$ at the downwind point of each element. Here k is the degree of the approximation polynomial used in an element. For the two-term fractional ODEs (3.52), the LDG methods retains optimal convergence order in L^2 norm, and superconvergence at the downwind point of each element as $k+1+\min\{k, \max\{\alpha, m\}\}$. We shall discuss the underlying mechanism of this superconvergence and illustrate the results of the analysis through a number of examples, including some going beyond the theoretical developments presented here.

3.2.1 *Local discontinuous Galerkin schemes*

The basic idea in the design of the LDG schemes is to rewrite the fractional ODEs as a system of the first order classical ODEs and a fractional integral. Since the integral operator naturally connects the discontinuous function, we need not add a penalty term or introduce a numerical fluxes for the integral equation. However, for the first order ODEs, upwind fluxes are used. In this subsection, we present the LDG schemes, prove numerical stability and discuss the underlying mechanism of superconvergence.

3.2.1.1 *LDG schemes*

We consider (3.51) and rewrite it as

$$\begin{cases} x_1(t) - \frac{dx_0(t)}{dt} = 0, \\ {}_0D_t^{-(1-\alpha)} x_1(t) = f(x,t), \quad \alpha \in (0,1], \\ x_0(0) = x_0, \end{cases} \tag{3.53}$$

where

$${}_0D_t^{-(1-\alpha)} x_1(t) = \frac{1}{\Gamma(1-\alpha)} \int_0^t (t-\xi)^{-\alpha} x_1(\xi) d\xi.$$

We consider a scheme for solving (3.53) in interval $[0,T]$. Given the nodes $0 = t_0 < t_1 < \cdots < t_{n-1} < t_n = T$, we define the mesh $\mathcal{T} = \{I_j = (t_{j-1}, t_j), j = 1, 2, \cdots, n\}$ and set $h_j := |I_j| = t_j - t_{j-1}$ and $h := \max_{j=1}^n h_j$. Associating with the mesh $\mathcal{T}$, we define the broken Sobolev spaces

$$L^2(\Omega, \mathcal{T}) := \{v : \Omega \to R \,|\, v|_{I_j} \in L^2(I_j), j = 1, 2, \cdots, n\}$$

and

$$H^1(\Omega, \mathcal{T}) := \{v : \Omega \to R \,|\, v|_{I_j} \in H^1(I_j), j = 1, 2, \cdots, n\}.$$

For a function $v \in H^1(\Omega, \mathcal{T})$, we denote the one-sided limits at the nodes t_j by

$$v^{\pm}(t_j) = v(t_j^{\pm}) := \lim_{t \to t_j^{\pm}} v(t).$$

Assume that the solutions belong to the corresponding spaces:

$$(x_0(t), x_1(t)) \in H^1(\Omega, \mathcal{T}) \times L^2(\Omega, \mathcal{T}).$$

We further define X_i as the approximation functions of x_i respectively, in the finite dimensional subspace $V \subset H^1(\Omega, \mathcal{T})$; and choose V to be the space of discontinuous, piecewise polynomial functions

$$V = \{v : \Omega \to R \,\big|\, v|_{I_j} \in \mathcal{P}^k(I_j),\, j = 1, 2, \cdots, n\},$$

where $\mathcal{P}^k(I_j)$ denotes the set of all polynomials of degree less than or equal to k on I_j. Using the upwind fluxes for the first order classical ODEs and discretizing the integral equation, we seek $X_i \in V$ such that for all $v_i \in V$, and $j = 1, 2, \cdots, n$, the following holds

$$\begin{cases} (X_1, v_0)_{I_j} + \left(X_0, \dfrac{dv_0}{dt}\right)_{I_j} - (X_0(t_j^-)v_0(t_j^-) - X_0(t_{j-1}^-)v_0(t_{j-1}^+)) = 0, \\ \left({}_0D_t^{-(1-\alpha)} X_1, v_1\right)_{I_j} = \left(f(X_0, t), v_1\right)_{I_j}, \\ X_0(t_0^-) = x_0. \end{cases} \tag{3.54}$$

Remark 3.5. If take $\alpha \in (0, 1]$ and $m = 1$ in (3.52), the scheme is

$$\begin{cases} (X_1, v_0)_{I_j} + \left(X_0, \dfrac{dv_0}{dt}\right)_{I_j} - (X_0(t_j^-)v_0(t_j^-) - X_0(t_{j-1}^-)v_0(t_{j-1}^+)) = 0, \\ \left({}_0D_t^{-(1-\alpha)} X_1 + X_1, v_1\right)_{I_j} = \left(f(X_0, t), v_0\right)_{I_j}, \\ X_0(t_0^-) = x_0. \end{cases}$$

The scheme is clearly consistent, i.e., the exact solutions of the corresponding models satisfy (3.54). Furthermore, since an upwind flux is used the solutions can be computed interval by interval and if $f(x, t)$ is a linear function in x, we just need to invert a small matrix in each interval to recover the solution.

3.2.1.2 *Numerical stability*

We consider the question of stability for the linear case of (3.54):

$$\begin{cases} {}_0^C D_t^\alpha x(t) = Ax(t) + B(t), \qquad \alpha \in (0, 1), \\ x(0) = x_0, \end{cases} \tag{3.55}$$

where A is a negative constant and $B(t)$ is sufficiently regular to ensure existence and uniqueness. The numerical scheme of (3.55) is to find $X_i \in V$ such that

$$\begin{cases} \left(X_0(t_j^-)v_0(t_j^-) - X_0(t_{j-1}^-)v_0(t_{j-1}^+)\right) - \left(X_1, v_0\right)_{I_j} - \left(X_0, \dfrac{dv_0}{dt}\right)_{I_j} = 0, \\ \left({}_0D_t^{-(1-\alpha)}X_1, v_1\right)_{I_j} = \left(AX_0 + B, v_1\right)_{I_j}, \\ X_0(t_0^-) = x_0, \end{cases} \tag{3.56}$$

holds for all $v_i \in V$.

First we present a lemma, its proof can be found in Sec. 4.3 of Chap. 5.

Lemma 3.2. *For $\beta > 0$ and $\alpha \in (0,1)$, it holds that*

$$\left({}_0D_t^{-\beta}u, v\right)_{L^2([0,t_j])} = \left(u, {}_tD_{t_j}^{-\beta}v\right)_{L^2([0,t_j])}; \tag{3.57}$$

$$\begin{aligned} \left({}_0D_t^{-(1-\alpha)}v, v\right)_{L^2([0,t_j])} &= \left({}_0D_t^{-\frac{1-\alpha}{2}}v, {}_tD_{t_j}^{-\frac{1-\alpha}{2}}v\right)_{L^2([0,t_j])} \\ &\geq \sin(\alpha\pi/2)\|{}_0D_t^{-\frac{1-\alpha}{2}}v\|^2_{L^2([0,t_j])}. \end{aligned} \tag{3.58}$$

Let $\widetilde{X}_i \in V$ be the approximate solution of X_i and denote $e_{X_i} := \widetilde{X}_i - X_i$ as the numerical errors. Stability of (3.56) is established in the following theorem.

Theorem 3.4. *Scheme (3.56) is L^∞ stable; and the numerical errors satisfy*

$$\begin{aligned} e^2_{X_0}(t_n^-) \leq e^2_{X_0}(t_0^-) &- \sum_{i=1}^{n}(\delta e_{X_0}(t_{i-1}))^2 \\ &+ \frac{2\sin(\alpha\pi/2)}{A}\|{}_0D_t^{-\frac{1-\alpha}{2}}e_{X_1}\|^2_{L^2([0,t_n])}, \end{aligned} \tag{3.59}$$

where $\delta e_{X_i}(t_{i-1}) = e_{X_i}(t_{i-1}^+) - e_{X_i}(t_{i-1}^-)$.

Since $A < 0$, the scheme is dissipative.

Proof. From (3.56), we recover the error equation

$$\begin{cases} \left(e_{X_0}(t_i^-)v_0(t_i^-) - e_{X_0}(t_{i-1}^-)v_0(t_{i-1}^+)\right) - \left(e_{X_1}, v_0\right)_{I_i} - \left(e_{X_0}, \frac{dv_0}{dt}\right)_{I_i} = 0, \\ -\frac{1}{A}\left({}_0D_t^{-(1-\alpha)}e_{X_1}, v_1\right)_{I_i} = -\left(e_{X_0}, v_1\right)_{I_i}, \end{cases} \tag{3.60}$$

for all $v_i \in V$. Taking $v_0 = e_{X_0}$, $v_1 = e_{X_1}$, and adding the two equations, we obtain

$$\left(e_{X_0}^2(t_i^-) - e_{X_0}(t_{i-1}^-)e_{X_0}(t_{i-1}^+)\right) - \frac{1}{2}\left(e_{X_0}^2(t_i^-) - e_{X_0}^2(t_{i-1}^+)\right)$$
$$- \frac{1}{A}\left({}_0D_t^{-(1-\alpha)} e_{X_1}, e_{X_1}\right)_{I_i} = 0.$$

Summing the equations for $i = 1, 2, \cdots, n$ leads to

$$e_{X_0}^2(t_n^-) - e_{X_0}^2(t_0^-) + \sum_{i=1}^{n}\left(e_{X_0}^2(t_{i-1}^-) - 2e_{X_0}(t_{i-1}^-)e_{X_0}(t_{i-1}^+) + e_{X_0}^2(t_{i-1}^+)\right)$$
$$-\frac{2}{A}\left({}_0D_t^{-(1-\alpha)} e_{X_1}, e_{X_1}\right)_{[0,I_n]} = 0,$$

and

$$e_{X_0}^2(t_n^-) = e_{X_0}^2(t_0^-) - \sum_{i=1}^{n}\left(e_{X_0}(t_{i-1}^+) - e_{X_0}(t_{i-1}^-)\right)^2$$
$$+ \frac{2}{A}\left({}_0D_t^{-(1-\alpha)} e_{X_1}, e_{X_1}\right)_{[0,I_n]}.$$

Using (3.58) of Lemma 3.2 yields the desired result. □

3.2.1.3 *Mechanism of superconvergence*

As we shall see shortly, the proposed LDG scheme is $k+1$ optimally convergent in the L^2 norm but superconvergent at downwind points. This is also known for the classic case where the downwind convergence is $2k+1$ [Delfour *et al.* (1981); Adjerid *et al.* (2002)]. However, for the fractional case, the order of the superconvergence depends on the order of the Caputo fractional derivatives. To understand this, let us again focus on the linear case of (3.54).

The error equation corresponding to (3.56) is

$$\begin{cases} \left(E_{X_0}(t_i^-)v_0(t_i^-) - E_{X_0}(t_{i-1}^-)v_0(t_{i-1}^+)\right) - \left(E_{X_1}, v_0\right)_{I_i} - \left(E_{X_0}, \dfrac{dv_0}{dt}\right)_{I_i} = 0, \\ -\dfrac{1}{A}\left({}_0D_t^{-(1-\alpha)} E_{X_1}, v_1\right)_{I_i} = -\left(E_{X_0}, v_1\right)_{I_i}, \end{cases} \tag{3.61}$$

where $E_{X_i} = x_i - X_i$. In (3.61), taking v_i to be continuous on the interval $[0, t_j]$ and summing the equations for $i = 1, 2, \cdots, n$ lead to

$$\begin{aligned} &E_{X_0}(t_n^-)v_0(t_n^-) - (E_{X_1}, v_0)_{[0,t_n]} - \left(E_{X_0}, \frac{dv_0}{dt}\right)_{[0,t_n]} \\ &- \frac{1}{A}({}_0D_t^{-(1-\alpha)} E_{X_1}, v_1)_{[0,t_n]} + (E_{X_0}, v_1)_{[0,t_n]} = 0. \end{aligned} \tag{3.62}$$

Following (3.57), we rewrite this as

$$\begin{aligned} &E_{X_0}(t_n^-)v_0(t_n^-) - (E_{X_1}, v_0)_{[0,t_n]} - \left(E_{X_0}, \frac{dv_0}{dt}\right)_{[0,t_n]} \\ &- \frac{1}{A}(E_{X_1}, {}_tD_{t_n}^{-(1-\alpha)}v_1)_{[0,t_n]} + (E_{X_0}, v_1)_{[0,t_n]} = 0. \end{aligned} \tag{3.63}$$

Rearranging the terms of (3.63) results in

$$E_{X_0}(t_n^-)v_0(t_n^-) - \left(E_{X_1}, v_0 + \frac{1}{A}{}_tD_{t_n}^{-(1-\alpha)}v_1\right)_{[0,t_n]} \tag{3.64}$$

$$- \left(E_{X_0}, \frac{dv_0}{dt} - v_1\right)_{[0,t_n]} = 0. \tag{3.65}$$

Solving $\tilde{v}_0 + \frac{1}{A}{}_tD_{t_n}^{-(1-\alpha)}\tilde{v}_1 = 0$ and $\frac{d\tilde{v}_0}{dt} - \tilde{v}_1 = 0$ for $t \in [0, t_n]$ with $\tilde{v}_0(t_n) = E_{X_0}(t_n^-)$, we get

$$\tilde{v}_0(t) = (E_{X_0}(t_n^-)/E_{\alpha,1}(-At_n^\alpha))E_{\alpha,1}(-At^\alpha), \tag{3.66}$$

where $E_{\alpha,1}$ is the Mittag-Leffler function. Taking v_i as the L^2 projection of $\tilde{v}_i$ onto $\mathcal{P}^k$, we recover that if α is an integer, there exists

$$\left\| v_0 + \frac{1}{A}{}_tD_{t_n}^{-(1-\alpha)}v_1 + \frac{dv_0}{dt} - v_1 \right\|_{L^2([0,t_n])} = O(h^k), \tag{3.67}$$

due to the regularity of the Mittag-Leffler function [Mainardi (2013)]. For the fractional case, the approximation [Guo and Heuer (2006)] yields

$$\left\| v_0 + \frac{1}{A}{}_tD_{t_n}^{-(1-\alpha)}v_1 + \frac{dv_0}{dt} - v_1 \right\|_{L^2([0,t_n])} = O(h^{\min\{k,\alpha\}}). \tag{3.68}$$

Combined with classic polynomial approximation results for E_{X_i} [Delfour *et al.* (1981)], it yields an order of convergence at the downwind point of $k + 1 + \min\{k, \alpha\}$.

3.2.2 *Numerical results*

Let us consider a few numerical examples to qualify the above analysis. All results assume that the corresponding analytical solutions are sufficiently regular. For showing the effectiveness of the LDG schemes and further confirming the predicted convergence orders, both linear and nonlinear cases are considered. Finally, we shall also consider the computation of the generalized Mittag-Leffler functions using the LDG schemes.

We use Newton's method to solve the nonlinear systems. The initial guess in the interval I_j ($j \geq 2$) is given as $X_i(t_{j-1}^-)$ or by extrapolating forward to the interval I_j. For the interval I_1, we use $X_0(t_0^-)$ ($= x_0$) as initial guess.

3.2.2.1 *Numerical results for (3.51) with $\alpha \in (0, 1]$*

We consider examples to confirm that the convergence order of (3.54) is $k + 1 + \alpha$ at downwind points and $k + 1$ in the L^2 sense, respectively, where k is the degree of the polynomial used in an element and $\alpha \in [0, 1)$. However, when $\alpha = 1$ the convergence order is $2k + 1$ at downwind point and still $k + 1$ in L^2 sense in agreement with classic theory [Delfour *et al.* (1981)].

Example 3.3. we consider the linear fractional ODE

$$ {}_0^C D_t^\alpha x(t) = -2x(t) + \frac{\Gamma(6)}{\Gamma(6-\alpha)} t^{5-\alpha} + 2t^5 + 2, \quad \alpha \in [0, 1] \tag{3.69}$$

for $t \in (0, 1)$, with the initial condition $x(0) = 1$ and the exact solution $x(t) = t^5 + 1$. Note that when $\alpha = 0$ this condition is still required for the form of (3.53).

Figure 3.2 displays convergence at the downwind point as well as in L^2 for $k = 1, 2, 3$, confirming optimal L^2 convergence and an order of convergence of $k + 1 + \alpha$ at the downwind point as predicted.

Example 3.4. We consider the nonlinear fractional ODE

$$ {}_0^C D_t^\alpha x(t) = -2x^2(t) + \frac{\Gamma(6)}{\Gamma(6-\alpha)} t^{5-\alpha} + 2t^{10} + 4t^5 + 2, \quad \alpha \in [0, 1] \tag{3.70}$$

for $t \in (0, 0.5)$, with the initial condition $x(0) = 1$, and the exact solution $x(t) = t^5 + 1$.

The results in Fig. 3.3 confirm that the optimal L^2 convergence and an order of convergence of $k + 1 + \alpha$ at the downwind point carries over to the nonlinear case.

3.2.2.2 *Calculating the generalized Mittag-Leffler function*

As a more general example, let us use the efficient and accurate solver for calculating the generalized Mittag-Leffler functions defined as

$$E_{\alpha,\beta}(At^\alpha) = \sum_{k=0}^{\infty} \frac{(At^\alpha)^k}{\Gamma(\alpha k + \beta)}, \quad \Re(\alpha) > 0. \tag{3.71}$$

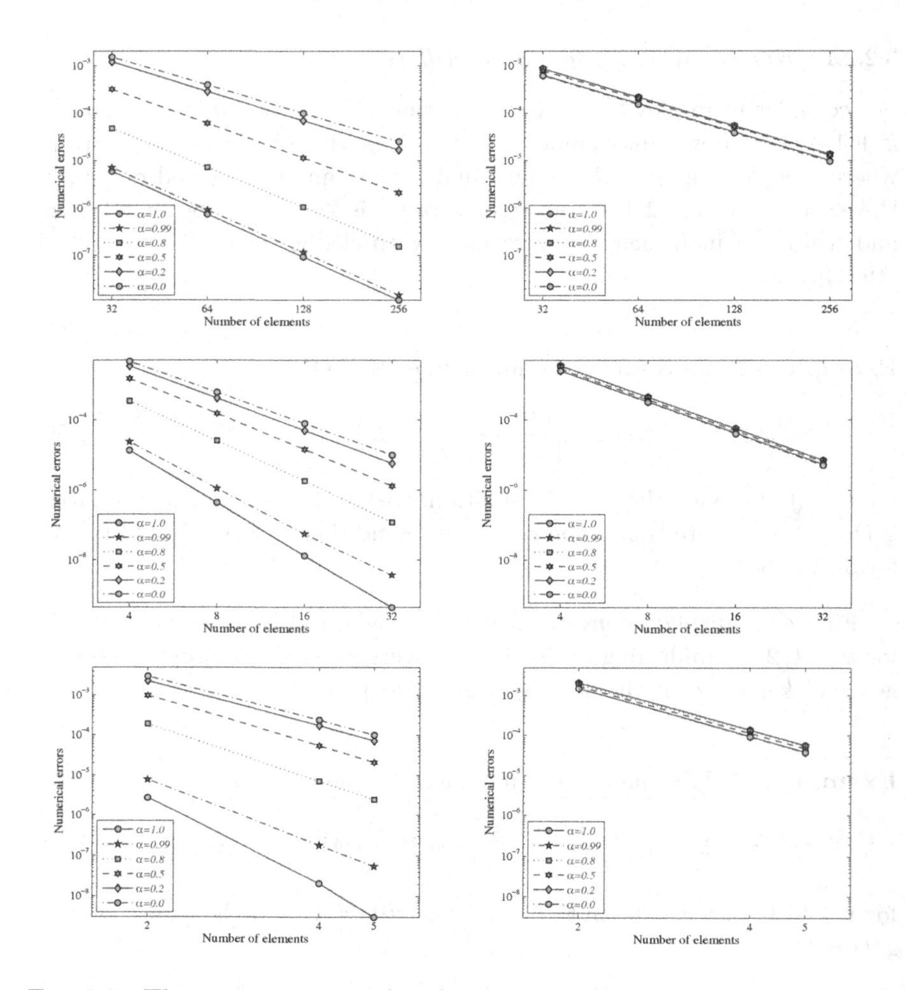

Fig. 3.2: The convergence of (3.54) for $k = 1$ (top row), $k = 2$ (middle row) and $k = 3$ (bottom row) in (3.69). In the left column we show the convergence at the downwind points while the right column displays L^2 convergence.

To build the relation between the Mittag-Leffler function and the fractional ODE consider

$$ {}_0^C D_t^\alpha x(t) = Ax(t), \quad x(0) = 1,\, x'(0) = 0, \cdots, x^{\lceil \alpha \rceil}(0) = 0. \tag{3.72} $$

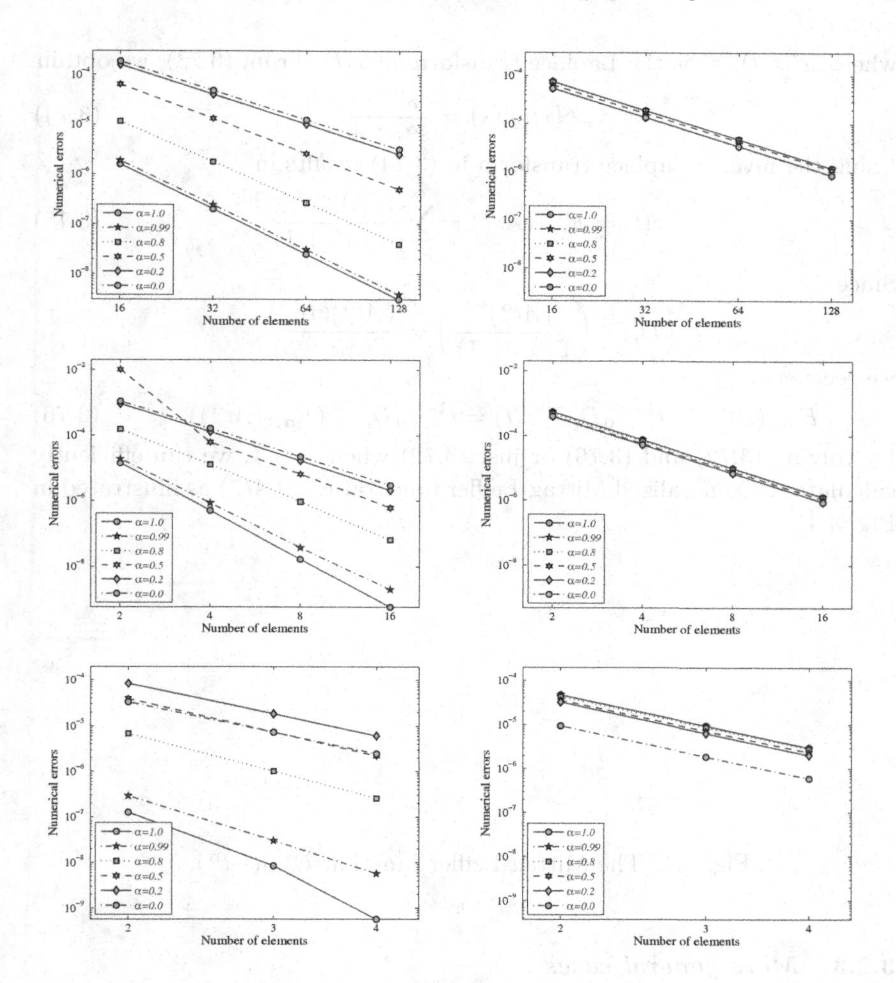

Fig. 3.3: The convergence of (3.54) for $k = 1$ (top row), $k = 2$ (middle row) and $k = 3$ (bottom row) in (3.70). In the left column we show the convergence at the downwind points while the right column displays L^2 convergence.

Taking the Laplace transform (see Appendix A) on both sides of the above equation, we recover

$$s^{\alpha}\mathscr{L}[x(t)](s) - s^{\alpha-1}x(0) = A\mathscr{L}[x(t)](s), \tag{3.73}$$

where $\mathscr{L}[x(t)](s)$ is the Laplace transform of $x(t)$. From (3.73), we obtain

$$\mathscr{L}[x(t)](s) = \frac{s^{\alpha-1}}{s^\alpha - A}. \tag{3.74}$$

Using the inverse Laplace transform in (3.74) results in

$$x(t) = E_{\alpha,1}(At^\alpha) = \sum_{k=0}^{\infty} \frac{(At^\alpha)^k}{\Gamma(\alpha k+1)}. \tag{3.75}$$

Since

$${}_0D_t^{1-\beta}\left(\frac{(At^\alpha)^k}{\Gamma(\alpha k+1)}\right) = \frac{(At^\alpha)^k t^{\beta-1}}{\Gamma(\alpha k+\beta)},$$

we recover

$$E_{\alpha,\beta}(At^\alpha) = t^{1-\beta}{}_0D_t^{1-\beta}x(t) = t^{1-\beta}{}_0D_t^{1-\beta}\left(E_{\alpha,1}(At^\alpha)\right). \tag{3.76}$$

By solving (3.72) and (3.76) or just (3.72) when $\beta = 1$, we can efficiently calculate the generalized Mittag-Leffler function $E_{\alpha,\beta}(At^\alpha)$ as illustrated in Fig. 3.4.

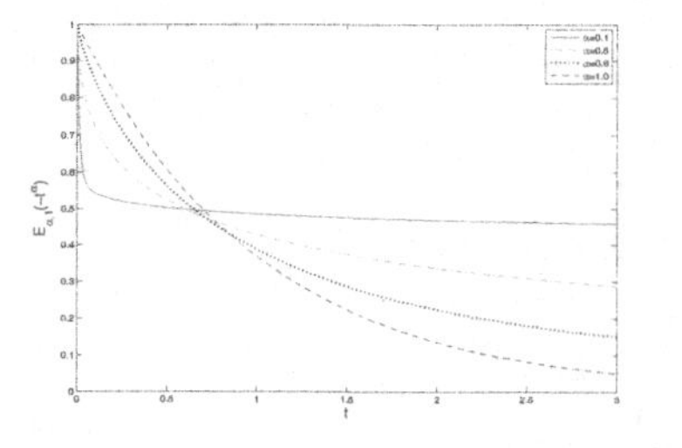

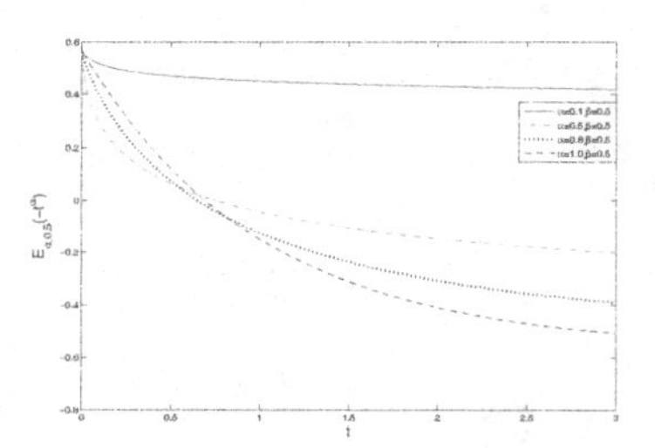

Fig. 3.4: The Mittag-Leffler function $E_{\alpha,\beta}(-t^\alpha)$.

3.2.3 *More general cases*

Let us now consider the generalized case, given as

$${}_a^CD_t^\alpha x(t) + d(t)\frac{d^m x(t)}{dt^m} = f(x,t), \tag{3.77}$$

where α in general is real, and m is a positive integer. We shall follow the same approach as previously, and consider the system of equations

$$\begin{cases} x_{i+1}(t) - \dfrac{dx_i(t)}{dt} = 0, i = 0, \ldots, \max(m,p) - 1, x_0(t) = x(t), \\ \\ {}_0D_t^{-(p-\alpha)}x_p(t) + d(t)x_m(t) = f(x_0,t), \;\; \alpha \in (0,1], \end{cases} \tag{3.78}$$

with appropriate given initial conditions $x_i(0)$. Here $p = \lceil \alpha \rceil$. For simplicity we assume $x_i(t) \in H^1(\Omega, \mathcal{T})$ except $x_m(t) \in L^2(\Omega, \mathcal{T})$. We will continue to use the upwind fluxes to seek X_i, such that for all $v_i \in V$, the following holds

$$\begin{cases} (X_{i+1}, v_i)_{I_j} + \left(X_i, \dfrac{dv_i}{dt}\right)_{I_j} - (X_i(t_j^-)v_i(t_j^-) - X_i(t_{j-1}^-)v_i(t_{j-1}^+)) = 0, \\ \quad i = 0, \ldots, \max(m, p) - 1, X_0(t) = x(t); \\ \\ \left({}_0D_t^{-(p-\alpha)} X_p, v_m\right)_{I_j} + (X_m, v_m)_{I_j} = (f(X_0, t), v_m)_{I_j}, \end{cases} \tag{3.79}$$

subject to the appropriate initial conditions.

The analysis of this scheme is generally similar to that of the previous one and will not be discussed further, although, as we shall illustrate shortly, there are details that remain open. A main difference is that the order of superconvergence at the downwind point changes to $k+1+\min\{k, \max\{\alpha, m\}\}$ and the impact of the fractional operator is thus eliminated by the linear classic operator as long as $m \geq \alpha$. However, for the case where $\lfloor \alpha \rfloor \geq k, m$, the situation is less clear.

Let us first consider a linear example to illustrate that the order of superconvergence $2k + 1$ at downwind points can also be obtained when α is an integer at the left end point of the interval, provided the initial condition is not overspecified.

Example 3.5. We consider

$${}_0^C D_t^\alpha x(t) + \frac{dx(t)}{dt} = -2x(t) + \frac{\Gamma(6)}{\Gamma(6-\alpha)} t^{5-\alpha} + 2t^5 + 5t^4 + 2 \quad \alpha \in [0, 1], \tag{3.80}$$

with the initial condition $x(0) = 1$ and the exact solution $x(t) = t^5 + 1$.

As expected, Fig. 3.5 confirms superconvergence of $k + 1 + \min\{k, \max\{\alpha, m\}\}$ at the downstream point.

Example 3.6. Let us consider a nonlinear problem, given for $t \in (0, 1), \alpha \in [1, 2]$, as

$$\begin{aligned} {}_0^C D_t^\alpha x(t) + \frac{d^3 x(t)}{dt^3} &= -2x^2(t) + \frac{\Gamma(6)}{\Gamma(6-\alpha)} t^{5-\alpha} + \frac{\Gamma(1)}{\Gamma(3-\alpha)} t^{2-\alpha} \\ &+ 2t^{10} + 2t^7 + 4t^6 + 4t^5 + 0.5t^4 + 2t^3 + 64t^2 + 4t + 2 \end{aligned} \tag{3.81}$$

with the initial condition $x(0) = 1$, $x'(0) = 1$, $x''(0) = 1$ and the exact solution $x(t) = t^5 + \frac{1}{2}t^2 + t + 1$.

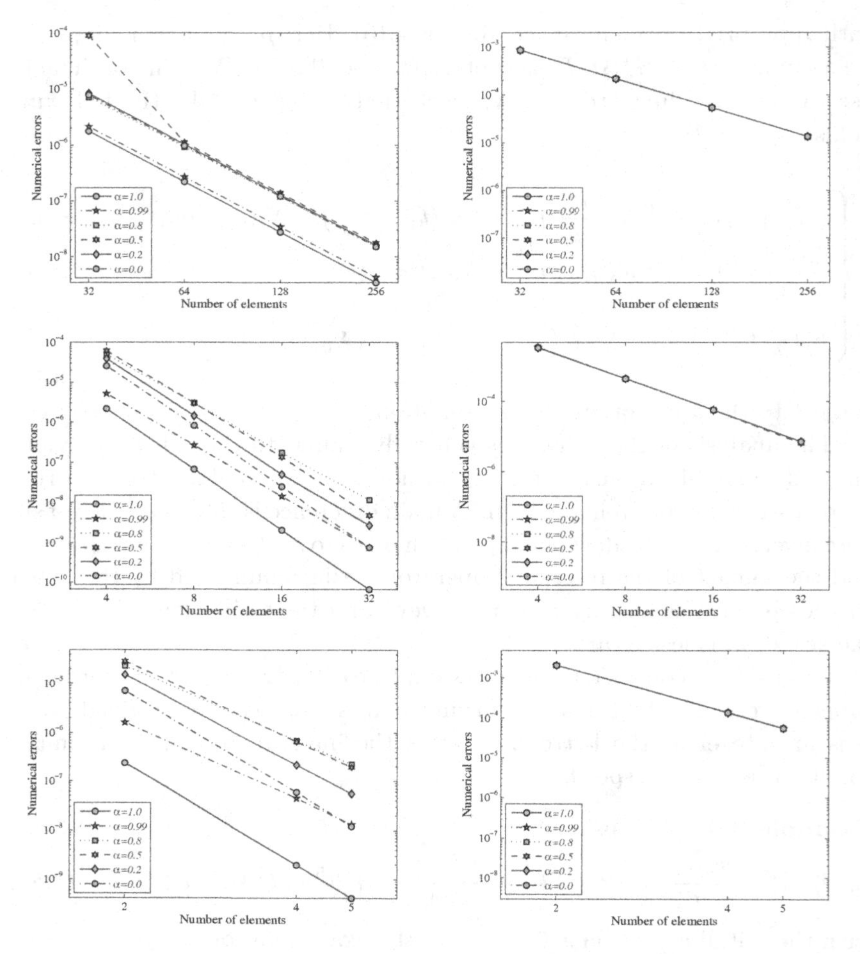

Fig. 3.5: The convergence of (3.80) for $k = 1$ (top row), $k = 2$ (middle row) and $k = 3$ (bottom row). In the left column we show the convergence at the downwind points while the right column displays L^2 convergence.

We note that in this case, $\alpha \in [1, 2]$ but $m = 3$ and, as shown in Fig. 3.6 super convergence of order $2k+1$ is maintained. However, if the assumption that $m \geq \lceil \alpha \rceil$ is violated, an α-dependent rate of convergence re-emerges as illustrated in the following example.

Example 3.7. Consider a nonlinear problem, given for $t \in [0, 1]$ and $\alpha \in [1, 2]$, as

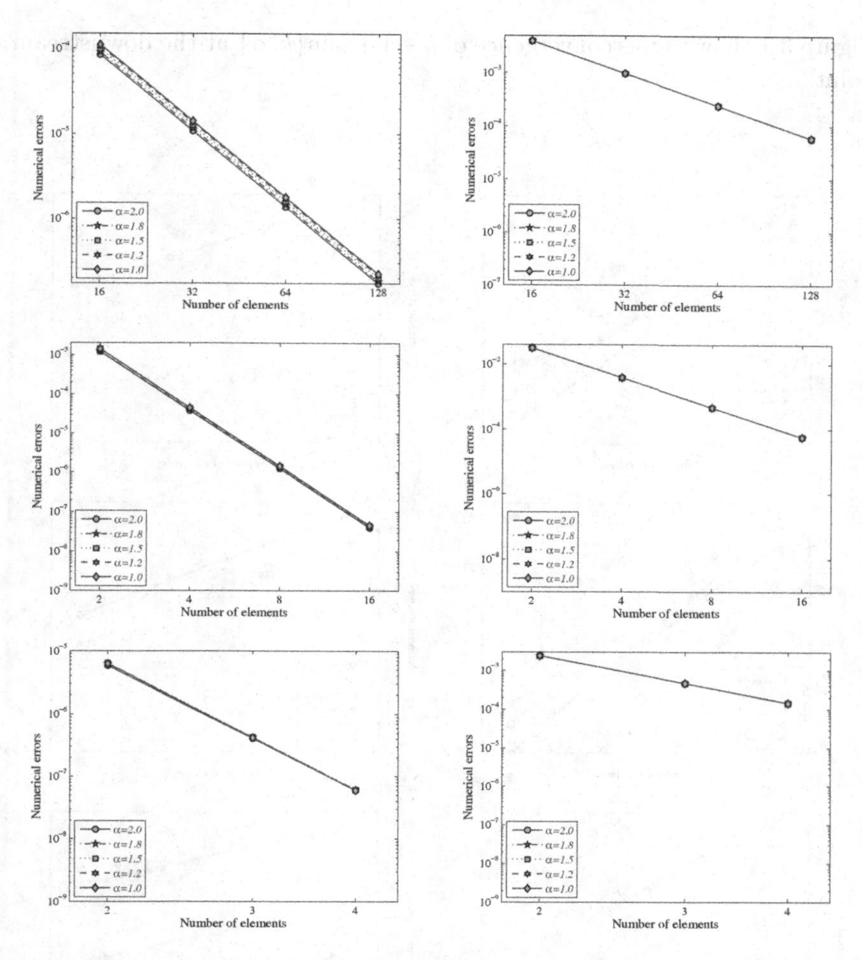

Fig. 3.6: The convergence of (3.79) for $k = 1$ (top row), $k = 2$ (middle row) and $k = 3$ (bottom row) in (3.81). In the left column we show the convergence at the downwind points while the right column displays L^2 convergence.

$$ {}_0^C D_t^{\alpha} x(t) + \frac{dx(t)}{dt} = -2x^2(t) + \frac{\Gamma(6)}{\Gamma(6-\alpha)} t^{5-\alpha} + 5t^4 + 1 + 2(t^5 + t + 1)^2 \tag{3.82} $$

with the initial condition $x(0) = 1$, $x'(0) = 1$ and the exact solution $x(t) = t^5 + t + 1$.

Figure 3.7 shows superconvergence of $k+1+\min\{k,\alpha\}$ at the downstream point.

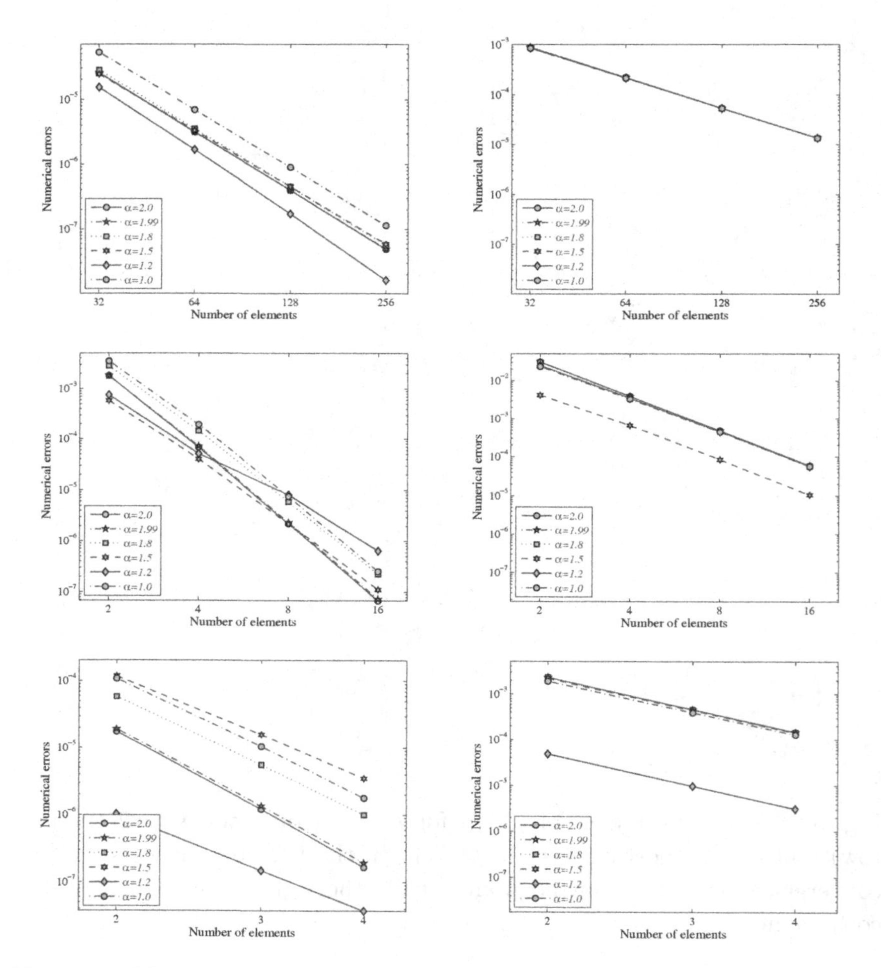

Fig. 3.7: The convergence of (3.79) for $k = 1$ (top row), $k = 2$ (middle row) and $k = 3$ (bottom row) in (3.82). In the left column we show the convergence at the downwind points while the right column displays L^2 convergence.

Example 3.8. As a final example, let us consider

$$ {}_0^C D_t^{\alpha} x(t) = -2x(t) + \frac{\Gamma(6)}{\Gamma(6-\alpha)} t^{5-\alpha} + 2t^5 + 2t + 2 \quad \alpha \in [1,2], \tag{3.83} $$

for $t \in (0,1)$, with the initial condition $x(0) = 1$, $x'(0) = 1$ and the exact solution $x(t) = t^5 + t + 1$.

In Fig. 3.8 we show the results for $k = 1$. Following the previous analysis, we would expect an order of convergence as $k + 1 + \min\{k, \alpha\}$ which in this case would be third order. However, the results in Fig. 3.8 highlight a reduction in the order of convergence at the endpoint as α approaches the value one. The mechanism for this is not fully understood but is likely associated with an over specification of the initial conditions in this singular limit. Increasing k recovers the expected convergence rate for all values of α.

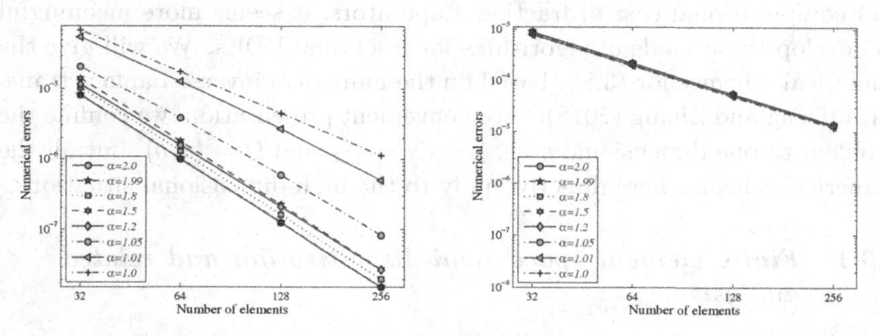

Fig. 3.8: The convergence of (3.79) for $k = 1$ in (3.83). On the left we show the convergence at the downwind points while the right figure displays L^2 convergence.

3.3 Numerical inverse Laplace transform method for the fractional subdiffusion equations

Now we discuss the exponential type integrator for solving the subdiffusion equation $(0 < \alpha < 1)$

$$ \begin{cases} \frac{\partial}{\partial t} u(x,t) - {}_0D_t^{1-\alpha} \Delta u(x,t) = f(x,t), & (x,t) \in \Omega \times (0,T], \\ u(x,0) = g(x), & x \in \Omega, \\ u(x,t) = 0, & x \in \partial\Omega, \end{cases} \tag{3.84} $$

where

$$ {}_0D_t^{1-\alpha} \Delta u(x,t) = \frac{1}{\Gamma(\alpha)} \frac{\partial}{\partial t} \int_0^t (t-\xi)^{\alpha-1} \Delta u(x,\xi) d\xi. $$

Acting operator ${}_0D_t^{-(1-\alpha)}$ on both sides of (3.84), and applying (B.26) and (B.19) in Appendix B, we have the following equivalent form, i.e.,

$$ {}_0^CD_t^\alpha u(x,t) + \Delta u(x,t) = f_1(x,t), \tag{3.85} $$

where $f_1(x,t) = {}_0D_t^{-(1-\alpha)} f(x,t) = \frac{1}{\Gamma(1-\alpha)} \int_0^t (t-\xi)^{-\alpha} f(x,\xi) d\xi$.

Model (3.84) describes the anomalous transport process with long-time memory, and its numerical solution has been extensively studied [Yuste and Acedo (2005); Ji and Sun (2015); Zhang *et al.* (2011); Zhuang *et al.* (2008)], where the finite difference methods are usually used to get the discretization in time. In fact, when $\alpha = 1$, both (3.84) and (3.85) reduce to the classic parabolic equation, and the exponential integrator schemes have been well developed for solving this special case [Gavrilyuk and Makarov (2005); Sheen *et al.* (2000, 2003)]. Because of the large storage requirement and computational cost of fractional operators, it seems more meaningful to develop these kinds of algorithms for fractional PDEs. We will give the numerical schemes for (3.85) based on the numerical inverse Laplace transform [Deng and Zhang (2018)]. For convenient presentation, we confine the problem to one dimensional case, i.e., $\Delta = \frac{d^2}{dx^2}$ and $\Omega = (a,b)$, but all the numerical schemes here directly apply to the high-dimensional situations.

3.3.1 *Finite element space semi-discretization and related analysis*

The Galerkin weak solution of (3.85) can be formulated as: find $u(\cdot,t) \in H_0^1(\Omega)$, such that for any $v \in H_0^1(\Omega)$

$$ \left({}_0^CD_t^\alpha u(\cdot,t), v\right) + (\nabla u(\cdot,t), \nabla v) = (f_1, v) \tag{3.86} $$

with $(u(\cdot,0),v) = (g,v)$, $\nabla := \frac{\partial}{\partial x}$. Partition Ω into $a = x_1 < \cdots < x_i < \cdots < x_N = b$ with $x_i = a + ih$, $h = \frac{b-a}{N}$ and define the finite element space

$$ S_h := \left\{ v \in C(\overline{\Omega}) : v(x)\,|_{I_i} \in P_l(I_i), v(a) = v(b) = 0 \right\}, \tag{3.87} $$

where $P_l(I_i)$ denotes the space of polynomials of degree no greater than $l (l \in \mathbb{N}^+)$ on $I_i = (x_{i-1}, x_i)$. We make finite element approximation for (3.86), i.e., find $u_h(t) \in S_h$ such that for any $v \in S_h$, there holds

$$ \left({}_0^CD_t^\alpha u_h(t), v\right) + (\nabla u_h(t), \nabla v) = (f_1, v), \tag{3.88} $$

where $u_h(0) = P_h g \in S_h$ with P_h denoting the L^2 projection operator.

For every t, define the Ritz projection $R_h : H_0^1(\Omega) \to S_h$ by

$$ \left(\nabla\left(R_h u(\cdot,t)\right), \nabla\eta\right) = (\nabla u(\cdot,t), \nabla\eta) \quad \forall \eta \in S_h. \tag{3.89} $$

Then for $u(\cdot,t) \in H_0^1(\Omega) \cap H^s(\Omega)\,(s \geq 1)$, one has

$$\begin{aligned}&\|u(\cdot,t) - R_h u(\cdot,t)\|_{L^2(\Omega)} + h\,\|u(\cdot,t) - R_h u(\cdot,t)\|_{H^1(\Omega)}\\ &\quad \leq ch^{\min\{l+1,s\}}\,\|u(\cdot,t)\|_{H^s(\Omega)}\,. \end{aligned} \tag{3.90}$$

Lemma 3.3. *Suppose that $\alpha \in (0,1)$, and $v(x) \in L^2(\mathbb{R})$ is a real function with compact support. Then*

$$\int_{\mathbb{R}} {}_{-\infty}D_s^{-\alpha} v(s)\, v(s) ds = \cos(\pi\alpha/2) \int_{\mathbb{R}} \left({}_{-\infty}D_s^{-\frac{\alpha}{2}} v(s)\right)^2 ds. \tag{3.91}$$

Proof. By Proposition B.8 and the Plancherel theorem (see Lemma A.16), it holds that

$$\begin{aligned}\int_{\mathbb{R}} {}_{-\infty}D_s^{-\alpha} v(s)\, v(s) ds &= \frac{1}{2\pi}\int_{\mathbb{R}} (-i\omega)^{-\alpha}\hat{v}(\omega)\,\overline{\hat{v}(\omega)} d\omega\\ &= \frac{1}{2\pi}\left(\int_0^\infty e^{\frac{\pi\alpha}{2}i}|\omega|^{-\alpha}\hat{v}(\omega)\,\overline{\hat{v}(\omega)}d\omega + \int_{-\infty}^0 e^{\frac{-\pi\alpha}{2}i}|\omega|^{-\alpha}\hat{v}(\omega)\,\overline{\hat{v}(\omega)}d\omega\right)\\ &= \frac{\cos(\pi\alpha/2)}{2\pi}\int_{\mathbb{R}} |\omega|^{-\alpha}\hat{v}(\omega)\,\overline{\hat{v}(\omega)}d\omega\\ &= \cos(\pi\alpha/2)\int_{\mathbb{R}}\left({}_{-\infty}D_s^{-\frac{\alpha}{2}} v(s)\right)^2 ds.\end{aligned}$$

Thus, we complete the proof. □

Theorem 3.5. *For $\alpha \in (0,1)$, the space semi-discrete scheme (3.88) is unconditionally stable, and there holds*

$$\|\nabla u_h(t)\|_{L^2(\Omega)}^2 \leq \|\nabla u_h(0)\|_{L^2(\Omega)}^2 + C_1\int_0^t \|f\|_{L^2(\Omega)}^2 ds, \tag{3.92}$$

$$\begin{aligned}\|u_h(t)\|_{L^2(\Omega)}^2 &\leq \|u_h(0)\|_{L^2(\Omega)}^2 + C_4 t^\alpha \|\nabla u_h(0)\|_{L^2(\Omega)}^2\\ &\quad + C_5 t^\alpha \int_0^t \|f\|_{L^2(\Omega)}^2 ds,\end{aligned} \tag{3.93}$$

where C_1 and C_5 are constants.

Proof. Let $v = \frac{\partial u_h}{\partial t} \in S_h$ in (3.88), i.e.,

$$\left({}_0^C D_t^\alpha u_h(t), \frac{\partial u_h(t)}{\partial t}\right) + \left(\nabla u_h(t), \nabla\left(\frac{\partial u_h(t)}{\partial t}\right)\right) = \left(f_1, \frac{\partial u_h(t)}{\partial t}\right). \tag{3.94}$$

It is easy to obtain

$$\int_0^t \left(\nabla u_h(s), \nabla\left(\frac{\partial u_h(s)}{\partial s}\right)\right) ds = \frac{1}{2}\left(\|\nabla u_h(t)\|_{L^2(\Omega)}^2 - \|\nabla u_h(0)\|_{L^2(\Omega)}^2\right). \tag{3.95}$$

Zero extending f and $\frac{\partial u_h(s)}{\partial s}$ outside of the interval $[0,t]$, and using the Plancherel theorem, one has

$$\begin{aligned}
&\int_0^t \left(f_1, \frac{\partial u_h(s)}{\partial s}\right) ds \\
&\leq \int_\Omega \left| \int_0^t \frac{\partial u_h(s)}{\partial s} \left({}_0D_s^{-(1-\alpha)} f\right) ds \right| dx \\
&\leq \frac{1}{2\pi} \int_\Omega \int_{-\infty}^{\infty} |-i\omega|^{-(1-\alpha)} |\mathscr{F}[f](\omega)| \left| \overline{\mathscr{F}[\frac{\partial u_h(s)}{\partial s}](\omega)} \right| d\omega dx \\
&\leq \frac{1}{\sin\left(\frac{\pi\alpha}{2}\right)} \sqrt{\int_0^t \left({}_0^C D_s^\alpha u_h(s), \frac{\partial u_h(s)}{\partial s}\right) ds} \sqrt{\int_0^t \left(f, {}_0D_s^{-(1-\alpha)} f\right) ds} \\
&\leq \frac{1}{2} \int_0^t \left({}_0^C D_s^\alpha u_h(s), \frac{\partial u_h(s)}{\partial s}\right) ds + \frac{1}{2\sin^2\left(\frac{\pi\alpha}{2}\right)} \int_0^t \left(f, {}_0D_s^{-(1-\alpha)} f\right) ds \\
&\leq \frac{1}{2} \int_0^t \left({}_0^C D_s^\alpha u_h(s), \frac{\partial u_h(s)}{\partial s}\right) ds + \frac{C_1}{2} \int_0^t \|f\|_{L^2(\Omega)}^2 ds, \qquad (3.96)
\end{aligned}$$

where the third step of above can be obtained by using the Hölder inequality and Lemma 3.3; and the last step of above can be obtained by using the Schwarz inequality and

$$\int_0^t \left({}_0D_s^{-(1-\alpha)} f\right)^2 ds \leq \int_{-\infty}^{\infty} \left(\frac{s^{-\alpha}\chi_{(0,t)}}{\Gamma(1-\alpha)} * f\right)^2 ds \leq \frac{t^{2(1-\alpha)}}{\Gamma^2(2-\alpha)} \int_0^t f^2 ds.$$

By zero extending $\frac{\partial u_h(s)}{\partial s}$ outside of the interval $[0,t]$ again and using Lemma 3.3, one has

$$\begin{aligned}
&\int_0^t \left({}_0^C D_s^\alpha u_h(s), \frac{\partial u_h(s)}{\partial s}\right) ds \\
&= \int_0^t \left({}_0D_s^{-(1-\alpha)} \frac{\partial u_h(s)}{\partial s}, \frac{\partial u_h(s)}{\partial s}\right) ds \\
&= \int_{-\infty}^{\infty} \left({}_{-\infty}D_s^{-(1-\alpha)} \frac{\partial u_h(s)}{\partial s}, \frac{\partial u_h(s)}{\partial s}\right) ds \\
&\geq \frac{\sin(\frac{\pi\alpha}{2})}{2\pi} \int_0^t \left\| {}_0D_s^{-\frac{1-\alpha}{2}} \frac{\partial u_h(s)}{\partial s} \right\|_{L^2(\Omega)}^2 ds. \qquad (3.97)
\end{aligned}$$

Noting that

$$\|u_h(t) - u_h(0)\|^2_{L^2(\Omega)} = \left\| {}_0D_t^{-\frac{1+\alpha}{2}} \left({}_0D_t^{-\frac{1-\alpha}{2}} \frac{\partial u_h(t)}{\partial t} \right) \right\|^2_{L^2(\Omega)}, \quad (3.98)$$

and by Hölder's inequality, one has

$$\begin{aligned}
&\|u_h(t) - u_h(0)\|^2_{L^2(\Omega)} \\
&\le \int_\Omega \left(\int_0^t \frac{(t-s)^{-(1-\alpha)}}{\Gamma^2(\frac{1+\alpha}{2})} ds \right) \left(\int_0^t \left({}_0D_t^{-\frac{1-\alpha}{2}} \frac{\partial u_h(t)}{\partial t} \right)^2 ds \right) dx \\
&\le C_3 t^\alpha \int_0^t \left\| {}_0D_t^{-\frac{1-\alpha}{2}} \frac{\partial u_h(t)}{\partial t} \right\|^2_{L^2(\Omega)} ds. \quad (3.99)
\end{aligned}$$

Now, combining all the previous results, it follows that

$$\begin{aligned}
\|u_h(t) - u_h(0)\|^2_{L^2(\Omega)} + C_4 t^\alpha \left(\|\nabla u_h(t)\|^2_{L^2(\Omega)} - \|\nabla u_h(0)\|^2_{L^2(\Omega)} \right) \\
\le C_5 t^\alpha \int_0^t \|f\|^2_{L^2(\Omega)} ds.
\end{aligned}$$

Thus, we complete the proof. □

Theorem 3.6. *Let $u(x,t)$ and $u_h(t)$ be the solutions of (3.86) and (3.88), respectively. Suppose that $u(\cdot,t) \in H_0^1(\Omega) \cap H^s(\Omega)\,(s \ge 1)$. Then*

$$\begin{aligned}
&\|u(\cdot,t) - u_h(t)\|_{L^2(\Omega)} \\
&\le C \left(\|u\|^2_{H^s(\Omega)} + \int_0^t (t-\xi)^{\alpha-1} \left\| {}_0^C D_\xi^\alpha u \right\|^2_{H^s(\Omega)} d\xi \right) h^{2\min\{l+1,s\}}, \quad (3.100)
\end{aligned}$$

where C is a constant independent of h.

Proof. Let $E_h(t) = u_h(t) - u(x,t) = \eta_h(t) - \theta_h(t)$, where $\eta_h(t) = u_h(t) - R_h u(x,t)$ and $\theta_h(t) = u(x,t) - R_h u(x,t)$. Then for any $v \in S_h$, it holds that

$$\left({}_0^C D_t^\alpha \eta_h(t), v \right) + (\nabla \eta_h(t), \nabla v) = \left({}_0^C D_t^\alpha \theta_h(t), v \right), \quad (3.101)$$

where $(\nabla \theta_h(t), \nabla v) = 0$ has been used. Take $v = \eta_h(t) \in S_h$ in (3.101). We first show that $\left({}_0^C D_t^\alpha \eta_h(t), \eta_h(t)\right) \geq \frac{1}{2} {}_0^C D_t^\alpha \left\|\eta_h(t)\right\|_{L^2(\Omega)}^2$. In fact,

$$\begin{aligned}
&\left({}_0^C D_t^\alpha \eta_h(t), \eta_h(t)\right) - \frac{1}{2} {}_0^C D_t^\alpha \left\|\eta_h(t)\right\|_{L^2(\Omega)}^2 \\
&= \frac{1}{\Gamma(1-\alpha)} \int_a^b \int_0^t \frac{[\eta_h(t) - \eta_h(\nu)]}{(t-\nu)^\alpha} \frac{\partial \eta_h(\nu)}{\partial \nu} d\nu dx \\
&= \frac{1}{\Gamma(1-\alpha)} \int_a^b \int_0^t \int_\nu^t \frac{\partial \eta_h(\xi)}{\partial \xi} \frac{\partial \eta_h(\nu)}{\partial \nu} (t-\nu)^{-\alpha} d\xi d\nu dx \\
&= \frac{1}{\Gamma(1-\alpha)} \int_a^b \int_0^t \frac{\partial \eta_h(\xi)}{\partial \xi} \int_0^\xi \frac{\partial \eta_h(\nu)}{\partial \nu} (t-\nu)^{-\alpha} d\nu d\xi dx \\
&= \frac{1}{2\Gamma(1-\alpha)} \int_a^b \int_0^t (t-\xi)^\alpha \frac{\partial}{\partial \xi} \left(\int_0^\xi \frac{\partial \eta_h(\nu)}{\partial \nu} (t-\nu)^{-\alpha} d\nu \right)^2 d\xi dx \\
&= \frac{1}{2\Gamma(1-\alpha)} \int_a^b \int_0^t (t-\xi)^{\alpha-1} \left(\int_0^\xi \frac{\partial \eta_h(\nu)}{\partial \nu} (t-\nu)^{-\alpha} d\nu \right)^2 d\xi dx \\
&\geq 0. \qquad (3.102)
\end{aligned}$$

In addition, it holds that

$$\left({}_0^C D_t^\alpha \theta_h(t), \eta_h(t)\right) \leq C \left\|\eta_h(t)\right\|_{L^2(\Omega)}^2 + \frac{1}{4C} \left\|{}_0^C D_t^\alpha \theta_h(t)\right\|_{L^2(\Omega)}^2, \qquad (3.103)$$

$$\left(\nabla \eta_h(t), \nabla \eta_h(t)\right) \geq C \left\|\eta_h(t)\right\|_{L^2(\Omega)}^2, \qquad (3.104)$$

where the first one is obtained by the Schwart and the Young inequalities, and the second one by the Poincaré inequality. Therefore,

$${}_0^C D_t^\alpha \left\|\eta_h(t)\right\|_{L^2(\Omega)}^2 \leq \frac{1}{2C} \left\|{}_0^C D_t^\alpha \theta_h(t)\right\|_{L^2(\Omega)}^2. \qquad (3.105)$$

Performing the operator ${}_0D_t^{-\alpha}$ on both sides of (3.105), and using $\eta_h(0) = 0$ and (B.29) yield

$$\begin{aligned}
\left\|\eta_h(t)\right\|_{L^2(\Omega)}^2 &\leq \frac{1}{2C\,\Gamma(\alpha)} \int_0^t (t-\xi)^{\alpha-1} \left\|{}_0^C D_\xi^\alpha \theta_h(\xi)\right\|_{L^2(\Omega)}^2 d\xi \\
&\leq C_1 h^{2\min\{l+1,s\}} \int_0^t (t-\xi)^{\alpha-1} \left\|{}_0^C D_\xi^\alpha u\right\|_{H^s(\Omega)}^2 d\xi. \qquad (3.106)
\end{aligned}$$

Therefore,

$$\begin{aligned}
&\left\|E_h(t)\right\|_{L^2(\Omega)}^2 \leq 2\left(\left\|\eta_h(t)\right\|_{L^2(\Omega)}^2 + \left\|\theta_h(t)\right\|_{L^2(\Omega)}^2\right) \\
&\leq C_2 \left(\left\|u\right\|_{H^s(\Omega)}^2 + \int_0^t (t-\xi)^{\alpha-1} \left\|{}_0^C D_\xi^\alpha u\right\|_{H^s(\Omega)}^2 d\xi \right) h^{2\min\{l+1,s\}}. \qquad (3.107)
\end{aligned}$$

Thus, we complete the proof. □

3.3.2 *Time discretization by numerical inverse Laplace transform*

Define discrete operator $\Delta_h : S_h \to S_h$ by

$$(\Delta_h \phi, \psi) = (\nabla\phi, \nabla\psi) \quad \forall \phi, \psi \in S_h.$$

Then (3.88) can be rewritten as

$$\begin{cases} {}_0^C D_t^\alpha u_h(t) + \Delta_h u_h(t) = P_h f_1(t), \\ u_h(0) = P_h g. \end{cases}$$

Applying the Laplace transform formula

$$\mathscr{L}\left[{}_0^C D_t^\alpha u_h(t)\right](z) = z^\alpha \mathscr{L}[u_h(t)](z) - z^{\alpha-1} u_h(0),$$

we formally obtain

$$(z^\alpha + \Delta_h)\mathscr{L}[u_h(t)](z) = z^{\alpha-1} u_h(0) + \mathscr{L}[P_h f_1(t)](z).$$

Assume that $H(\Delta_h, z) := (z^\alpha + \Delta_h)^{-1}\left(z^{\alpha-1} P_h g + \mathscr{L}[P_h f_1(t)](z)\right)$ is analytic on the right of a contour Γ. By inverse Laplace transform one has

$$u_h(t) = \frac{1}{2\pi i}\int_\Gamma e^{zt} H(\Delta_h, z) dz.$$

In order to effectively calculate the above integral, one usually choose the contour Γ such that the integrand function $e^{zt} H(\Delta_h, z)$ decays exponentially towards both ends of the contour. Here, we will consider a slightly different way to approximate $u_h(t)$.

By the Laplace transform formula

$$\mathscr{L}\left[t^{n\alpha+\beta-1} E_{\alpha,\beta}(st^\alpha)\right](z) = \frac{z^{\alpha-\beta}}{(z^\alpha - s)^{n+1}} \tag{3.108}$$

of the Mittag-Leffler function

$$E_{\alpha,\beta}(t) = \sum_{k=0}^{\infty} \frac{t^k}{\Gamma(\alpha k + \beta)} \quad (t \in \mathbb{C}, \alpha, \beta > 0),$$

it is easy to obtain that

$$\begin{aligned} u_h(t) &= E_{\alpha,1}(-t^\alpha \Delta_h) P_h g \\ &\quad + \int_0^t (t-s)^{\alpha-1} E_{\alpha,\alpha}(-(t-s)^\alpha \Delta_h) P_h f_1 ds. \end{aligned} \tag{3.109}$$

Note that for $\nu > 0$ and $\beta > 0$, there exists

$$\begin{aligned} &\int_a^t (t-s)^{\beta-1} E_{\alpha,\beta}(-q(t-s)^\alpha)(t-a)^{\nu-1} ds \\ &\quad = \Gamma(\nu)(t-a)^{\beta+\nu-1} E_{\alpha,\beta+\nu}\left(-q(t-a)^\alpha\right). \end{aligned} \tag{3.110}$$

Therefore, if P_hf has the form $(t^{\nu_1-1}g_1(x)+t^{\nu_2-1}g_2(x)+...)$, then one can remove the second integral symbol in (3.109) exactly. Otherwise, if P_hf is piecewise smooth w.r.t. t, then the piecewise polynomial interpolation can be used, i.e., letting $0\le t_0<\cdots<t_{M-1}<t_M=t$ be the partition of $[0,t]$, and

$$h_k(x,s)=\sum_{l=0}^{m_k}c_{m_k,l}^L(s-t_k)^l=\sum_{l=0}^{m_k}c_{m_k,l}^R(s-t_{k+1})^l \tag{3.111}$$

be the m_k degree interpolation of P_hf_1 in interval $[t_k,t_{k+1}]$, then

$$\begin{aligned}
&\int_0^t(t-s)^{\alpha-1}E_{\alpha,\alpha}\left(-(t-s)^\alpha\Delta_h\right)P_hf_1ds\\
&\approx\sum_{k=0}^{M-1}\left(\int_{t_k}^t-\int_{t_{k+1}}^t\right)(t-s)^{\alpha-1}E_{\alpha,\alpha}\left(-(t-s)^\alpha\Delta_h\right)h_k(x,s)ds\\
&\approx\sum_{k=0}^{M-1}\sum_{l=0}^{\max\{m_k,m_{k-1}\}}\Gamma(l+1)(t-t_k)^{\alpha+l}E_{\alpha,\alpha+l+1}(-(t-t_k)^\alpha\Delta_h)\\
&\times\left(c_{m_k,l}^L-c_{m_{k-1},l}^R\right)+\sum_{l=0}^{m_0}\Gamma(l+1)t^{\alpha+l}E_{\alpha,\alpha+l+1}(-t^\alpha\Delta_h)c_{m_0,l}^L,
\end{aligned}\tag{3.112}$$

where we take $c_{m_k,l}^L=0$ for $l>m_k$ and $c_{m_{k-1},l}^R=0$ for $l>m_{k-1}$, respectively. Therefore, the problem can be attributed to how to handle

$$\Pi^{\alpha,\beta}v=t^{\beta-1}E_{\alpha,\beta}\left(-t^\alpha\Delta_h\right)v \tag{3.113}$$

effectively. Following the work by [Garrappa (2015); Schmelzer and Trefethen (2007); Trefethen *et al.* (2006)], two kinds of rational approximations based on the inverse Laplace transform can be adopted here.

The first one is Caratheéodory-Fejér (CF) approximation. Note that $-\Delta_h$ generates an analytic semigroup $e^{-t\Delta_h}$ on S_h, and it satisfies the resolvent estimate

$$\left\|(z+\Delta_h)^{-1}\right\|_{L^2(\Omega)}\le C|z|^{-1}\quad z\in\Sigma_\delta:=\{z\in\mathbb{C}:|argz|\le\pi-\delta\} \tag{3.114}$$

for any $\delta\in(0,\frac{\pi}{2})$. Then when $\alpha\in(0,1)$ and $z\in\Sigma_\delta$

$$\left\|z^{\alpha-\beta}\left(z^\alpha+t^\alpha\Delta_h\right)^{-1}v\right\|_{L^2(\Omega)}\le C|z|^{-\beta}\left\|v\right\|_{L^2(\Omega)}. \tag{3.115}$$

Therefore, $z^{\alpha-\beta}\left(z^\alpha+t^\alpha\Delta_h\right)^{-1}v$ is analytic in $\mathbb{C}/(-\infty,0]$, and it tends to zero uniformly as $|z|\to\infty$. By (3.108) and the Cauchy Theorem, it follows that

$$\begin{aligned}
\Pi^{\alpha,\beta}v&=\frac{1}{2\pi i}\int_{\Gamma_1}e^{st}s^{\alpha-\beta}\left(s^\alpha+\Delta_h\right)^{-1}vds\\
&=\frac{t^{\beta-1}}{2\pi i}\int_{\Gamma_1}e^zz^{\alpha-\beta}\left(z^\alpha+t^\alpha\Delta_h\right)^{-1}vdz,
\end{aligned}\tag{3.116}$$

where the path Γ_1 is a deformed Bromwich contour enclosing the negative axis in the anticlockwise sense. By [Schmelzer and Trefethen (2007)], we can replace e^z with the (N_1-1, N_1) type CF rational approximation $r_{N_1,N_1-1} = \sum_{k=1}^{N_1} \frac{c_k}{z-z_k}$ to obtain an approximation with the form

$$\begin{aligned}\Pi_{N_1}^{\alpha,\beta} v &= \frac{t^{\beta-1}}{2\pi i} \int_{\Gamma_1} \left(\sum_{k=1}^{N_1} \frac{c_k}{z - z_k} \right) z^{\alpha-\beta} \left(z^\alpha + t^\alpha \Delta_h\right)^{-1} v dz \\ &= -t^{\beta-1} \sum_{k=1}^{N_1} c_k z_k^{\alpha-\beta} \left(z_k^\alpha + t^\alpha \Delta_h\right)^{-1} v, \end{aligned} \tag{3.117}$$

where the poles $\{z_k\}$ and $\{c_k\}$ just need to be computed one time. To implement (3.117), it seems that we need to solve N_1 elliptic problems, i.e., find $\tilde{v}(z_k) = (z_k^\alpha + t^\alpha \Delta_h)^{-1} v \in S_h$, such that

$$((z_k^\alpha + t^\alpha \Delta_h)\tilde{v}(z_k), \psi) = (v, \psi) \qquad \forall \psi \in S_h, \tag{3.118}$$

in fact, half of the computation cost can be reduced, because of the complex conjugate nature of $\{c_k, z_k\}$ [Trefethen *et al.* (2006)]. Moreover, they can be calculated in parallel.

By (3.115) and Theorem 5.2 in [Schmelzer and Trefethen (2007)], the theoretical convergence rate of the CF approximation is geometric. Our numerical experiments also show that it always gives excellent results for $\beta \leq 4$ while N equals 14 to 16. However, for the big β, the CF approximation is distorted due to the fact that $z^{\alpha-\beta}(z^\alpha + \Delta_h)^{-1}$ decays so fast that the left-most nodes make a negligible contribution (the same happens when approximating the Gamma function (see Schmelzer and Trefethen, 2007, Fig. 4.3)). One can overcome this by fine-tuning the integral depending on α, β. Here we choose Γ_1 as the simplest parabolic contour (PC) $z(p) = \sigma(ip + 1)^2$ with $\sigma > 0$ and $p \in \mathbb{R}$ in [Garrappa (2015)]. Then

$$\Pi^{\alpha,\beta} v = \frac{t^{\beta-1}}{2\pi i} \int_{-\infty}^{\infty} z'(p) e^{z(p)} z^{\alpha-\beta} (z^\alpha(p) + t^\alpha \Delta_h)^{-1} v dp. \tag{3.119}$$

The fast decay of $|e^{z(p)}|$ for $|p| \to \infty$ allows one to produce an approximation of (3.119) by the truncated trapezoidal formula as

$$\begin{aligned}\Pi_{N_1}^{\alpha,\beta} v &= \frac{t^{\beta-1}\tau_1}{2\pi i} \sum_{k=-N_1}^{N_1} z_k' e^{z_k} z_k^{\alpha-\beta} \left(z_k^\alpha + t^\alpha \Delta_h\right)^{-1} v \\ &= \frac{t^{\beta-1}\tau_1}{2\pi i} \sum_{k=0}^{N_1} \nu_k z_k' e^{z_k} z_k^{\alpha-\beta} (z_k^\alpha + t^\alpha \Delta_h)^{-1} v, \end{aligned} \tag{3.120}$$

where $z_k = z(p_k), z'_k = z'(p_k), p_k = k\tau_1$, and $\nu_k = \begin{cases} 1, & k = 0, \\ 2, & k \geq 1. \end{cases}$

Lemma 3.4. *For any given $\sigma > 0$, the PC approximation $\Pi_{N_1}^{\alpha,\beta} v$ is stable for $\beta \geq \frac{1}{2}$ and $\tau_1 \leq 1$.*

Proof. By (3.120), it follows that

$$\begin{aligned}
\left\| \Pi_{N_1}^{\alpha,\beta} v \right\|_{L^2(\Omega)} &\leq C \frac{t^{\beta-1}\tau_1}{2\pi} \sum_{k=-N_1}^{N_1} |z'_k e^{z_k}| |z_k|^{-\beta} \|v\|_{L^2(\Omega)} \\
&= C \frac{t^{\beta-1}\tau_1}{2\pi} e^{\sigma} \sigma^{-\beta+1} \left(1 + 2\sum_{k=1}^{N_1} e^{-\sigma p_k^2} (1 + p_k^2)^{-\beta+\frac{1}{2}} \right) \|v\|_{L^2(\Omega)} \\
&\leq C \frac{t^{\beta-1}}{2\pi} e^{\sigma} \sigma^{-\beta+1} \left(\tau_1 + 2\int_0^{\infty} e^{-\sigma p^2} (1 + p^2)^{-\beta+\frac{1}{2}} dp \right) \|v\|_{L^2(\Omega)}
\end{aligned}$$

which completes the proof. □

In order to determine the optimal parameters τ_1, σ, and N_1, we extend the function $z(p)$ defined for $p \in \mathbb{R}$ as an analytic function in a strip $Y = \{p = \xi + i\eta, a_- < \eta < a_+\}$. It is easy to check that the neighbourhood $Z = \{z(p), p \in Y\}$ of the contour Γ_1 lies in $\mathbb{C}/(-\infty, 0]$ for any $0 < a_+ < 1$ and $a_- < 0$. Then according to the convergence result of the trapezoidal rule for the integral over the real line (see the proof by residue calculus of Theorem 5.1 in [Trefethen and Weideman (2014), Section 5.1]), for $\beta \geq \frac{1}{2}$, it holds that

$$\left\| \Pi^{\alpha,\beta} v - \Pi_{N_1}^{\alpha,\beta} v \right\|_{L^2(\Omega)} \leq \frac{M_-}{e^{-2\pi a_-/\tau_1} - 1} + \frac{M_+}{e^{2\pi a_+/\tau_1 - 1}} + TE, \tag{3.121}$$

where

$$\begin{aligned}
M_{\pm} &\leq C \frac{t^{\beta-1}}{2\pi} \int_{-\infty}^{\infty} \left| z'(\zeta + ia_{\pm}) e^{z(\zeta + ia_{\pm})} \right| |z(\zeta + ia_{\pm})|^{-\beta} \|v\|_{L^2(\Omega)} d\zeta \\
&\leq C \frac{t^{\beta-1}\sigma^{-\beta+1}}{\pi} e^{\sigma(1-a_{\pm})^2} \Big(\left((1 - a_{\pm})^2\right)^{-\beta+\frac{1}{2}} \int_0^1 e^{-\sigma\zeta^2} d\zeta \\
&\quad + \int_1^{\infty} e^{-\sigma\zeta^2} d\zeta \Big) \|v\|_{L^2(\Omega)} \\
&\leq C \frac{t^{\beta-1}\sigma^{-\beta+1}}{\pi} e^{\sigma(1-a_{\pm})^2} \left((1 - a_{\pm})^{-2\beta+1} + \frac{\sqrt{\pi}}{2\sigma^{1/2}} \right) \|v\|_{L^2(\Omega)},
\end{aligned}$$

and

$$
\begin{aligned}
TE &= C\frac{t^{\beta-1}\tau_1}{\pi}\sum_{k\geq N_1+1}|z_k' e^{z_k}|\,|z_k|^{-\beta}\,\|v\|_{L^2(\Omega)}\\
&\leq C\frac{t^{\beta-1}}{\pi}e^{\sigma}\sigma^{-\beta+1}\int_{N_1\tau_1}^{\infty}e^{-\sigma p^2}(1+p^2)^{-\beta+\frac{1}{2}}\,dp\,\|v\|_{L^2(\Omega)}\\
&\leq C\frac{t^{\beta-1}}{2\sqrt{\pi}}e^{\sigma}\sigma^{-\beta+1/2}e^{-\sigma(N_1\tau_1)^2}\,\|v\|_{L^2(\Omega)}\,.
\end{aligned}
$$

For the sufficient small τ_1, by choosing $a_- = 1 - \frac{\pi}{\sigma\tau_1} < 0$, we have

$$
\left(e^{\sigma(1-a_-)^2+\frac{2\pi a_-}{\tau_1}}\right)_{\min} = e^{-\frac{\pi^2}{\sigma\tau_1^2}+\frac{2\pi}{\tau_1}}.
$$

To determine the parameters (i.e., σ, τ_1, N_1), one can balance the orders of magnitude of three error terms in (3.121), i.e.,

$$
e^{-\frac{\pi^2}{\sigma\tau_1^2}+\frac{2\pi}{\tau_1}} \simeq e^{-\frac{2\pi a_+}{\tau_1}}(1-a_+)^{-2\beta+1} \simeq e^{\sigma-\sigma(N_1\tau_1)^2},
$$

where the factor $(1-a_+)^{-2\beta+1}$ can not be directly discarded since $(1-a_+)^{-2\beta+1} \to +\infty$ when $a_+ \to 1$. Note that

$$
e^{-\frac{2\pi a_+}{\tau_1}}(1-a_+)^{-2\beta+1} = e^{-\frac{2\pi a_+}{\tau_1}}(1-a_+)^{-2(\beta+1)+3}. \tag{3.122}
$$

Then the algorithms developed in [Garrappa (2015), Subsection 3.2.2] for computing the Mittag-Leffler function can be used here by replacing the corresponding β with $\beta+1$; and they produce good numerical results for all $\beta \geq \frac{1}{2}$.

Remark 3.6. In the two kinds of approximations, the quadrature points $\{z_k\}$ are independent of the time, so they can be pre-computed and stored.

3.3.3 *Numerical results*

We give a numerical experiment to assess the computational performance and effectiveness of the numerical schemes. The codes to generate the quadrature points z_k (c_k) in the CF and PC schemes are obtained by modifying the codes given in [Trefethen *et al.* (2006); Garrappa (2015)], respectively; we choose $N_1 = 16$ for the CF scheme, and parameters of the PC scheme are adaptively produced by code itself.

Example 3.9. Consider

$$
{}_0^C D_t^{\alpha} u(x,t) = \frac{1}{\pi^2}\frac{\partial u(x,t)}{\partial x} + \left(\frac{\Gamma(\gamma+1)t^{\gamma-\alpha}}{\Gamma(\gamma-\alpha+1)} + t^{\gamma} + 1\right) sin(\pi x) \tag{3.123}
$$

for $(x,t) \in (0,1)\times(0,1]$, with the exact solution $u(x,t) = (t^{\gamma}+1)\sin(\pi x)$.

The numerical simulations at $t = 1$ are listed in Tables 3.8 and 3.9, where the quadratic elements are used for space discretizations, and the first and quadratic Lagrange interpolation polynomials are, respectively, used for the time interpolation of $P_h f$. For comparison, the finite element scheme with the $2-\alpha$ order time discretization (i.e., the L1 approximation, see (5.51) of Chap. 5.2 for details)

$$
\begin{aligned}
&{}_0D_t^\alpha u(x,t)\,|_{t=t_m}\\
&= \frac{\tau^{-\alpha}}{\Gamma(2-\alpha)}\left[\sum_{j=0}^{m-1}\left(Q_j^\alpha - Q_{j-1}^\alpha\right)u(x,t_{m-j}) - Q_{m-1}^\alpha u(x,t_0)\right] + \mathcal{O}(\tau^{2-\alpha})
\end{aligned}
$$

is also provided in the last two columns. Here $Q_j^\alpha = (j+1)^{1-\alpha} - j^{1-\alpha}$ for $j = 0, 1, \cdots, m-1$ and $Q_{-1}^\alpha := 0$.

Table 3.8: The numerical performance of Example 3.9 with the first interpolation in time. Here '$(\mathrm{CF}, 2^J)$' denotes that the CF scheme and 2^J equidistant partitions in space and time are used, being similar for the other ones.

(α,γ)	J	(CF, 2^J) L2-Err	rate	(PC, 2^J) L2-Err	rate	(L1, 2^J) L2-Err	rate
	7	7.7672e-06	-	9.2477e-06	-	4.4775e-04	-
$(0.6, 0.6)$	8	2.5756e-06	1.5925	3.0289e-06	1.6103	2.2033e-04	1.0230
	9	8.5323e-07	1.5939	9.9381e-07	1.6077	1.0890e-04	1.0166
	7	2.1538e-05	-	2.1529e-05	-	1.9084e-03	-
$(0.8, 3)$	8	5.3885e-06	1.9989	5.3876e-06	1.9986	8.3345e-04	1.1952
	9	1.3476e-06	1.9995	1.3477e-06	1.9992	3.6348e-04	1.1972

Table 3.9: The numerical performance of Example 3.9 with the quadratic interpolation in time.

(α,γ)	J	(CF, 2^J) L2-Err	rate	(PC, 2^J) L2-Err	rate	(L1, 2^J) L2-Err	rate
	7	2.7777e-06	-	9.6335e-07	-	4.4775e-04	-
$(0.6, 0.6)$	8	9.1590e-07	1.6006	3.1656e-07	1.6056	2.2033e-04	1.0230
	9	3.0423e-07	1.5900	1.0359e-07	1.6116	1.0890e-04	1.0166
	7	1.2024e-07	-	1.2024e-07	-	1.9084e-03	-
$(0.8, 3)$	8	1.5030e-08	3.0000	1.5030e-08	3.0000	8.3345e-04	1.1952
	9	1.8795e-09	2.9994	1.8788e-09	3.0000	3.6348e-04	1.1972

Chapter 4

High order finite difference methods for the space fractional PDEs

Nowadays, many numerical methods have been proposed to solve the fractional PDEs, such as the finite difference methods [Çelik and Duman (2012); Meerschaert and Tadjeran (2004, 2006); Yang *et al.* (2010)], the finite element methods [Ervin and Roop (2006); Deng (2008); Wang and Zhang (2015); Jin *et al.* (2015a)], and the spectral methods [Li and Xu (2009, 2010); Zayernouri and Karniadakis (2014b)], etc. Among them, the finite difference methods are the most popular and dominant methods. They are usually formulated by using the equivalence of the Riemann-Liouville and Grünwald-Letnikov fractional derivatives [Podlubny (1999)]. However, the difference scheme based on the standard Grünwald-Letnikov formula for time dependent problems is unstable [Meerschaert and Tadjeran (2004)]. To overcome this problem, Meerschaert and Tadjeran in [Meerschaert and Tadjeran (2004, 2006); Meerschaert *et al.* (2006)] present the shifted Grünwald-Letnikov formula to approximate the fractional advection-dispersion flow or the purely dispersion equations, and then the extrapolation techniques in space are used to increase the spatial convergence rate to the second-order [Tadjeran *et al.* (2006)]. Later, Ortigueira [Ortigueira (2006)] gives the "fractional centred derivative" to approximate the Riesz fractional derivative with second order accuracy, and this method is used by Çelik and Duman in [Çelik and Duman (2012)] to approximate the fractional diffusion equation with the Riesz fractional derivative in a finite domain. On the other hand, the second order approximations to Riemann-Liouville fractional derivatives based on the piecewise linear interpolation are also studied. For example, Sousa and Li propose a second order discretization for Riemann-Liouville fractional derivative and established an unconditionally stable weighted average finite difference method for the one-dimensional fractional diffusion equation [Sousa and Li (2015)], and the results in the two-dimensional two-sided space fractional convection diffusion equation in finite domain can be found in [Chen and Deng (2014b)].

In fact, in numerically solving fractional PDEs, besides a little big complex numerical analysis, one of the biggest challenges comes from the computational cost caused by the nonlocal properties of fractional operators. High order scheme is a natural idea to release the challenge of cost. Comparing the first order schemes, the high order schemes for fractional operators do not increase computational cost but may greatly improve the accuracy. The reason is that both the derived differential matrices corresponding to the high-order schemes and low-order schemes are full and have the same structure, the computational cost for solving the matrix algebraic equations is almost the same [Chen *et al.* (2014)].

The outline of this chapter is as follows. In Sec. 4.1, we describe some basic approximations to the (tempered) Riemann-Liouville (R-L) fractional derivatives, which will be used to develop the high order schemes for solving the space fractional PDEs. For the sake of self-containedness, in Sec. 4.2, we give the numerical schemes of using the first order shifted Grünwald-Letnikov formula to solve the fractional advection-diffusion equations with variable coefficients, and the detailed theoretical analysis is also presented. In Sec. 4.3 and Sec. 4.4, we introduce the weighted and shifted Grünwald difference (WSGD) operators and the quasi-compact WSGD operators, which have the second and third approximation accuracy, respectively. By weighting and shifting Lubich's second order discretization, a class of effective fourth order approximation schemes are provided in Sec. 4.5, and the detailed stability and convergence analysis is also presented. Finally, in Sec. 4.6 we extend the proposed algorithms to the tempered fractional derivatives, and the second order approximations are provided to solve the fractional substantial differential equation with truncated Lévy flights.

4.1 Basic approximations to the fractional derivatives and some results on matrix analysis

In this section, we first give some basic approximation formulae to the (tempered) R-L fractional derivatives, and then some lemmas on matrix analysis. They will be used in the subsequent sections.

4.1.1 *Basic approximations to the Riemann-Liouville fractional derivatives*

We first discuss the Grünwald-Letnikov (G-L) approximation, and then more general Lubich's fractional backward difference approximations to the R-L fractional derivatives.

For any $\beta \geq 0$, the left and right R-L fractional derivatives ${}_{-\infty}D_x^{\beta}u$ and ${}_xD_{\infty}^{\beta}u$ on $\mathbb{R}$ of order β are, respectively, defined by

$$ {}_{-\infty}D_x^{\beta}u(x) := \frac{d^n}{dx^n}\left({}_{-\infty}D_x^{-(n-\beta)}u(x)\right) \tag{4.1}$$

and

$$ {}_xD_{\infty}^{\beta}u(x) := (-1)^n\frac{d^n}{dx^n}\left({}_xD_{\infty}^{-(n-\beta)}u(x)\right) \tag{4.2}$$

where $n = [\beta]+1$, $[\beta]$ is the integer part of β, and ${}_{-\infty}D_x^{-(n-\beta)}u(x)$ and ${}_xD_{\infty}^{-(n-\beta)}u(x)$ denote the left and right R-L fractional integrals, respectively, i.e.,

$$ {}_{-\infty}D_x^{-(n-\beta)}u(x) = \frac{1}{\Gamma(n-\beta)}\int_{-\infty}^{x}(x-\xi)^{n-\beta-1}u(\xi)d\xi, \tag{4.3}$$

and

$$ {}_xD_{\infty}^{-(n-\beta)}u(x) = \frac{1}{\Gamma(n-\beta)}\int_{x}^{\infty}(\xi-x)^{n-\beta-1}u(\xi)d\xi. \tag{4.4}$$

Obviously, if $u(x) = 0$ for $x < a$, we have

$$ {}_{-\infty}D_x^{\beta}u(x) = {}_aD_x^{\beta}u(x) := \frac{d^n}{dx^n}\left({}_aD_x^{-(n-\beta)}u(x)\right) \quad \text{for } x > a, \tag{4.5}$$

and if $u(x) = 0$ for $x > b$, we have

$$ {}_xD_{\infty}^{\beta}u(x) = {}_xD_b^{\beta}u(x) := (-1)^n\frac{d^n}{dx^n}\left({}_xD_b^{-(n-\beta)}u(x)\right) \quad \text{for } x < b, \tag{4.6}$$

where

$$ {}_aD_x^{-(n-\beta)}u(x) = \frac{1}{\Gamma(n-\beta)}\int_{a}^{x}(x-\xi)^{n-\beta-1}u(\xi)d\xi, \tag{4.7}$$

$$ {}_xD_b^{-(n-\beta)}u(x) = \frac{1}{\Gamma(n-\beta)}\int_{x}^{b}(\xi-x)^{n-\beta-1}u(\xi)d\xi. \tag{4.8}$$

For the detailed definitions and properties of the fractional derivatives and integrals, see Appendix B.

4.1.1.1 *The standard G-L approximations to the R-L fractional derivatives*

It is well known that a function $u(x)$ which is differentiable up to order $n\,(n \in \mathbb{N})$ admits the formula

$$u^{(n)}(x) = \lim_{h\to 0} \frac{1}{h^n} \sum_{k=0}^{\infty} (-1)^k \binom{n}{k} u(x-kh),$$

where $\binom{n}{k}$ is the special case of

$$\binom{\beta}{\gamma} := \frac{\Gamma(\beta+1)}{\Gamma(\gamma+1)\Gamma(\beta-\gamma+1)}$$

with $\beta = n$ and $\gamma = k$, having the properties (see Samko *et al.*, 1993, p. 15) that $\binom{n}{k} = 0$ for $k > n$, and

$$\left|\binom{\beta}{\gamma}\right| \leq \frac{c}{\gamma^{1+\beta}} \tag{4.9}$$

for any fixed $\beta\,(\neq -1, -2, \cdots)$ and real $\gamma \to +\infty$.

Replacing the order of derivative n by $\beta > 0$, leads to the so-called left G-L fractional derivative

$${}^{G}_{-\infty}D^{\beta}_{x}u(x) := \lim_{h\to 0} \frac{1}{h^\beta} \sum_{k=0}^{\infty} (-1)^k \binom{\beta}{k} u(x-kh). \tag{4.10}$$

Obviously, for a function $u(x)$ satisfying $u(x) = 0$ for $x < a$, one has

$${}^{G}_{-\infty}D^{\beta}_{x}u(x) = {}^{G}_{a}D^{\beta}_{x}u(x) := \lim_{h\to 0} \frac{1}{h^\beta} \sum_{k=0}^{\left[\frac{x-a}{h}\right]} (-1)^k \binom{\beta}{k} u(x-kh), \ x > a. \tag{4.11}$$

Similarly, one can define the right G-L fractional derivative

$${}^{G}_{x}D^{\beta}_{\infty}u(x) := \lim_{h\to 0} \frac{1}{h^\beta} \sum_{k=0}^{\infty} (-1)^k \binom{\beta}{k} u(x+kh), \tag{4.12}$$

and when $u(x) = 0$ for $x > b$, we have

$${}^{G}_{x}D^{\beta}_{\infty}u(x) = {}^{G}_{x}D^{\beta}_{b}u(x) := \lim_{h\to 0} \frac{1}{h^\beta} \sum_{k=0}^{\left[\frac{b-x}{h}\right]} (-1)^k \binom{\beta}{k} u(x+kh), \ x < b. \tag{4.13}$$

Lemma 4.1 ([Diethelm (2010); Podlubny (1999)]). *Let $n-1 < \beta \leq n, n \in \mathbb{N}$ and $u(x) \in C^n[a,b]$. Then for $x \in (a,b]$,*

$${}_{a}D^{\beta}_{x}u(x) = {}^{G}_{a}D^{\beta}_{x}u(x). \tag{4.14}$$

The G-L definition of fractional derivative is important for our purposes because it allows to estimate the R-L derivative numerically in a simple and efficient way:

Lemma 4.2 ([Weilbee (2005); Podlubny (1999)]). *Let $\beta \geq 0, n = \lceil \beta \rceil$, $u(x) \in C^n[a,b]$, and $h = \frac{x-a}{N}$. Then the finite G-L differential operator*

$$\frac{1}{h^\beta}\sum_{k=0}^{N}(-1)^k\binom{\beta}{k}u(x-kh) \tag{4.15}$$

yields a first order approximation for the R-L differential operator ${}_aD_x^\beta$ if and only if $u(a) = 0$, i.e.,

$${}_aD_x^\beta u(x) = \frac{1}{h^\beta}\sum_{k=0}^{N}(-1)^k\binom{\beta}{k}u(x-kh) + \mathcal{O}(h). \tag{4.16}$$

4.1.1.2 *The shifted G-L approximations to the R-L fractional derivatives*

Unfortunately, Meerschaert and Tadjeran (see Meerschaert and Tadjeran, 2004, 2006) show that the difference discretization with (4.16) results in unstable numerical methods for fractional diffusion equation of order β $(1 < \beta < 2)$, even for some implicit methods that are well known to be stable for integer derivative. Therefore, the shifted G-L difference operators

$${}_LA_{h,p}^\beta u(x) := \frac{1}{h^\beta}\sum_{k=0}^{\infty}(-1)^k\binom{\beta}{k}u\left(x-(k-p)h\right) \tag{4.17}$$

and

$${}_RA_{h,p}^\beta u(x) := \frac{1}{h^\beta}\sum_{k=0}^{\infty}(-1)^k\binom{\beta}{k}u\left(x+(k-p)h\right) \tag{4.18}$$

are introduced.

To discuss the approximation properties of these shifted G-L operators, for $\beta > 0$, we define the set

$$\mathscr{S}^{\beta+m}(\mathbb{R}) = \left\{u(x) \in L^1(\mathbb{R}) : \int_{-\infty}^{\infty}(1+|\omega|)^{\beta+m}|\hat{u}(\omega)|d\omega < \infty\right\}. \tag{4.19}$$

We first give some characteristics of $\mathscr{S}^{\beta+m}(\mathbb{R})$.

Lemma 4.3. *Let $u(x) \in \mathscr{S}^{\beta+m}(\mathbb{R})$. Then for each $k = 0, 1, \cdots, m + [\beta]$, $u^{(k)}(x)$ is bounded and uniformly continuous, and $\lim_{|x|\to\infty} u^{(k)}(x) = 0$.*

Proof. Since $u(x), \hat{u}(\omega) \in L^1(\mathbb{R})$, we have

$$u(x) = \frac{1}{2\pi} \int_{\mathbb{R}} e^{i\omega x} \hat{u}(\omega) d\omega. \tag{4.20}$$

Note that for $n = 1, 2, \cdots, [\beta] + m$, $|e^{i\omega x}(i\omega)^n \hat{u}(\omega)| \leq (1 + |w|)^{\beta+m} |\hat{u}(\omega)|$, which implies the uniform convergence of $\int_{\mathbb{R}} e^{i\omega x}(i\omega)^n \hat{u}(\omega) d\omega$ with respect to the parameter x. Then the differentiation on the right side of (4.20) can be performed on its integration, in fact, this action can be successively proceeded, so

$$u^{(k)}(x) = \frac{i^k}{2\pi} \int_{\mathbb{R}} e^{i\omega x} w^k \hat{u}(\omega) d\omega. \tag{4.21}$$

By the Riemann-Lebesgue Lemma (see Appendix A), one has $u^{(k)}(x) \to 0$ for $|x| \to \infty$, and obviously,

$$\left| u^{(k)}(x) \right| \leq \frac{1}{2\pi} \int_{\mathbb{R}} \left| e^{i\omega x} w^k \hat{u}(\omega) \right| d\omega < \infty. \tag{4.22}$$

In the following, we will prove the continuity. For $x_1, x_2 \in \mathbb{R}$,

$$\left| u^{(k)}(x_1) - u^{(k)}(x_2) \right| \leq \frac{1}{2\pi} \int_{\mathbb{R}} \left| \omega^k \hat{u}(\omega) \right| \left| e^{i\omega(x_1 - x_2)} - 1 \right| d\omega. \tag{4.23}$$

By $\int_{\mathbb{R}} |w^k \hat{u}(\omega)| d\omega < \infty$, one has

$$\lim_{H \to \infty} \int_{\{\omega : |\omega| > H\}} |\omega^k \hat{u}(\omega)| d\omega = 0. \tag{4.24}$$

Therefore, for any $\epsilon > 0$, there exists $H_1 > 0$ such that

$$\begin{aligned} &\frac{1}{2\pi} \int_{\{\omega : |\omega| > H_1\}} |\omega^k \hat{u}(\omega)| |e^{i\omega(x_1 - x_2)} - 1| d\omega \\ &\qquad \leq \frac{1}{\pi} \int_{\{\omega : |\omega| > H_1\}} |\omega^k \hat{u}(\omega)| d\omega < \frac{\epsilon}{2}. \end{aligned} \tag{4.25}$$

Choosing $\sigma = \frac{\pi \epsilon}{H_1 C_0}$ with $C_0 = \int_{\mathbb{R}} |w^k \hat{u}(\omega)| d\omega < \infty$, then for $|x_1 - x_2| \leq \sigma$, one has

$$\begin{aligned} &\frac{1}{2\pi} \int_{-H_1}^{H_1} |\omega^k \hat{u}(\omega)| \left| e^{-i\omega(x_1 - x_2)} - 1 \right| d\omega \\ &\qquad \leq \frac{1}{2\pi} \int_{-H_1}^{H_1} |\omega^k \hat{u}(\omega)| \left| \omega(x_1 - x_2) \right| d\omega \leq \frac{\epsilon}{2}, \end{aligned} \tag{4.26}$$

where the inequality $\left| e^{iz} - 1 \right| \leq |z|$, $z \in \mathbb{R}$ has been used. Combining (4.25) and (4.26) leads to the desired result. □

Remark 4.1. In fact, using the knowledge of harmonic analysis, a more precise result can be given as: if $u(x) \in \mathscr{S}^{\beta+m}(\mathbb{R})$, then $u \in C^{n,\mu}(\mathbb{R})$, where $n = [\beta] + m$ and $\mu = \beta - [\beta]$.

Lemma 4.4. *Let $\beta > 0$. If $u(x)$ and all of its derivatives up to $m + [\beta] + 2$ exist and belong to $L^1(\mathbb{R})$, then $u \in \mathscr{S}^{\beta+m}(\mathbb{R})$.*

Proof. Since $u(x)$ and all of its derivatives up to $m + [\beta] + 2$ belong to $L^1(\mathbb{R})$, by Lemma A.15, there exists $M \geq 1$ such that for $|\omega| \geq M$

$$|\hat{u}(\omega)| \leq \frac{C_1}{|\omega|^{m+[\beta]+2}} \leq \frac{2C_1}{1+|\omega|^{m+[\beta]+2}}.$$

In addition, by $|\hat{u}(\omega)| \leq \int_{\mathbb{R}} |u(x)|dx$ one has

$$|\hat{u}(\omega)| \leq \frac{C_2}{1+|\omega|^{m+[\beta]+2}} \quad \text{for } |\omega| \leq M,$$

where $C_2 = \left(1 + M^{m+[\beta]+2}\right) \int_{\mathbb{R}} |u(x)|dx$. Therefore,

$$(1+|\omega|)^{\beta+m}|\hat{u}(\omega)| \leq 2^{\beta+m} \max\{2C_1, C_2\} \frac{1+|\omega|^{\beta+m}}{1+|\omega|^{m+[\beta]+2}} \in L^1(\mathbb{R}).$$

Thus, we complete the proof. □

Theorem 4.1. *Suppose that $u \in \mathscr{S}^{\beta+1}(\mathbb{R})$ and $u(x)$ have compact support. Then*

$$_LA_{h,p}^{\beta}u(x) := {}_{-\infty}D_x^{\beta}u(x) + \mathcal{O}(h) \tag{4.27}$$

and

$$_RA_{h,p}^{\beta}u(x) := {}_xD_{\infty}^{\beta}u(x) + \mathcal{O}(h) \tag{4.28}$$

uniformly in $x \in \mathbb{R}$ as $h \to 0$.

Proof. Here we only give the proof of (4.28), being similar for (4.27).

Making use of Lemma 4.3, it holds that $u \in C^{1+[\beta]}(\mathbb{R})$ and $\|u\|_{L^\infty(\mathbb{R})} \leq \|\hat{u}(\omega)\|_{L^\infty(\mathbb{R})}$. Since $u(x)$ has compact support, by the Heine-Borel theorem, there exists $r \in (0, +\infty)$ such that $\operatorname{supp} u(x) = 0$ for $|x| \geq r$; combining with (B.21), we have

$$\mathscr{F}\left[{}_xD_{\infty}^{\beta}u(x)\right](\omega) = (-i\omega)^{\beta}\hat{u}(\omega). \tag{4.29}$$

By (4.9), we have

$$\sum_{k=0}^{\infty}\left|(-1)^k\binom{\beta}{k}\right| < \infty, \tag{4.30}$$

then the series (4.18) converges absolutely and uniformly for each bounded function $u(x)$. Hence

$$\left\| {}_R A_{h,p}^{\beta} u(x) \right\|_{L^1(\mathbb{R})} \leq \sum_{k=0}^{\infty} \left| \binom{\beta}{k} \right| \| u(x + (k-p)h) \|_{L^1(\mathbb{R})} < \infty \quad (4.31)$$

and therefore

$$\begin{aligned} \mathscr{F}[{}_R A_{h,p}^{\beta} u(x)](\omega) &= \frac{1}{h^\beta} e^{-i\omega p h} \sum_{k=0}^{\infty} (-1)^k \binom{\beta}{k} e^{ikh\omega} \hat{u}(\omega) \\ &= \frac{1}{h^\beta} e^{-iph\omega} \left(1 - e^{ih\omega}\right)^\beta \hat{u}(\omega) \\ &= (-i\omega)^\beta \left(\frac{1 - e^{i\omega h}}{-i\omega h} \right)^\beta e^{-iph\omega} \hat{u}(\omega), \end{aligned} \quad (4.32)$$

where the properties $\widehat{u(x-c)}(\omega) = e^{-ic\omega}\hat{u}(\omega)$ and $(1+z)^\beta = \sum_{k=0}^{\infty} \binom{\beta}{k} z^k$ for $|z| \leq 1$ have been used.

Denote $W_{\beta,p}(z) = e^{pz} \left(\frac{1-e^{-z}}{z} \right)^\beta$. Since $W_{\beta,p}(z)$ is analytic in some neighborhood of the origin we have the power series expansion $W_{\beta,p}(z) = \sum_{l=0}^{\infty} a_l z^l$, which converges absolutely for all $|z| \leq \mathcal{X}$ and some $\mathcal{X} > 0$. Next we will show that if $|\mathrm{Re}(z)|$ is bounded (i.e., $|\mathrm{Re}(z)| \leq M$ for some constant M), then there exists a constant $C_1 > 0$ such that

$$\left| W_{\beta,p}(z) - \sum_{l=0}^{n-1} a_l z^l \right| \leq C_1 |z|^n, \quad n \in \mathbb{N}^+ \quad (4.33)$$

uniformly in $z \in \mathbb{C}$. In fact, for $|z| \leq \mathcal{X}$ we have

$$\begin{aligned} &\left| W_{\beta,p}(z) - \sum_{l=0}^{n-1} a_l z^l \right| \\ &\qquad \leq \left| \sum_{l=n}^{\infty} a_l z^l \right| \leq |z|^n \sum_{l=n}^{\infty} |a_l| |z|^{l-n} \leq C_2 |z|^n, \end{aligned}$$

where $C_2 = \mathcal{X}^{-n} \sum_{l=0}^{\infty} |a_l| \mathcal{X}^l < \infty$; while for $|z| > \mathcal{X}$ we have

$$|W_{\beta,p}(z)| \leq \left| \left(\frac{1-e^{-z}}{z} \right)^\beta e^{pz} \right| \leq \frac{e^{|p|M}(1+e^M)^\beta}{\mathcal{X}^\beta} \leq C_3 |z|^n,$$

where $C_3 = \frac{e^{|p|M}(1+e^M)^\beta}{\mathcal{X}^{\beta+n}} < \infty$, and

$$\left| \sum_{l=0}^{n-1} a_l z^l \right| \leq |z|^n \sum_{l=0}^{n-1} |a_l| |z|^{l-n} \leq C_4 |z|^n, \quad (4.34)$$

where $C_4 = \sum_{l=0}^{n-1} |a_l| \mathcal{X}^{l-n} < \infty$. Now if we set $C_1 = \max\{C_2, C_3 + C_4\}$ then (4.33) holds for all $z \in \mathbb{C}$.

Note that $a_0 = 1, a_1 = p - \frac{\beta}{2}, a_2 = \frac{1}{24}\left(\beta + 3\beta^2 - 12\beta p + 12p^2\right), a_3 = \frac{1}{48}\left(\beta^3 + (1-6p)\beta^2 + 2p(6p-1)\beta - 8p^3\right), \cdots$. In view of (4.32) we can write

$$\begin{aligned}\mathscr{F}[{}_RA_{h,p}^{\beta}u(x)](\omega) &= (-i\omega)^{\beta}u(x) + (-i\omega)^{\beta}\left(W_{\alpha,p}(-i\omega h) - 1\right)\hat{u}(\omega)] \\ &= \mathscr{F}\left[{}_xD_{\infty}^{\beta}u(x)\right](\omega) + \hat{\phi}(h,\omega), \qquad (4.35)\end{aligned}$$

where $\hat{\phi}(h,\omega) := (-i\omega)^{\beta}\left(W_{\beta,p}(-i\omega h) - 1\right)\hat{u}(\omega)$. Using (4.33), it follows that $\left|\hat{\phi}(h,\omega)\right| \leq |\omega|^{\beta}C_1|h\omega||\hat{u}(\omega)|$. By

$$C_5 = \int_{-\infty}^{\infty}(1+|\omega|)^{\beta+1}|\hat{u}(\omega)|d\omega < \infty$$

and the inverse Fourier transform, we finally obtain

$$|\phi(h,x)| \leq \left|\frac{1}{2\pi}\int_{-\infty}^{\infty}e^{-i\omega x}\hat{\phi}(h,\omega)d\omega\right| \leq \frac{C_1C_5}{2\pi}h \qquad (4.36)$$

uniformly in $x \in \mathbb{R}$. Thus, we complete the proof. □

Similar to the proof of Theorem 4.1, we have

Corollary 4.1. *If $u(x) \in \mathscr{S}^{\beta+m}(\mathbb{R})$ and $u(x)$ has compact support, then*

$${}_LA_{h,p}^{\beta}u(x) = {}_{-\infty}D_x^{\beta}u(x) + \sum_{l=1}^{m-1} a_l\, {}_{-\infty}D_x^{\beta+l}u(x)h^l + \mathcal{O}(h^m) \qquad (4.37)$$

and

$${}_RA_{h,p}^{\beta}u(x) = {}_xD_{\infty}^{\beta}u(x) + \sum_{l=1}^{m-1} a_l\, {}_xD_{\infty}^{\beta+l}u(x)h^l + \mathcal{O}(h^m). \qquad (4.38)$$

Since $a_1 = p - \frac{\beta}{2}$, it is easy to see that when $u \in \mathscr{S}^{\beta+2}(\mathbb{R})$, a second-order approximation can be obtained by choosing $p = \frac{\beta}{2}$. However, when using it to solve the fractional PDEs, we have to use the interpolation to make the computation performed on grid $\{x_n\}$. Usually, the optimal integer p should be the one that makes p minimum. It is obviously that $p = 0$ for $0 < \beta \leq 1$ and $p = 1$ for $1 < \beta \leq 2$ are optimal.

Remark 4.2. If a function $u(x)$ satisfies $u(x) = 0$ for $x \leq a$, and we only consider the approximation on finite interval (a, b), then

$${}_LA_{h,p}^{\beta}u(x) = \frac{1}{h^{\beta}}\sum_{k=0}^{[(x-a)/h+p]}(-1)^k\binom{\beta}{k}u\left(x-(k-p)h\right), \quad x \in (a,b). \qquad (4.39)$$

Assuming $u(x) \in C^{[\beta]+m+2}[a,b]$ with $u^{(k)}(a) = 0, k = 0, 1, \cdots, [\beta]+m+2$, we introduce a function $u^*(x) : \mathbb{R} \to \mathbb{R}$ such that

$$u^*(x) = \begin{cases} u(x), & x \in [a,b], \\ v(x), & x \in (b,b'], \\ 0, & \text{otherwise}, \end{cases} \tag{4.40}$$

where $v(x)$ is the smooth extension of $u(x)$, satisfying $v^{(k)}(b') = 0$ for $k = 0, 1, \cdots, m + [\beta] + 2$. Then by Lemma 4.4 and Corollary 4.1, for $x \in (a, b]$, one has

$$\begin{aligned} {}_LA^{\beta}_{h,p}u(x) &= {}_{-\infty}D^{\beta}_{x}u^*(x) + \sum_{l=1}^{m-1} a_l \, {}_{-\infty}D^{\beta+l}_{x}u^*(x)h^l + \mathcal{O}(h^m) \\ &= {}_aD^{\beta}_{x}u(x) + \sum_{l=1}^{m-1} a_l \, {}_aD^{\beta+l}_{x}u(x)h^l + \mathcal{O}(h^m). \end{aligned} \tag{4.41}$$

The conditions required by (4.41) are far stronger than the ones in Corollary 4.2 ($p = 0$), even in Theorem 4.1, which are caused by the way of proof, i.e., instead of proving (4.41) directly on the given interval $[a, b]$ [Podlubny (1999); Weilbee (2005)], we have extended it to the whole real axis and then used the Fourier transform, where the global smoothness is needed. However, we will see that this Fourier analysis method provides useful guidances to construct the high order discretizations and carried out error analysis.

4.1.1.3 *Properties of the coefficients of the G-L approximation*

In fact, the coefficients $g^{\beta}_k = (-1)^k \binom{\beta}{k}, k = 0, 1, \cdots$ in (4.17) and (4.18) are the coefficients of Taylor expansion of the function $\delta^{\beta,1}(z) := (1-z)^{\beta}$, i.e.,

$$(1-z)^{\beta} = \sum_{k=0}^{\infty} g^{\beta}_k z^k, \quad |z| \le 1. \tag{4.42}$$

Therefore, we usually call $(1-z)^{\beta}$ the generating function of coefficients g^{β}_k, which is the βth power of the generating function of the first order backward multi-step method for the first order differential equation. It is easy to check that the coefficients g^{β}_k have the following properties.

Proposition 4.1. *The coefficients $g^{\beta}_k, k = 0, 1, \cdots$ satisfy*

(1) With small random error g^{β}_k can be obtained by

$$g^{\beta}_0 = 1, \; g^{\beta}_1 = -\beta, \; g^{\beta}_k = \left(1 - \frac{\beta+1}{k}\right) g^{\beta}_{k-1}, \; k = 1, 2, \cdots. \tag{4.43}$$

(2) If $0 < \beta < 1$, *then*

$$g_k^\beta < 0\,(k > 1),\ \sum_{k=0}^{\infty} g_k^\beta = 0,\ \ \sum_{k=0}^{N} g_k^\beta > 0\,(\forall N \in \mathbb{N}). \tag{4.44}$$

(3) If $1 < \beta < 2$, *then*

$$g_1^\beta < 0,\ g_k^\beta > 0\,(k \neq 1),\ \sum_{k=0}^{\infty} g_k^\beta = 0,\ \ \sum_{k=0}^{N} g_k^\beta < 0\,(\forall N \in \mathbb{N}). \tag{4.45}$$

Proposition 4.1 is very important in performing the stability and convergence analyses of the numerical schemes for fractional PDEs.

4.1.1.4 *Shifted Lubich's fractional difference approximations to the R-L fractional derivatives*

We have already seen that the function $(1-z)$ generates the coefficients for the first order approximation of the first order derivative, and its βth power, the function $(1-z)^\beta$, generates the coefficients of the first order approximation of the βth order derivative. In fact, using the generating function $\delta^K(z) := \sum_{l=1}^{K} \frac{1}{l}(1-z)^l$ of Kth order ($K = 1, \cdots, 6$) backward multi-step method for the first order derivative, Lubich [Lubich (1986)] has developed the corresponding Kth order fractional multi-step method (i.e., taking $p = 0$ in (4.47)) for the βth order derivative with the generating function

$$\delta^{\beta,K}(z) := \left(\sum_{l=1}^{K} \frac{1}{l}(1-z)^l\right)^\beta = \sum_{k=0}^{\infty} g_k^{\beta,K} z^k. \tag{4.46}$$

However, as Meerschaert et al. have pointed out, for the time dependent equations the non-shifted difference schemes are unstable. Therefore, here we define and check the shifted Lubich fractional difference operators

$$_L A_p^\beta u(x) := \frac{1}{h^\beta} \sum_{k=0}^{\infty} g_k^{\beta,K} u\left(x - (k-p)h\right),\ \ K = 1, 2, \cdots, 6. \tag{4.47}$$

We have the following results.

Theorem 4.2. *Let* $u \in \mathscr{S}^{\beta+1}(\mathbb{R})$ *and* $u(x)$ *have compact support. Then*

$$_{-\infty}D_x^\beta u(x) = {}_L A_p^\beta u(x) + \mathcal{O}(h) \tag{4.48}$$

uniformly in $x \in \mathbb{R}$ *as* $h \to 0$. *In particular, if* $p = 0$ *and* $u \in \mathscr{S}^{\beta+K}(\mathbb{R})$, *then*

$$_{-\infty}D_x^\beta u(x) = {}_L A_p^\beta u(x) + \mathcal{O}(h^K). \tag{4.49}$$

Proof. We only give the proof for $K = 2$; for the other cases, see [Chen and Deng (2014a)].

$$\begin{aligned}
&\mathscr{F}\left[{}_L A_p^\beta u\right](\omega) \\
&= h^{-\beta} e^{i\omega ph} \sum_{k=0}^{\infty} g_k^{\beta,K} \left(e^{-i\omega h}\right)^k \widehat{u}(\omega) \\
&= (i\omega)^\beta \left[e^{i\omega ph} \left(\frac{1-e^{-i\omega h}}{i\omega h} \right)^\beta \right] \left(1 + \frac{1}{2}\left(1 - e^{-i\omega h}\right) \right)^\beta \widehat{u}(\omega) \\
&= (zh^{-1})^\beta e^{pz} \left(\frac{1-e^{-z}}{z} \right)^\beta \left(1 + \frac{1}{2}\left(1-e^{-z}\right) \right)^\beta \widehat{u}(\omega), \qquad (4.50)
\end{aligned}$$

with $z = i\omega h$.

Note that $e^{pz}\left(\frac{1-e^{-z}}{z}\right)^\beta \left(1 + \frac{1}{2}\left(1-e^{-z}\right)\right)^\beta$ is analytic in some neighborhood of the origin, which can be expanded as power series for $|z| \le \mathcal{X}$ and some $\mathcal{X} > 0$. In fact, by

$$\begin{aligned}
&e^{pz}\left(\frac{1-e^{-z}}{z}\right)^\beta \\
&= \Big[1 + \left(p - \frac{\beta}{2}\right) z + \left(\frac{1}{2}p^2 - \frac{\beta}{2}p + \frac{3\beta^2+\beta}{24}\right) z^2 \\
&\quad + \left(\frac{1}{6}p^3 - \frac{\beta}{4}p^2 + \frac{3\beta^2+\beta}{24}p - \frac{\beta^3+\beta^2}{48}\right) z^3 + \cdots \Big],
\end{aligned}$$

and

$$\begin{aligned}
&\left(1 + \frac{1}{2}\left(1-e^{-z}\right)\right)^\beta \\
&= \left[1 + \frac{\beta}{2} z + \frac{\beta(\beta-3)}{8} z^2 + \frac{\beta(\beta^2-9\beta+12)}{48} z^3 + \cdots \right],
\end{aligned}$$

it holds that

$$\begin{aligned}
&e^{pz}\left(\frac{1-e^{-z}}{z}\right)^\beta \left(1 + \frac{1}{2}\left(1-e^{-z}\right)\right)^\beta \\
&= 1 + pz + \frac{3p^2-2\beta}{6} z^2 + \frac{2p^3+\beta(3-4p)}{12} z^3 + \cdots.
\end{aligned}$$

Since $|\mathrm{Re}(z)|$ is bounded, similar to the proof of Theorem 4.1, we have

$$\begin{aligned}
&\Bigg| e^{pz}\left(\frac{1-e^{-z}}{z}\right)^\beta \left(1 + \frac{1}{2}\left(1-e^{-z}\right)\right)^\beta \\
&\quad - \left(1 + pz + \frac{3p^2-2\beta}{6} z^2 + \frac{2p^3+\beta(3-4p)}{12} z^3\right) \Bigg| \le C|z|^4, \qquad (4.51)
\end{aligned}$$

where C is a constant independent of z.

Note that

$$\mathscr{F}\left[{}_LA_p^\beta u\right](\omega) = \mathscr{F}({}_{-\infty}D_x^\beta u(x)) + \widehat{\phi}(\omega, h),$$

where

$$\widehat{\phi}(\omega, h) = (zh^{-1})^\beta \left(pz + \frac{3p^2 - 2\beta}{6} z^2 + \frac{2p^3 + \beta(3 - 4p)}{12} z^3 + \cdots \right) \widehat{u}(\omega).$$

Then using

$$\begin{aligned} |\widehat{\phi}(\omega, h)| &\le \widetilde{c}|i\omega|^{\beta+1}|\widehat{u}(\omega)| \cdot h, \qquad p \ne 0, \\ |\widehat{\phi}(\omega, h)| &\le c|i\omega|^{\beta+2}|\widehat{u}(\omega)| \cdot h^2, \qquad p = 0, \end{aligned}$$

and

$$|\phi(x, h)| \le \frac{1}{2\pi} \int_{\mathbb{R}} |\widehat{\phi}(\omega, h)| dx,$$

we complete the proof. □

Similarly, if one defines

$$_RA_p^\beta u(x) = \frac{1}{h^\beta} \sum_{k=0}^{\infty} g_k^{\beta,K} u(x + (k - p)h), \tag{4.52}$$

then ${}_RA_p^\beta u(x) \to {}_xD_\infty^\beta u(x)$ has the same approximation results as the ones in Theorem 4.2. We will use them to construct the high order approximation schemes in Sec. 4.5.

One of the difficulties in the shifted Lubich fractional difference approximations is in evaluating the weights $g_k^{\beta,K}$, $k = 0, 1, 2, \cdots$. The following lemma states that they can still be computed with the linear complexity.

Lemma 4.5. *Let*

$$\delta^K(z) = \sum_{l=1}^{K} \frac{1}{l}(1 - z)^l = \sum_{l=0}^{K} w_l z^l.$$

Then $g_0^{\beta,K} = w_0^\beta$ *and*

$$g_k^{\beta,K} = \frac{1}{kw_0} \sum_{m=\max\{0,k-K\}}^{k-1} \left[(k - m)\beta - m\right] g_m^{\beta,K} w_{k-m}, \quad k \ge 1.$$

Proof. Obviously, $g_0^{\beta,L} = \delta^{\beta,K}(0) = w_0^\beta$ and $\sum_{k=0}^{\infty} g_k^{\beta,K} = \delta^{\beta,K}(1) = 0$. Note that $\delta^{\beta,K}(z) = (\delta^K(z))^\beta$. By

$$\frac{d\delta^{\beta,K}(z)}{dz} = \beta \frac{d\delta^K(z)}{dz} \left(\delta^K(z)\right)^{\beta-1},$$

one has

$$\frac{d\delta^{\beta,K}(z)}{dz} \delta^K(z) = \beta \frac{d\delta^K(z)}{dz} \delta^{\beta,K}(z),$$

which means

$$\left(\sum_{m=1}^{\infty} k g_m^{\beta,K} z^{m-1}\right) \left(\sum_{l=0}^{K} w_l z^l\right) = \beta \left(\sum_{l=1}^{K} l w_l z^{l-1}\right) \left(\sum_{m=0}^{\infty} g_m^{\beta,K} z^m\right).$$

Checking the coefficient of z^{k-1} $(k \geq 1)$, one has

$$k g_k^{\beta,K} + \sum_{m=0}^{k-2} w_{k-m-1}(m+1) g_{m+1}^{\beta,K} = \beta \sum_{m=0}^{k-1} (k-m) w_{k-m} g_m^{\beta,K}.$$

Therefore,

$$g_k^{\beta,K} = \frac{1}{k w_0} \sum_{m=0}^{k-1} \left[(k-m)\beta - m\right] g_m^{\beta,K} w_{k-m} \text{ for } k \geq 1.$$

Combining with $w_l = 0$ for $l > K$, one completes the proof. □

For the special case $K = 2$, by Lemma 4.5, there exist

$$g_0^{\beta,2} = \left(\frac{3}{2}\right)^\beta, \quad g_1^{\beta,2} = -\left(\frac{3}{2}\right)^\beta \frac{4\beta}{3}, \tag{4.53}$$

$$g_k^{\beta,2} = \frac{1}{3}\left(\frac{2\beta+2}{k} - 1\right) g_{k-2}^{\beta,2} + \frac{4}{3}\left(1 - \frac{\beta+1}{k}\right) g_{k-1}^{\beta,2}, k \geq 2. \tag{4.54}$$

4.1.2 *Basic approximations to the tempered fractional derivatives*

The tempered R-L fractional derivatives for $1 < \beta < 2$ are defined by the inverse Fourier formulae (see Appendix B):

$${}_{-\infty}D_x^{\beta,\lambda} u(x) = \mathscr{F}^{-1}[\left((\lambda + i\omega)^\beta - i\omega\beta\lambda^{\beta-1} - \lambda^\beta\right) \hat{u}(\omega)](x), \tag{4.55}$$

$${}_x D_\infty^{\beta,\lambda} u(x) = \mathscr{F}^{-1}[\left((\lambda - i\omega)^\beta + i\omega\beta\lambda^{\beta-1} - \lambda^\beta\right) \hat{u}(\omega)](x); \tag{4.56}$$

and if $u(x) \in \mathscr{S}^{\beta+1}(\mathbb{R})$, similar to the proof of Theorem 5.1 of [Sabzikar *et al.* (2015)], it holds that

$$
\begin{aligned}
{}_{-\infty}\mathbb{D}_x^{\beta,\lambda}u(x) &= \lim_{h\to 0}\frac{1}{h^\beta}\sum_{j=0}^{\infty} g_j^\beta e^{-\lambda jh}u(x-jh),\\
{}_x\mathbb{D}_\infty^{\beta,\lambda}u(x) &= \lim_{h\to 0}\frac{1}{h^\beta}\sum_{j=0}^{\infty} g_j^\beta e^{-\lambda jh}u(x+jh),
\end{aligned}
$$

where

$$
{}_{-\infty}\mathbb{D}_x^{\beta,\lambda}u(x) := \mathscr{F}^{-1}[(\lambda+i\omega)^\beta](x), \tag{4.57}
$$

$$
{}_x\mathbb{D}_\infty^{\beta,\lambda}u(x) := \mathscr{F}^{-1}[(\lambda-i\omega)^\beta](x). \tag{4.58}
$$

Moreover, if $u(x) \in H^\beta(\mathbb{R})$, one can also relate the tempered R-L fractional derivatives to the R-L fractional derivatives by

$$
{}_{-\infty}D_x^{\beta,\lambda}u(x) = {}_{-\infty}\mathbb{D}_x^{\beta,\lambda}u(x) - \beta\lambda^{\beta-1}\frac{du(x)}{dx} - \lambda^\beta u(x), \tag{4.59}
$$

$$
{}_xD_\infty^{\beta,\lambda}u(x) = {}_x\mathbb{D}_\infty^{\beta,\lambda}u(x) + \beta\lambda^{\beta-1}\frac{du(x)}{dx} - \lambda^\beta u(x) \tag{4.60}
$$

with

$$
{}_{-\infty}\mathbb{D}_x^{\beta,\lambda}u(x) = e^{-\lambda x}\,{}_{-\infty}D_x^\beta\left(e^{\lambda x}u(x)\right), \tag{4.61}
$$

$$
{}_x\mathbb{D}_\infty^{\beta,\lambda}u(x) = e^{\lambda x}\,{}_xD_\infty^\beta\left(e^{-\lambda x}u(x)\right). \tag{4.62}
$$

In the following, we will extend the shifted G-L and Lubich approximations introduced above to the tempered R-L fractional derivatives. Define

$$
{}_LA_{h,p}^{\beta,\lambda}u(x) := \frac{1}{h^\beta}\sum_{k=0}^{\infty} g_k^\beta e^{-(k-p)h\lambda}u(x-(k-p)h) \tag{4.63}
$$

and

$$
\Lambda_{h,p}^{\beta,\lambda} = \frac{e^{ph\lambda}}{h^\beta}\left(1-e^{-h\lambda}\right)^\beta. \tag{4.64}
$$

Theorem 4.3. *Let $u(x) \in \mathscr{S}^{\beta+1}(\mathbb{R})$ and $u(x)$ have compact support. Then*

$$
{}_LA_{h,p}^{\beta,\lambda}u(x) = {}_{-\infty}\mathbb{D}_x^{\beta,\lambda}u(x) + \mathcal{O}(h) \tag{4.65}
$$

uniformly in $x \in \mathbb{R}$ for $h \to 0$. In addition, it always holds that

$$
\Lambda_{h,p}^{\beta,\lambda} = \lambda^\beta + \mathcal{O}(h). \tag{4.66}
$$

Proof. Let $W_{\beta,p}(z) = e^{pz}\left(\frac{1-e^{-z}}{z}\right)^{\beta}$. From the proof of Theorem 4.1, we know that if $|\mathrm{Re}(z)|$ is bounded, there exists a constant $C_1 > 0$ such that

$$\left|W_{\beta,p}(z) - \sum_{l=0}^{n-1} a_l z^l\right| \le C_1 |z|^n, \quad n \in \mathbb{N}^+ \tag{4.67}$$

uniformly in $z \in \mathbb{C}$, where $\sum_{l=0}^{\infty} a_l z^l$ is the Taylor expansion of $W_{\beta,p}(z)$ with $a_0 = 1, a_1 = p - \frac{\beta}{2}$.

Since

$$\begin{aligned}\mathscr{F}\left[{}_L A_{h,p}^{\beta,\lambda} u\right](\omega) &= \frac{1}{h^\beta}\sum_{k=0}^{\infty} g_k^\beta e^{-(k-p)h(\lambda+i\omega)}\hat{u}(\omega)\\ &= e^{ph(\lambda+i\omega)}\left(\frac{1-e^{-h(\lambda+i\omega)}}{h}\right)^{\beta}\hat{u}(\omega)\\ &= (\lambda+i\omega)^\beta W_{\beta,p}\left((\lambda+i\omega)h\right)\hat{u}(\omega).\end{aligned}$$

and

$$\Lambda_{h,p}^{\beta,\lambda} = \lambda^\beta W_{\beta,p}(\lambda h). \tag{4.68}$$

Therefore

$$\begin{aligned}&\left|{}_L A_{h,p}^{\beta,\lambda} u(x) - {}_{-\infty}\mathbb{D}_x^{\beta,\lambda} u(x)\right|\\ &\le \frac{1}{2\pi}\int_{\mathbb{R}}\left|\mathscr{F}\left[{}_L A_{h,p}^{\beta,\lambda} u\right]\hat{u}(\omega) - (\lambda+i\omega)^\beta\hat{u}(\omega)\right| dw\\ &\le C\int_{\mathbb{R}}(\lambda+|\omega|)^{\beta+1}|\hat{u}(\omega)|d\omega\, h\\ &\le C\max\{1,\lambda^{\beta+1}\}\int_{\mathbb{R}}(1+|\omega|)^{\beta+1}|\hat{u}(\omega)|d\omega\, h\end{aligned}$$

and

$$\left|\lambda^\beta - \Lambda_{h,p}^{\beta,\lambda}\right| \le C\lambda^{\beta+1}h. \tag{4.69}$$

Thus, we complete the proof. □

Remark 4.3.

(1) The approximation of λ^β is necessary to ensure the stability of the numerical schemes (see Sec. 4.6).
(2) Similarly, we can define

$${}_R A_{h,p}^{\beta,\lambda} u(x) := \frac{1}{h^\beta}\sum_{k=0}^{\infty} g_k^\beta e^{-(k-p)h\lambda} u(x+(k-p)h); \tag{4.70}$$

if $u(x) \in \mathscr{F}^{\beta+1}(\mathbb{R})$ and $u(x)$ has compact support, it also holds that

$${}_R A_{h,p}^{\beta,\lambda} u(x) = {}_x\mathbb{D}_{\infty}^{\beta,\lambda} u(x) + \mathcal{O}(h). \tag{4.71}$$

Equation (4.63) may be derived directly from the G-L approximation of the R-L derivative. In fact, replacing $u(x)$ in (4.17) and (4.27) with $e^{\lambda x}u(x)$ leads to

$$
\begin{aligned}
{}_{-\infty}\mathbb{D}_x^{\beta,\lambda}u(x) &\approx \frac{e^{-\lambda x}}{h^\beta}\sum_{k=0}^{\infty} g_k^\beta e^{\lambda h(x-(k-p))}u(x-(k-p)h) \\
&= \frac{1}{h^\beta}\sum_{k=0}^{\infty} g_k^\beta e^{-\lambda h(k-p)}u(x-(k-p)h). \qquad (4.72)
\end{aligned}
$$

Therefore, if we replace the G-L coefficients g_k^β with the Lubich coefficients $g_k^{\beta,K}$, we can expect that

Theorem 4.4.

(1) For any tempered parameter $\lambda \geq 0$, it always holds that

$$
\lambda^\beta = \frac{e^{p\lambda h}}{h^\beta}\left(\sum_{l=1}^{K}\frac{1}{l}(1-e^{-\lambda h})^l\right)^\beta + \begin{cases}\mathcal{O}(h) \text{ for } p=0\\ \mathcal{O}(h^K) \text{ for } p\neq 0.\end{cases}
$$

(2) Let $u(x) \in \mathscr{S}^{\beta+1}(\mathbb{R})$ (or $u(x) \in \mathscr{F}^{\beta+2}(\mathbb{R})$) and $u(x)$ have compact support. Then

$$
{}_{-\infty}\mathbb{D}_x^{\beta}u(x) = \frac{1}{h^\beta}\sum_{k=0}^{\infty} g_k^{\beta,K} e^{-\lambda h(k-p)}uh(x-(k-p)h) + \begin{cases}\mathcal{O}(h) \text{ for } p=0\\ \mathcal{O}(h^K) \text{ for } p\neq 0\end{cases}
$$

uniformly in $x \in \mathbb{R}$ as $h \to 0$.

Proof. One easily prove the results by replacing the z in (4.50) with $(\lambda + i\omega)h$ and λh, respectively. □

4.1.3 *Some results on matrix analysis*

In the following, we give some results on matrix analysis that will be used later in this book. The reader can refer to [Quarteroni *et al.* (2007); Chan and Jin (2007); Laub (2005)] for the detailed proofs.

Definition 4.1. A matrix $A \in \mathbb{R}^{n\times n}$ is said to be positive definite in $\mathbb{R}^n$ if $(Ax, x) > 0$, $\forall x \in \mathbb{R}^n$, $x \neq 0$.

Lemma 4.6. *A real matrix A of order n is positive definite if and only if its symmetric part $H = \frac{A+A^{\mathrm{T}}}{2}$ is positive definite; H is positive definite if and only if the eigenvalues of H are positive.*

Lemma 4.7. *If $A \in \mathbb{C}^{n\times n}$, $H = \frac{A+A^*}{2}$ is the Hermitian part of A, and A^* is the conjugate transpose of A, then for any eigenvalue λ of A, there exists*

$$\lambda_{\min}(H) \le \mathrm{Re}(\lambda) \le \lambda_{\max}(H),$$

where $\mathrm{Re}(\lambda)$ represents the real part of λ, and $\lambda_{\min}(H)$, $\lambda_{\max}(H)$ are the minimum and maximum of the eigenvalues of H.

Definition 4.2. Let T_n be a Toeplitz matrix with the following form

$$T_n = \begin{pmatrix} t_0 & t_{-1} & \cdots & t_{2-n} & t_{1-n} \\ t_1 & t_0 & t_{-1} & \cdots & t_{2-n} \\ \vdots & t_1 & t_0 & \ddots & \vdots \\ t_{n-2} & \cdots & \ddots & \ddots & t_{-1} \\ t_{n-1} & t_{n-2} & \cdots & t_1 & t_0 \end{pmatrix}. \tag{4.73}$$

If we assume that the diagonals $\{t_k\}_{k=-n+1}^{n-1}$ are the Fourier coefficients of a function $f(x)$, i.e.,

$$t_k = \frac{1}{2\pi}\int_{-\pi}^{\pi} f(x)e^{-ikx}dx,$$

then $f(x)$ is called the generating function of matrix T_n.

We note that $f(x)$ is 2π-periodic continuous and satisfies

- When f is real-valued $\rightleftharpoons$ T_n are Hermitian for all n.
- When f is real-valued and even $\rightleftharpoons$ T_n are real symmetric for all n.

Lemma 4.8. *Let T_n be the Toeplitz matrix given by (4.73) with $f(x)$ being a 2π-periodic continuous real-valued function. Denote $\lambda_{\min}(T_n)$ and $\lambda_{\max}(T_n)$ as the smallest and largest eigenvalues of T_n, respectively. Then we have*

$$f_{\min} \le \lambda_{\min}(T_n) \le \lambda_{\max}(T_n) \le f_{\max},$$

where $f_{\min}$, $f_{\max}$ denote the minimum and maximum values of $f(x)$, respectively. Moreover, if $f_{\min} < f_{\max}$, then all eigenvalues of T_n satisfy

$$f_{\min} < \lambda(T_n) < f_{\max},$$

for all $n > 0$. Note that if $f_{\min} \ge 0$, then T_n is positive definite.

The following results on the Kronecker product will be used in performing the stability and convergence analyses of the high-dimensional model.

Lemma 4.9. *Let $A \in \mathbb{R}^{n\times n}$ have eigenvalues $\{\lambda_i\}_{i=1}^{n}$, and $B \in \mathbb{R}^{m\times m}$ have eigenvalues $\{\mu_j\}_{j=1}^{m}$. Then the mn eigenvalues of $A \otimes B$, which represents the Kronecker product of matrices A and B, are*

$$\lambda_1\mu_1, \ldots, \lambda_1\mu_m, \lambda_2\mu_1, \ldots, \lambda_2\mu_m, \ldots, \lambda_n\mu_1, \ldots, \lambda_n\mu_m.$$

Lemma 4.10. *Let $A \in \mathbb{R}^{m\times n}, B \in \mathbb{R}^{r\times s}, C \in \mathbb{R}^{n\times p}, D \in \mathbb{R}^{s\times t}$. Then*

$$(A \otimes B)(C \otimes D) = AC \otimes BD \quad (\in \mathbb{R}^{mr\times pt}). \tag{4.74}$$

Moreover, if $A, B \in \mathbb{R}^{n\times n}$, I is a unit matrix of order n, then matrices $I \otimes A$ and $B \otimes I$ commute.

Lemma 4.11. *For all A and B, $(A \otimes B)^{\mathrm{T}} = A^{\mathrm{T}} \otimes B^{\mathrm{T}}$; and if A and B are invertible, then $(A \otimes B)^{-1} = A^{-1} \otimes B^{-1}$.*

4.2 First order difference scheme for the variable coefficients advection-diffusion equations

In this section, we mainly introduce the first order shifted Grünwald difference schemes for solving the following fractional advection-diffusion equation

$$\begin{aligned}\frac{\partial u(x,t)}{\partial t} &= c_+(x,t)\,{}_{-\infty}D_x^{\beta}u(x,t) + c_-(x,t)\,{}_xD_{\infty}^{\beta}u(x,t)\\ &\quad - v(x,t)\frac{\partial u(x,t)}{\partial x} + f(x,t)\end{aligned} \tag{4.75}$$

in a finite domain $\Omega = (a,b)$, $0 < t \leq T$, with initial-boundary conditions

$$u(x,0) = g(x), \quad x \in \Omega, \tag{4.76}$$

and

$$u(x,t) = 0, \quad x \in \mathbb{R}\backslash\Omega, \ 0 \leq t \leq T. \tag{4.77}$$

Here, we consider the case $1 < \beta < 2$, where the parameter β is the fractional order (fractor) of the space derivatives. The function $f(x,t)$ is a source/sink term. The functions $c_+(x,t),\ c_-(x,t) \geq 0$ with $c_+(x,t) + c_-(x,t) \neq 0$ are the diffusion coefficients. We also assume $v(x,t) \geq 0$ so that the flow is from left to right. If we let $\beta = 2$ in (4.75), the left and right derivatives become localized and equal, then (4.75) becomes the classic advection-diffusion equation

$$\frac{\partial u(x,t)}{\partial t} = (c_-(x,t) + c_+(x,t))\frac{\partial^2 u(x,t)}{\partial x^2} - v(x,t)\frac{\partial u(x,t)}{\partial x} + f(x,t),$$

and the corresponding initial-boundary conditions can be given as

$$u(x,0) = g(x), \ x \in \Omega \ \text{ and } \ u(a,t) = u(b,t) = 0. \tag{4.78}$$

For the sake of completeness, we also extend the proposed finite difference scheme to the two-dimensional space fractional model in the concluding subsection. More specifically, we consider the model

$$\begin{aligned}\frac{\partial u(x,y,t)}{\partial t} &= d_+(x,y,t)_{-\infty}D_x^{\beta}u(x,y,t) + d_-(x,y,t)_xD_{\infty}^{\beta}u(x,y,t)\\ &+ e_+(x,y,t)_{-\infty}D_y^{\beta_1}u(x,y,t) + e_-(x,y,t)_xD_{\infty}^{\beta_1}u(x,y,t)\\ &- v_1(x,y,t)\frac{\partial u(x,y,t)}{\partial x} - v_2(x,y,t)\frac{\partial u(x,y,t)}{\partial y} + f(x,y,t), \qquad (4.79)\end{aligned}$$

with the initial-boundary conditions

$$u(x,y,0) = u_0(x,y), \quad (x,y) \in \Omega, \qquad (4.80)$$

$$u(x,y,t) = 0, \quad (x,y,t) \in \mathbb{R}^2\backslash\Omega \times [0,T], \qquad (4.81)$$

where $\Omega = (a,b)\times(c,d)$, ${}_{-\infty}D_x^{\beta}$, ${}_xD_{\infty}^{\beta}$ $(1<\beta,\beta_1\le 2)$ and ${}_{-\infty}D_y^{\beta_1}$, ${}_yD_{\infty}^{\beta_1}$ are R-L fractional operators, and all the variable coefficients are non-negative.

4.2.1 *Derivation of the numerical scheme*

Define $t_n = n\tau, n = 0,1,\cdots,M$ to be the integration time with $\tau = T/M$, and $x_i = a+ih, i = 0,\cdots,N$ to be the grid in spatial dimension with $h = (b-a)/N$. Meanwhile, define $c_{-,i}^n = c_-(x_i,t_n), c_{+,i}^n = c_+(x_i,t_n)$, $v_i^n = v(x_i,t_n)$ and $f_i^n = f(x_i,t_n)$. To obtain a stable difference scheme, we use the shifted Grünwald formulae with $p=1$ (see (4.17) and (4.18))

$$_LA_{h,p}^{\beta}u(x) := \frac{1}{h^{\beta}}\sum_{k=0}^{\infty} g_k^{\beta}u\left(x-(k-p)h\right), \qquad (4.82)$$

$$_RA_{h,p}^{\beta}u(x) := \frac{1}{h^{\beta}}\sum_{k=0}^{\infty} g_k^{\beta}ku\left(x+(k-p)h\right) \qquad (4.83)$$

to approximate the R-L fractional derivatives ${}_{-\infty}D_x^{\beta}u(x)$ and ${}_xD_{\infty}^{\beta}u(x)$, respectively. Lemma 4.1 shows that the truncation errors are $\mathcal{O}(h)$. Then, using $\frac{u(x_i,t)-u(x_{i-1},t)}{h}$ to approximate the first order derivative $\frac{\partial u(x_i,t)}{\partial x}$ yields the implicit Euler finite difference scheme as follows:

$$\begin{aligned}\frac{U_i^n - U_i^{n-1}}{\tau} &= \frac{1}{h^{\beta}}\left(\sum_{k=0}^{i+1} g_k^{\beta}c_{+,i}^nU_{i-k+1}^n + \sum_{k=0}^{N-i+1} g_k^{\beta}c_{-,i}^nU_{i+k-1}^n\right)\\ &- v_i^n\frac{U_i^n - U_{i-1}^n}{h} + f_i^n, \qquad (4.84)\end{aligned}$$

where $i = 1, 2, \cdots, N$, $n = 1, 2, \cdots, M$, and U_i^n denotes the numerical approximation of u_i^n. It is easy to check that this difference scheme is consistent, and the local truncation errors are $\mathcal{O}(h + \tau)$.

Define $(N-1) \times (N-1)$ matrices

$$A_{\beta,N} = \begin{bmatrix} g_1^\beta & g_0^\beta & & & \\ g_2^\beta & g_1^\beta & g_0^\beta & & \\ \vdots & g_2^\beta & g_1^\beta & \ddots & \\ g_{N-2}^\beta & \cdots & \ddots & \ddots & g_0^\beta \\ g_{N-1}^\beta & g_{N-2}^\beta & \cdots & g_2^\beta & g_1^\beta \end{bmatrix}, \quad B_N = \begin{bmatrix} 1 & & & & \\ -1 & 1 & & & \\ \vdots & \ddots & \ddots & \ddots & \\ 0 & \cdots & -1 & 1 & 0 \\ 0 & 0 & \cdots & -1 & 1 \end{bmatrix}, \tag{4.85}$$

$$C_+^n = \operatorname{diag}\left(c_{+,1}^n, c_{+,2}^n, \cdots, c_{+,N-1}^n\right), \tag{4.86}$$

$$C_-^n = \operatorname{diag}\left(c_{-,1}^n, c_{-,2}^n, \cdots, c_{-,N-1}^n\right), \tag{4.87}$$

$$V^n = \operatorname{diag}\left(v_1^n, v_2^n, \cdots, v_{N-1}^n\right), \tag{4.88}$$

and vectors

$$U^n = \left(U_1^n, U_2^n, \cdots, U_{N-1}^n\right)^{\mathrm{T}}, \quad F^n = \left(f_1^n, f_2^n, \cdots, f_{N-1}^n\right)^{\mathrm{T}}, \tag{4.89}$$

where $g_k^\beta = (-1)^k \binom{\beta}{k}$ satisfies Proposition 4.1. Then (4.84) can be rewritten as the matrix vector form

$$\left(I - \frac{\tau}{h^\beta}\left(C_+^n A_{\beta,N} + C_-^n A_{\beta,N}^{\mathrm{T}}\right) + \frac{\tau}{h} V^n B_N\right) U^n = U^{n-1} + \tau F^n, \tag{4.90}$$

for $n = 1, 2, \cdots, M$.

4.2.2 *Stability and convergence analyses*

Since $u(x,t) = 0$ for $x \in \mathbb{R}\backslash\Omega$, we introduce the grid function space $\mathcal{V} = \{v | v = (v_0, v_1, \cdots, v_N), v_0 = v_N = 0\}$ on the mesh $\{x_i, 0 \leq i \leq N\}$. For any $v, w \in \mathcal{V}$, we define the discrete L^2 inner product and norms:

$$(v, w) = h \sum_{i=1}^{N-1} v_i \overline{w_i}, \quad \|v\| = \sqrt{(v,v)}; \quad \|v\|_\infty = \max_{1 \leq i \leq N-1} |v_i|. \tag{4.91}$$

In the following, we first prove the stability and convergence in maximum norm sense.

Lemma 4.12. *Let $A_{\beta,N}, B_N, C_-^n, C_+^n, V^n$ and U^n be given as above. Then*

$$\left\|\left(I - \frac{\tau}{h^\beta}\left(C_+^n A_{\beta,N} + C_-^n A_{\beta,N}^{\mathrm{T}}\right) + \frac{\tau}{h} V^n B_N\right) U^n\right\|_\infty \geq \|U^n\|_\infty. \tag{4.92}$$

Proof. Let $\|U^n\|_\infty = |U_p^n|$ for $1 \le p \le N-1$. For $1 < \beta < 2$, it follows that (see Proposition 4.1)

$$g_1^\beta < 0,\ g_k^\beta > 0\,(k \neq 1),\ \sum_{k=0}^{K} g_k^\beta < 0\,(\forall K \in \mathbb{N}), \tag{4.93}$$

which results in

$$\begin{aligned} &U_p^n\left[\frac{1}{h^\beta}\left(\sum_{k=0}^{p+1} g_k^\beta c_{+,p}^n U_{p-k+1}^n + \sum_{k=0}^{N-p+1} g_k^\beta c_{-,p}^n U_{p+k-1}^n\right)\right.\\ &\qquad\left.-\frac{v_p^n}{h}\frac{U_p^n - U_{p-1}^n}{h}\right]\\ &\le \frac{1}{h^\beta}\left(\sum_{k=0}^{p+1} g_k^\beta c_{+,p}^n \left(U_p^n\right)^2 + \sum_{k=0}^{N-p+1} g_k^\beta c_{-,p}^n \left(U_p^n\right)^2\right)\\ &\qquad-\frac{v_p^n}{h}\frac{\left(U_p^n\right)^2 - \left(U_p^n\right)^2}{h} \le 0. \end{aligned}$$

Then

$$\begin{aligned} &\left|U_p^n - \tau\left[\frac{1}{h^\beta}\left(\sum_{k=0}^{p+1} g_k^\beta c_{+,p}^n U_{p-k+1}^n + \sum_{k=0}^{N-p+1} g_k^\beta c_{-,p}^n U_{p+k-1}^n\right)\right.\right.\\ &\qquad\left.\left.-\frac{v_p^n}{h}\frac{U_p^n - U_{p-1}^n}{h}\right]\right| \ge |U_p^n| = \|U^n\|_\infty, \end{aligned}$$

which means that

$$\left\|\left(I - \frac{\tau}{h^\beta}\left(C_+^n A_{\beta,N} + C_-^n A_{\beta,N}^{\mathrm{T}}\right) + \frac{\tau}{h} V^n B_N\right) U^n\right\|_\infty \ge \|U^n\|_\infty.$$

□

Theorem 4.5. *For $1 < \beta < 2$, the discrete scheme (4.90) is unconditionally stable with maximum norm, i.e., for $1 \le n \le M$, it holds that*

$$\|U^n\|_\infty \le \left\|U^0\right\|_\infty + \sum_{k=1}^{n} \tau \left\|F^k\right\|_\infty. \tag{4.94}$$

Moreover, one has

$$\|E^n\|_\infty \le c(\tau + h), \tag{4.95}$$

where $E^n = \left(u_1^n - U_1^n, \cdots, u_i^n - U_i^n, \cdots, u_{N-1}^n - U_{N-1}^n\right)^{\mathrm{T}}$, and c is a positive constant depending on T, the exact solution $u(x,t)$, but not depending on τ and h.

Proof. Starting from (4.90) and using Lemma 4.12, one has

$$\|U^n\|_\infty \leq \left\|U^{n-1} + \tau F^n\right\|_\infty \leq \left\|U^{n-1}\right\|_\infty + \tau \left\|F^n\right\|_\infty.$$

Replacing n by k and then adding k from 1 to n, one arrives at the desired stability results.

Since the error equation can be given as

$$\left(I - \frac{\tau}{h^\beta}\left(C_+^n A_{\beta,N} + C_-^n A_{\beta,N}^{\mathrm{T}}\right) + \frac{\tau}{h} V^n B_N\right) E^n = E^{n-1} + \tau\,\mathcal{E}^n \tag{4.96}$$

with $E^0 = \mathbf{0}$, where

$$\mathcal{E}^n = \left(\varepsilon_1^n, \cdots, \varepsilon_i^n, \cdots, \varepsilon_{N-1}^n\right)^{\mathrm{T}} \quad \text{with} \quad \varepsilon_i^n \leq \tilde{c}(\tau + h).$$

Then

$$\|E^n\|_\infty \leq \sum_{k=1}^{n} \tau \left\|\mathcal{E}^k\right\|_\infty \leq c_1 T(\tau + h).$$

□

Remark 4.4. Making use of (4.93), it is easy to prove that the matrix $\frac{\tau}{h^\beta}\left(C_+^n A_{\beta,N} + C_-^n A_{\beta,N}^{\mathrm{T}}\right) - \frac{\tau}{h} V^n B_N$ is strictly row diagonally dominant, and the entries in the main diagonal are negative. According to the Greschgorin theorem [Isaacson and Keller (1996)], the real parts of all the eigenvalues are negative. Therefore, the spectral radius of the system matrix $\left(I - \frac{\tau}{h^\beta}\left(C_+^n A_{\beta,N} + C_-^n A_{\beta,N}^{\mathrm{T}}\right) + \frac{\tau}{h} V^n B_N\right)^{-1}$ is less than one. In this case, we also say that the difference scheme (4.90) is stable.

If $c_-(x,t), c_+(x,t)$ and $v(x,t)$ are all non-negative constants, we can also give the stability and convergence analysis in L^2 norm.

Lemma 4.13. *Let $1 < \beta < 2$. Then one has the coercivity*

$$-\frac{1}{h^\beta}\left(A_{\beta,N} U^n, U^n\right) \geq -\frac{2}{(b-a)^\beta \Gamma(1-\beta)} \|U^n\|^2 > 0. \tag{4.97}$$

Proof.

$$\begin{aligned}
\left(A_{\beta,N} U^n, U^n\right) &= h \sum_{i=1}^{N-1}\left(\sum_{k=0}^{i} g_k^\beta U_{i-k+1}^n\right) U_i^n \\
&= h \sum_{k=0}^{N-1} g_k^\beta \sum_{i=k}^{N-1} U_{i-k+1}^n u_i^n \\
&= h\left(g_1^\beta \sum_{i=1}^{N-1} (U_i^n)^2 + \sum_{k=0,k\neq 1}^{N-1} g_k^\beta \sum_{i=k}^{N-1} \frac{\left(U_{i-k+1}^n\right)^2 + (U_i^n)^2}{2}\right) \\
&\leq \left(\sum_{k=0}^{M} g_k^\beta\right) \|U^n\|^2 .
\end{aligned}$$

Considering the mth coefficient of the Taylor expansion of $(1-z)^{\beta-1} = (1-z)^{-1}(1-z)^{\beta}$, one has

$$\sum_{k=0}^{m} g_k^{\beta} = (-1)^m \binom{\beta-1}{m} = \frac{\Gamma(m+1-\beta)}{\Gamma(1-\beta)\Gamma(m+1)}.$$

According to asymptotic expansion of the derivation of two Gamma functions [Samko *et al.* (1993)]

$$\frac{\Gamma(x+c_1)}{\Gamma(x+c_2)} = x^{c_1-c_2} + x^{c_1-c_2}\mathcal{O}(x^{-1}), \quad |x| \to +\infty,$$

it holds that

$$m^{\beta} \sum_{k=0}^{m} g_k^{\beta} \to \frac{1}{\Gamma(1-\beta)}, \quad m \to +\infty.$$

On the other hand, using the fact that $f(x) = (1-x)^{\beta} - (1-\beta x), \beta > 1$ is strictly monotonically increasing in the interval $(0,1]$, we have

$$\frac{\sum_{k=0}^{m} g_k^{\beta}}{\sum_{k=0}^{m-1} g_k^{\beta}} = \frac{m-\beta}{m} < \left(1 - \frac{1}{m}\right)^{\beta}, \quad m \geq 1,$$

which means that the sequence $m^{\beta} \sum_{k=0}^{m} g_k^{\beta}$ is strictly monotonically increasing w.r.t. m. Therefore,

$$\sum_{k=0}^{m} g_k^{\beta} < \frac{1}{m^{\beta}\Gamma(1-\beta)}.$$

Then

$$\begin{aligned} -\frac{1}{h^{\beta}} (A_{\beta,N}U^n, U^n) &\geq -\frac{1}{h^{\beta}} \left(\sum_{k=0}^{M} g_k^{\beta} \right) \|U^n\|^2 \\ &\geq -\frac{2}{(b-a)^{\beta}\Gamma(1-\beta)} \|U^n\|^2 . \end{aligned}$$

Thus, we complete the proof. □

Theorem 4.6. *If $c_-(x,t), c_+(x,t)$ and $v(x,t)$ are all constants, then the discrete scheme (4.90) is unconditionally stable in L^2 norm, i.e., for $1 \leq n \leq M$, it holds that*

$$\|U^n\| \leq \left\|U^0\right\| + \tau \sum_{k=0}^{n} \|F^n\| . \tag{4.98}$$

In addition, one has

$$\|E^n\| \leq c(\tau + h). \tag{4.99}$$

Proof. Taking an inner product of (4.90) with U^n, it yields that

$$(U^n, U^n) - \frac{\tau}{h^\beta}\left(\left(C_+^n A_{\beta,N} + C_-^n A_{\beta,N}^{\mathrm{T}} - V^n B_N\right) U^n, U^n\right) \\ = \left(U^{n-1}, U^n\right) + \tau\left(F^n, U^n\right).$$

Note that

$$-\frac{1}{h^\beta}\left(A_{\beta,N}U^n, U^n\right) = -\frac{1}{h^\beta}\left(U^n, A_{\beta,N}^{\mathrm{T}}U^n\right) \geq 0,$$

and

$$\left(B_N U^n, U^n\right) = \frac{1}{2}\left(U^n\right)^{\mathrm{T}}\left(B_N + B_N{}^{\mathrm{T}}\right)U^n \geq 0. \tag{4.100}$$

By the Cauchy-Schwarz inequality, one has

$$\|U^n\| \leq \left\|U^{n-1}\right\| + \tau\left\|F^n\right\|.$$

Replacing n by k and then adding k from 1 to n, it yields the desired stability result (4.98). Starting from the error equation (4.96) and using the same proof skills above, one can give the convergence estimate easily. □

4.2.3 *Extending the proposed numerical scheme to the two dimensional space fractional model*

Now we establish the first order difference scheme for problem (4.79)–(4.81). Let N_x, N_y and M be positive integers and

$$h_x = \frac{b-a}{N_x}, \quad h_y = \frac{d-c}{N_y}, \quad \tau = \frac{T}{M} \tag{4.101}$$

the sizes of the spatial grids and time step, respectively. We define the spatial and temporal partitions $x_i = a + ih_x$ for $i = 0, 1, \cdots, N_x$, $y_j = c + jh_y$ for $j = 0, 1, \cdots, N_y$ and $t_n = n\tau$ for $n = 0, 1, \cdots, M$ and also write $u_{i,j}^n = u(x_i, y_j, t_n)$, $d_{\pm,i,j}^n = d_\pm(x_i, y_j, t_n)$, $e_{\pm,i,j}^n = e_\pm(x_i, y_j, t_n)$, $v_{1,i,j}^n = v_1(x_i, y_j, t_n)$, $v_{2,i,j}^n = v_2(x_i, y_j, t_n)$ and $f_{i,j}^n = f(x_i, y_j, t_n)$. Similar to (4.84), the difference scheme of (4.79) can be given as

$$\begin{aligned} \frac{U_{i,j}^n - U_{i,j}^{n-1}}{\tau} &= \frac{1}{h_x^\beta}\left(\sum_{k=0}^{i+1} g_k^\beta d_{+,i,j}^n U_{i-k+1,j}^n + \sum_{k=0}^{N_x-i+1} g_k^\beta d_{-,i,j}^n U_{i+k-1,j}^n\right) \\ &+ \frac{1}{h_y^{\beta_1}}\left(\sum_{k=0}^{j+1} g_k^{\beta_1} e_{+,i,j}^n U_{i,j-k+1}^n + \sum_{k=0}^{N_y-j+1} g_k^{\beta_1} e_{-,i,j}^n U_{i,j+k-1}^n\right) \\ &- v_{1,i,j}^n \frac{U_{i,j}^n - U_{i-1,j}^n}{h_x} - v_{2,i,j}^n \frac{U_{i,j}^n - U_{i,j-1}^n}{h_y} + f_{i,j}^n, \end{aligned} \tag{4.102}$$

where $U_{i,j}^n$ denotes the numerical approximation of $u_{i,j}^n$. It is easy to see that the local truncation errors are $\mathcal{O}(\tau + h_x + h_y)$.

Define diagonal matrix

$$D_+^n = \mathrm{diag}\Big(d_{+,1,1}^n, d_{+,2,1}^n, \cdots, d_{+,N_x-1,1}^n, d_{+,1,2}^n, \cdots, d_{+,N_x-1,2}^n,$$
$$d_{+,N_x-1,N_y-1}^n, \cdots, d_{+,N_x-1,N_y-1}^n\Big). \tag{4.103}$$

Similarly, we can define diagonal matrices $D_-^n, E_+^n, E_-^n, V_1^n, V_2^n$, for example E_-^n can be defined by replacing the 'D' by 'E', '+' by '−', 'd' by 'e' in (4.103). Let

$$\begin{aligned}
U^n =&(U_{1,1}^n, U_{2,1}^n, \cdots, U_{N_x-1,1}^n, U_{1,2}^n, U_{2,2}^n, \cdots, U_{N_x-1,2}^n,\\
&\cdots, U_{1,N_y-1}^n, U_{2,N_y-1}^n, \cdots, U_{N_x-1,N_y-1}^n)^{\mathrm{T}},\\
F^n =&(f_{1,1}^n, f_{2,1}^n, \cdots, f_{N_x-1,1}^n, f_{1,2}^n, f_{2,2}^n, \cdots, f_{N_x-1,2}^n,\\
&\cdots, f_{1,N_y-1}^n, f_{2,N_y-1}^n, \cdots, f_{N_x-1,N_y-1}^n)^{\mathrm{T}}.
\end{aligned}$$

Then the matrix-vector form of (4.102) can be given as

$$(I - \mathcal{A}_x - \mathcal{A}_y + \mathcal{B})\, U^n = U^{n-1} + \tau\, F^n, \tag{4.104}$$

where

$$\begin{aligned}
\mathcal{A}_x &= \frac{\tau}{h_x^\beta}\left(D_+^n(I_y \otimes A_{\beta,N_x}) + D_-^n(I_y \otimes A_{\beta,N_x}^{\mathrm{T}})\right),\\
\mathcal{A}_y &= \frac{\tau}{h_y^{\beta_1}}\Big(E_+^n(A_{\beta_1,N_y} \otimes I_x) + E_-^n(A_{\beta_1,N_y}^{\mathrm{T}} \otimes I_x)\Big),\\
\mathcal{B} &= \frac{\tau V_1^n}{h_x}(I_y \otimes B_{N_x}) + \frac{\tau V_2^n}{h_y}\left(B_{N_y} \otimes I_x\right), \quad I = I_y \otimes I_x,
\end{aligned} \tag{4.105}$$

with $\otimes$ denoting the Kronecker product, I_x and I_y being $(N_x-1)\times(N_x-1)$ and $(N_y-1)\times(N_y-1)$ unit matrices, respectively, and A_{β,N_x} and A_{β_1,N_y}, B_{N_x} and B_{N_y} corresponding to matrices $A_{\beta,N}$ and B being defined in (4.85), respectively.

Define the sets of the index of the interior and boundary mesh grid points in domain $[a,b]\times[c,d]$, respectively, as

$$\begin{aligned}
&\Lambda_h = \{(i,j) : 1 \le i \le N_x - 1, 1 \le j \le N_y - 1\},\\
&\partial\Lambda_h = \{(i,j) : i = 0, N_x; 0 \le j \le N_y\} \cup \{(i,j) : 1 \le i \le N_x - 1; j = 0, N_y\}.
\end{aligned}$$

We introduce the grid function space

$$\mathcal{V} = \{v : v = \{v_{i,j}\} \text{ is a grid function in } \Lambda_h \cup \partial\Lambda_h \text{ and } v_{i,j} = 0 \text{ on } \partial\Lambda_h\}.$$

For any $v, w \in \mathcal{V}$, we define the discrete L^2 inner product and norms

$$(v,w) = h_x h_y \sum_{i=1}^{N_x-1} \sum_{j=1}^{N_y-1} v_{i,j}\overline{w_{i,j}}, \quad \|v\| = \sqrt{(v,v)}, \quad \|v\|_\infty = \max_{(i,j)\in\Lambda_h} |v_{i,j}|. \tag{4.106}$$

Theorem 4.7. *For $1 < \beta,\ \beta_1 < 2$, the discrete scheme (4.102) is unconditionally stable under maximum norm, i.e., for $1 \le n \le M$, it holds that*

$$\|U^n\|_\infty \le \|U^0\|_\infty + \sum_{k=1}^{n} \tau \|F^k\|_\infty. \tag{4.107}$$

Moreover, one has

$$\|E^n\|_\infty \le c\,(\tau + h_x + h_y), \tag{4.108}$$

where

$$E^n = \left(u_{1,1}^n - U_{1,1}^n, \cdots, u_{1,N_y-1}^n - U_{1,N_y-1}^n, \cdots, u_{N_x-1,N_y-1}^n - U_{N_x-1,N_y-1}^n\right)^{\mathrm{T}},$$

and c is a positive constant depending on T, the exact solution $u(x,y,t)$ but not depending on τ, h_x, and h_y.

Proof. Suppose that $|U_{i,j}^n| = \|U^n\|_\infty$. Then

$$\begin{aligned}
&U_{i,j}^n \Bigg[\frac{1}{h_x^\beta} \left(\sum_{k=0}^{i+1} g_k^\beta d_{+,i,j}^n U_{i-k+1,j}^n + \sum_{k=0}^{N_x-i+1} g_k^\beta d_{-,i,j}^n U_{i+k-1,j}^n \right) \\
&\quad + \frac{1}{h_y^{\beta_1}} \left(\sum_{k=0}^{j+1} g_k^{\beta_1} e_{+,i,j}^n U_{i,j-k+1}^n + \sum_{k=0}^{N_y-j+1} g_k^{\beta_1} e_{-,i,j}^n U_{i,j+k-1}^n \right) \\
&\quad - v_{1,i,j}^n \frac{U_{i,j}^n - U_{i-1,j}^n}{h_x} - v_{2,i,j}^n \frac{U_{i,j}^n - U_{i,j-1}^n}{h_y} \Bigg] \\
&\le \frac{1}{h_x^\beta} \left(\sum_{k=0}^{i+1} g_k^\beta d_{+,i,j}^n \left(U_{i,j}^n\right)^2 + \sum_{k=0}^{N_x-i+1} g_k^\beta d_{-,i,j}^n \left(U_{i,j}^n\right)^2 \right) \\
&\quad + \frac{1}{h_y^{\beta_1}} \left(\sum_{k=0}^{j+1} g_k^{\beta_1} e_{+,i,j}^n \left(U_{i,j}^n\right)^2 + \sum_{k=0}^{N_y-j+1} g_k^{\beta_1} e_{-,i,j}^n \left(U_{i,j}^n\right)^2 \right) \\
&\quad - v_{1,i,j}^n \frac{\left(U_{i,j}^n\right)^2 - \left(U_{i,j}^n\right)^2}{h_x} - v_{2,i,j}^n \frac{\left(U_{i,j}^n\right)^2 - \left(U_{i,j}^n\right)^2}{h_y} \le 0,
\end{aligned}$$

which means that

$$\left\| \left(I - \mathcal{A}_x - \mathcal{A}_y + \mathcal{B}\right) U^n \right\|_\infty \ge \|U^n\|_\infty. \tag{4.109}$$

The subsequent proofs are similar to Theorem 4.5. □

Theorem 4.8. *If $d_\pm(x,y,t)$, $e_\pm(x,y,t)$, $v_1(x,y,t)$ and $v_2(x,y,t)$ are all constants, then the discrete scheme (4.102) is unconditionally stable in L^2 norm, i.e., for $1 \le n \le M$, it holds that*

$$\|U^n\| \le \left\|U^0\right\| + \tau \sum_{k=0}^{n} \|F^n\| . \tag{4.110}$$

In addition, one has

$$\|E^n\| \le c(\tau + h_x + h_y). \tag{4.111}$$

Proof. By Lemma 4.11, it holds that

$$\begin{aligned}
\frac{\mathcal{A}_x + \mathcal{A}_x^{\mathrm{T}}}{2} &= \frac{\left(D_+^n + D_-^n\right)\tau}{2h_x^\beta} I_y \otimes \left(A_{\beta,N_x} + A_{\beta,N_x}^{\mathrm{T}}\right), \\
\frac{\mathcal{A}_y + \mathcal{A}_y^{\mathrm{T}}}{2} &= \frac{\left(E_+^n + E_-^n\right)\tau}{2h_y^{\beta_1}} \left(A_{\beta_1,N_y} + A_{\beta_1,N_y}^{\mathrm{T}}\right) \otimes I_x, \\
\frac{\mathcal{B} + \mathcal{B}}{2} &= \frac{V_1^n}{2h_x} I_y \otimes \left(B_{N_x} + B_{N_x}^{\mathrm{T}}\right) + \frac{V_2^n}{2h_y}\left(B_{N_y} + B_{N_y}^{\mathrm{T}}\right) \otimes I_x.
\end{aligned}$$

Then from Lemma 4.13, (4.100), and Lemmas 4.6 and 4.9, the eigenvalues of $-\frac{\mathcal{A}_x+\mathcal{A}_x^{\mathrm{T}}}{2}$, $-\frac{\mathcal{A}_y+\mathcal{A}_y^{\mathrm{T}}}{2}$ and $\frac{\mathcal{B}+\mathcal{B}}{2}$ are all positive. Therefore, using Lemma 4.6 again, one has

$$\left(\left(-\mathcal{A}_x - \mathcal{A}_y + \mathcal{B}\right) U^n, U^n\right) \ge 0. \tag{4.112}$$

The subsequent proofs are similar to Theorem 4.6. □

4.2.4 *Numerical results*

In the following, the numerical experiments are carried out to assess the computational performance and effectiveness of the numerical schemes. All numerical experiments are run in MATLAB 7.11 (R2010b) on a PC with Intel(R) Core (TM)i7-4510U 2.6 GHz processor and 8.0 GB RAM.

Example 4.1. Consider (4.75) with $c_+(x,t) = \Gamma(1.2)x^{1.8}$, $c_-(x,t) = \Gamma(1.2)(2-x)^{1.8}$, $v(x,t) = 2x$, and

$$\begin{aligned}
f(x,t) = -32\Bigg(& x^2 + (2-x)^2 - 2.5(x^3 + (2-x)^3) \\
&+ \frac{25}{22}(x^4 + (2-x)^4) - \left(x^3 - x^2\right)(x-2)\Bigg) - 4x^2(2-x)^2.
\end{aligned}$$

For $(x,t) \in [0,2] \times (0,0.5)$, the exact solution is $u(x,t) = 4e^{-t}x^2(2-x)^2$. We list the numerical results in Tables 4.1 and 4.2, where 'Gauss' denotes the Gaussian elimination method, the 'PGMRES (C_F)' and 'PGMRES(S)' denote the left preconditioned GMRES iteration with the Strang and T. Chan type preconditioners (see Appendix C), respectively, performed as Algorithm C.2, 'Iter' denotes the average iterations of all the time levels $t_n, n = 1, 2, \cdots, M$. For each t_n, the initial iteration vector is chosen as U^{n-1}, the stopping criterion is $\frac{r^m}{r_0} \leq 1e-8$ with $r(k)$ being the residual vector of linear systems after k iterations, and the preconditioning matrix is chosen as the circulant matrix (more specifically, the Strang or the T. Chan preconditioner) generated from Toeplitz matrix

$$I - \frac{\tau}{h^{1.8}}\left(\frac{\sum_{i=1}^{N-1} c_+(x_i,t_n)}{N-1}A_{\beta,N} + \frac{\sum_{i=1}^{N-1} c_-(x_i,t_n)}{N-1}A_{\beta,N}^{\mathrm{T}}\right)$$
$$+ \frac{\tau}{h}\frac{\sum_{i=1}^{N-1} v(x_i,t_n)}{N-1}B_N.$$

Table 4.1: The discrete maximum errors and convergence rates for Example 4.1 at $t = 0.5$ with $\tau = h$.

	Gauss		PGMRES(S)		PGMRES(C_F)	
N	$\Vert u^M - U^M\Vert_\infty$	rate	$\Vert u^M - U^M\Vert_\infty$	rate	$\Vert u^M - U^M\Vert_\infty$	rate
2^9	1.6637e-03	-	1.6637e-03	-	1.6638e-03	-
2^{10}	8.3458e-04	0.99	8.3458e-04	0.99	8.3467e-04	0.99
2^{11}	4.1797e-04	1.00	4.1774e-04	1.00	4.1798e-04	1.00

Table 4.2: The iterative numbers and computational time corresponding to Table 4.1.

	Gauss	PGMRES(S)		PGMRES(C_F)	
N	CPU(s)	Iter	CPU(s)	Iter	CPU(s)
2^9	8.2257	6.0	3.69	8.0	3.28
2^{10}	1.0415e+02	5.0	18.14	7.0	18.81
2^{11}	1.5857e+03	4.0	107.09	6.0	103.58

Example 4.2. Consider variable coefficients fractional diffusion equation

$$\frac{\partial u(x,y,t)}{\partial t} = d_+(x,y)\,{}_{-\infty}D_x^{\beta}u(x,y,t) + e_+(x,y)\,{}_{-\infty}D_y^{\beta_1}u(x,y,t) + f \quad (4.113)$$

in $(0,1)^2 \times (0,1]$, with $d_+(x,y) = x^\beta y$, $e_+(x,y) = xy^{\beta_1}$ and

$$u(x,y,t) = 0, \quad (x,y,t) \in \mathbb{R}^2 \backslash (0,1)^2 \times [0,1], \tag{4.114}$$

$$u(x,y,0) = (x^3 - x^2)(y^2 - y). \tag{4.115}$$

The source term $f(x,y,t)$ is derived from the exact solution $u(x,y,t) = e^{-t}(x^3 - x^2)(y^2 - y)$, $(x,y,t) \in [0,1]^2 \times [0,1]$.

The numerical performances of the Gaussian elimination method and the iteration methods for solving (4.104) are listed in Tables 4.3 and 4.4, where the preconditioned matrix is chosen as the block circulant circulant block (BCCB) matrix

$$I_y \otimes I_x - \frac{\tau}{h_x^\beta} \frac{\sum_{i=1,j=1}^{N_x,N_y} d_+(x_i,y_j)}{N_x N_y} \left(I_y \otimes C_F\left(A_{\beta,N_x}\right)\right)$$

$$- \frac{\tau}{h_x^{\beta_1}} \frac{\sum_{i=1,j=1}^{N_x,N_y} e_+(x_i,y_j)}{N_x N_y} \left(C_F\left(A_{\beta_1,N_y}\right) \otimes I_x\right).$$

Here $C_F\left(A_{\beta_1,N_x}\right)$ denotes the T. Chan circulant matrix generated from A_{β_1,N_x}, being similar for $C_F\left(A_{\beta_1,N_y}\right)$.

Table 4.3: The discrete L^2 errors and convergence rates for Example 4.2 at $t = 1$, with $(\beta, \beta_1) = (1.3, 1.8)$ and $\tau = h_x = h_y$.

	Gauss		GMRES		PGMRES	
N_x	$\|u^M - U^M\|$	rate	$\|u^M - U^M\|$	rate	$\|u^M - U^M\|$	rate
2^6	1.0934e-04	-	1.0934e-04	-	1.0934e-04	-
2^7	5.5259e-05	0.98	5.5259e-05	0.98	5.5261e-05	0.98
2^8	$\cdots$	$\cdots$	2.7779e-05	0.99	2.7787e-05	0.99

Table 4.4: The iterative numbers and computation time corresponding to Table 4.3.

	Gauss	GMRES		PGMRES	
N_x	CPU(s)	Iter	CPU(s)	Iter	CPU(s)
2^6	70.58	75.0	10.96	39.0	5.260
2^7	8.4197e+03	100.0	137.51	45.0	50.81
2^8	$>$ 8 hour	126.0	1.8347e+03	46.7	448.67

Finally, we add a $\mathcal{O}(\tau^2)$ perturbation term to the left hand side of (4.102) and develop the finite difference alternating direction implicit (ADI) scheme

$$\begin{cases} (1-\tau\delta_x)\, V_{i,j}^n = U_{i,j}^{n-1} + \tau f_{i,j}^n, \\ (1-\tau\delta_y)\, U_{i,j}^n = V_{i,j}^n, \end{cases} \tag{4.116}$$

where

$$\delta_x v_{i,j}^n := \frac{1}{h_x^{\beta}} \sum_{k=0}^{i+1} g_k^{\beta} d_{+,i,j}^n v_{i-k+1,j}^n,$$

$$\delta_y v_{i,j}^n := \frac{1}{h_y^{\beta_1}} \sum_{k=0}^{j+1} g_k^{\beta_1} e_{+,i,j}^n v_{i,j-k+1}^n.$$

The simulation results for (4.116) are listed in Table 4.5, which shows that the convergence rate is also $\mathcal{O}(\tau + h_x + h_y)$.

Table 4.5: Numerical simulations for Example 4.2 approximated by the ADI scheme (4.116) at $t=1$ with $\tau = h_x = h_y$. Here 'CPU(s)' denotes the computational time solved by the Gaussian elimination method.

	$(\beta,\beta_1)=(1.3,1.8)$			$(\beta,\beta_1)=(1.5,1.5)$		
N_x	$\Vert u^M - U^M\Vert_\infty$	rate	CPU(s)	$\Vert u^M - U^M\Vert$	rate	CPU(s)
2^6	4.9714e-04	-	1.27	2.1335e-04	-	1.24
2^7	2.5359e-04	0.97	15.16	1.0870e-04	0.97	15.43
2^8	1.2810e-04	0.99	263.17	5.4882e-05	0.99	282.36

4.3 Second order schemes for space fractional diffusion equations

In this section, we first provide a class of second order schemes to approximate the (untempered) R-L fractional derivative via combining the distinct shifted G-L formulae with their corresponding weights, and then use them to solve the fractional diffusion equations. Comparing with the extrapolation method, they look more flexible, and have the lower computation cost. The detailed stability and convergence analysis is presented. Numerical examples illustrate the effectiveness of the schemes and confirm the theoretical estimations.

4.3.1 Second order approximations for the Riemann-Liouville fractional derivatives

Theorem 4.9 ([Tian *et al.* (2015)]). *Suppose that* $u(x) \in \mathscr{S}^{\beta+2}(\mathbb{R})$ *and* $u(x)$ *has compact support. Define the WSGD operator by*

$$_L\mathcal{D}^{\beta}_{h,p,q}u(x) := \frac{\beta-2q}{2(p-q)}\,{}_LA^{\beta}_{h,p}u(x) + \frac{2p-\beta}{2(p-q)}\,{}_LA^{\beta}_{h,q}u(x), \tag{4.117}$$

then we have

$$_L\mathcal{D}^{\beta}_{h,p,q}u(x) = {}_{-\infty}D^{\beta}_{x}u(x) + O(h^2) \tag{4.118}$$

uniformly for $x \in \mathbb{R}$, *where*

$$_LA^{\beta}_{h,p}u(x) = \frac{1}{h^{\beta}}\sum_{k=0}^{\infty} g^{\beta}_{k} u\left(x-(k-p)h\right)$$

is given by (4.17).

Proof. By the definition of $_LA^{\beta}_{h,p}$, we assume the WSGD operator with the form

$$_L\mathcal{D}^{\beta}_{h,p,q}u(x) = \frac{\lambda_1}{h^{\beta}}\sum_{k=0}^{\infty} g^{(\beta)}_{k} u(x-(k-p)h) + \frac{\lambda_2}{h^{\beta}}\sum_{k=0}^{\infty} g^{(\beta)}_{k} u(x-(k-q)h). \tag{4.119}$$

Taking Fourier transform on (4.119), we obtain

$$\begin{aligned}\mathscr{F}[{}_L\mathcal{D}^{\beta}_{h,p,q}u](\omega) &= \frac{1}{h^{\beta}}\sum_{k=0}^{\infty} g^{(\beta)}_{k}\Big(\lambda_1 \mathrm{e}^{-i\omega(k-p)h} + \lambda_2 \mathrm{e}^{-i\omega(k-q)h}\Big)\hat{u}(\omega)\\ &= \frac{1}{h^{\beta}}\Big(\lambda_1(1-\mathrm{e}^{-i\omega h})^{\beta}\mathrm{e}^{i\omega hp} + \lambda_2(1-\mathrm{e}^{-i\omega h})^{\beta}\mathrm{e}^{i\omega hq}\Big)\hat{u}(\omega)\\ &= (i\omega)^{\beta}\Big(\lambda_1 W_p(i\omega h) + \lambda_2 W_q(i\omega h)\Big)\hat{u}(\omega),\end{aligned} \tag{4.120}$$

where for $r = p, q$ (see the proof of Theorem 4.1),

$$\begin{aligned}W_r(z) &= \Big(\frac{1-\mathrm{e}^{-z}}{z}\Big)^{\beta}\mathrm{e}^{rz}\\ &= 1 + \Big(r-\frac{\beta}{2}\Big)z + \frac{1}{24}\left(\beta+3\beta^2-12\beta r+12r^2\right)z^2 + \cdots.\end{aligned} \tag{4.121}$$

In order to have second order accuracy, coefficients λ_1 and λ_2 should satisfy

$$\begin{cases}\lambda_1+\lambda_2 = 1,\\ (p-\frac{\beta}{2})\lambda_1 + (q-\frac{\beta}{2})\lambda_2 = 0,\end{cases}$$

which indicates that $\lambda_1 = \frac{\beta-2q}{2(p-q)}$ and $\lambda_2 = \frac{2p-\beta}{2(p-q)}$.

Denoting $\hat{\phi}(\omega, h) = \mathscr{F}[{}_L\mathcal{D}^{\beta}_{h,p,q}u](\omega) - \mathscr{F}[{}_{-\infty}D^{\beta}_{x}u](\omega)$, then from (4.120) and (4.33) there exists

$$|\hat{\phi}(\omega, h)| \leq Ch^2|i\omega|^{\beta+2}|\hat{u}(\omega)|. \tag{4.122}$$

With the condition $u(x) \in \mathscr{S}^{\beta+2}(\mathbb{R})$, it yields

$$\left|{}_L\mathcal{D}^{\beta}_{h,p,q}u - {}_{-\infty}D^{\beta}_{x}u\right| \leq \frac{1}{2\pi}\int_{\mathbb{R}} |\hat{\phi}(\omega, h)| = O(h^2). \tag{4.123}$$

Thus, we complete the proof. □

Similarly, we can prove that

Theorem 4.10. *Let $u \in \mathscr{S}^{\beta+2}(\mathbb{R})$ and $u(x)$ have compact support, and define the WSGD operator by*

$$ {}_R\mathcal{D}^{\beta}_{h,p,q}u(x) := \frac{\beta-2q}{2(p-q)}{}_RA^{\beta}_{h,p}u(x) + \frac{2p-\beta}{2(p-q)}{}_RA^{\beta}_{h,q}u(x). \tag{4.124}$$

Then

$$ {}_R\mathcal{D}^{\beta}_{h,p,q}u(x) = {}_xD^{\beta}_{\infty}u(x) + O(h^2)$$

uniformly for $x \in \mathbb{R}$, where

$$ {}_RA^{\beta}_{h,p}u(x) = \frac{1}{h^{\beta}}\sum_{k=0}^{\infty} g^{\beta}_{k}u\left(x + (k-p)h\right) \tag{4.125}$$

is given by (4.18).

Remark 4.5. If $p, q \in \mathbb{N}$, then $|p|, |q|$ are the numbers of the points located on the right/left hand of the point x used for evaluating the β order left/right R-L fractional derivatives at x. Thus, if $u(x)$ has compact support in (a, b), it may be better to choose the integers p, q such that $|p| \leq 1, |q| \leq 1$. Since for $(p, q) = (0, -1)$, with the same proof process of Theorem 4.11 below, the approximation method turns out to be unstable for time dependent problems. Therefore, two sets of (p, q) can be selected to establish the difference scheme for fractional diffusion equations, that is $(1, 0)$, $(1, -1)$, and the corresponding weights in (4.117) and (4.124) are $(\frac{\beta}{2}, \frac{2-\beta}{2})$ and $(\frac{2+\beta}{4}, \frac{2-\beta}{4})$, respectively.

In fact, if $u(x) = 0$ for $x \in \mathbb{R}\backslash\Omega$ with $\Omega = (a, b)$, the simplified forms of the discrete approximations in Theorems 4.9 and 4.10 on grid points

$\{x_i = a + ih, h = (b-a)/n, i = 1, \ldots, n-1\}$ with $(p,q) = (1,0)$, $(1,-1)$ reduce to

$$\begin{aligned} {}_L\mathcal{D}^{\beta}_{h,p,q} u(x_i) &= \frac{1}{h^\beta} \sum_{k=0}^{i+1} w_k^{(\beta)} u(x_{i-k+1}), \\ {}_R\mathcal{D}^{\beta}_{h,p,q} u(x_i) &= \frac{1}{h^\beta} \sum_{k=0}^{N-i+1} w_k^{(\beta)} u(x_{i+k-1}), \end{aligned} \tag{4.126}$$

where

$$\begin{cases} (p,q) = (1,0), & w_0^{(\beta)} = \dfrac{\beta}{2} g_0^{(\beta)}, \ w_k^{(\beta)} = \dfrac{\beta}{2} g_k^{(\beta)} + \dfrac{2-\beta}{2} g_{k-1}^{(\beta)}, \ k \geq 1; \\ (p,q) = (1,-1), & w_0^{(\beta)} = \dfrac{2+\beta}{4} g_0^{(\beta)}, \ w_1^{(\beta)} = \dfrac{2+\beta}{4} g_1^{(\beta)}, \\ & w_k^{(\beta)} = \dfrac{2+\beta}{4} g_k^{(\beta)} + \dfrac{2-\beta}{4} g_{k-2}^{(\beta)}, \ k \geq 2. \end{cases} \tag{4.127}$$

Obviously, the WSGD operators (4.117) and (4.124) are the centered difference approximations of second order derivative when (p,q) equals to $(1,0)$ or $(1,-1)$, and $\beta = 2$.

Theorem 4.11. *Let*

$$A = \begin{bmatrix} w_1^{(\beta)} & w_0^{(\beta)} & & & \\ w_2^{(\beta)} & w_1^{(\beta)} & w_0^{(\beta)} & & \\ \vdots & w_2^{(\beta)} & w_1^{(\beta)} & \ddots & \\ w_{n-2}^{(\beta)} & \cdots & \ddots & \ddots & w_0^{(\beta)} \\ w_{n-1}^{(\beta)} & w_{n-2}^{(\beta)} & \cdots & w_2^{(\beta)} & w_1^{(\beta)} \end{bmatrix}, \tag{4.128}$$

where the diagonals $\{w_k^{(\beta)}\}_{k=0}^{n-1}$ *are the coefficients given in (4.126) corresponding to* $(p,q) = (1,0)$ *or* $(1,-1)$. *Then any eigenvalue* λ *of* A *satisfies*

(1) $\mathrm{Re}(\lambda) \equiv 0$, *for* $(p,q) = (1,0)$, $\beta = 1$;
(2) $\mathrm{Re}(\lambda) < 0$, *for* $(p,q) = (1,0)$, $1 < \beta \leq 2$;
(3) $\mathrm{Re}(\lambda) < 0$, *for* $(p,q) = (1,-1)$, $1 < \beta \leq 2$.

Moreover, when $1 < \beta \leq 2$, *matrix* A *is negative definite, and the real parts of the eigenvalues of matrix* $c_1 A + c_2 A^{\mathrm{T}}$ *are less than zero, where* $c_1, c_2 \geq 0, c_1^2 + c_2^2 \neq 0$.

Proof. Let $H = \frac{A+A^{\mathrm{T}}}{2}$ be the symmetric part of matrix A. Since the generating functions of A and A^{T} are

$$f_A(x) = \sum_{k=0}^{\infty} w_k^{(\beta)} \mathrm{e}^{i(k-1)x}, \quad f_{A^{\mathrm{T}}}(x) = \sum_{k=0}^{\infty} w_k^{(\beta)} \mathrm{e}^{-i(k-1)x},$$

respectively, then $f(\beta;x)=\frac{f_A(x)+f_{A^{\mathrm{T}}}(x)}{2}$ is the generating function of H, and $f(\beta;x)$ is a 2π periodic continuous real-valued function on $[-\pi,\pi]$.

Case $(p,q)=(1,0)$: with the corresponding coefficients $w_k^{(\beta)}$ given by (4.127), then

$$
\begin{aligned}
f(\beta;x) =& \frac{1}{2}\Big(\sum_{k=0}^{\infty} w_k^{(\beta)} \mathrm{e}^{i(k-1)x} + \sum_{k=0}^{\infty} w_k^{(\beta)} \mathrm{e}^{-i(k-1)x}\Big) \\
=& \frac{1}{2}\Big(\frac{\beta}{2}\mathrm{e}^{-ix}\sum_{k=0}^{\infty} g_k^{(\beta)} \mathrm{e}^{ikx} + \frac{2-\beta}{2}\sum_{k=0}^{\infty} g_k^{(\beta)} \mathrm{e}^{ikx} \\
&+ \frac{\beta}{2}\mathrm{e}^{ix}\sum_{k=0}^{\infty} g_k^{(\beta)} \mathrm{e}^{-ikx} + \frac{2-\beta}{2}\sum_{k=0}^{\infty} g_k^{(\beta)} \mathrm{e}^{-ikx}\Big) \\
=& \frac{\beta}{4}\Big(\mathrm{e}^{-ix}(1-\mathrm{e}^{ix})^{\beta} + \mathrm{e}^{ix}(1-\mathrm{e}^{-ix})^{\beta}\Big) \\
&+ \frac{2-\beta}{4}\Big((1-\mathrm{e}^{ix})^{\beta} + (1-\mathrm{e}^{-ix})^{\beta}\Big).
\end{aligned}
$$

Next we check $f(\beta;x)\le 0$ for $1<\beta\le 2$. Since $f(\beta;x)$ is a real-valued and even function, we just consider its principal value on $[0,\pi]$. By the formula

$$
\mathrm{e}^{i\theta}-\mathrm{e}^{i\phi} = 2i\sin\Big(\frac{\theta-\phi}{2}\Big)\mathrm{e}^{\frac{i(\theta+\phi)}{2}},
$$

we obtain

$$
f(\beta;x) = \Big(2\sin\Big(\frac{x}{2}\Big)\Big)^{\beta}\Big(\frac{\beta}{2}\cos\Big(\frac{\beta}{2}(x-\pi)-x\Big) + \frac{2-\beta}{2}\cos\Big(\frac{\beta}{2}(x-\pi)\Big)\Big). \tag{4.129}
$$

Denote

$$
g(\beta;x) = \frac{\beta}{2}\cos\Big(\frac{\beta}{2}(x-\pi)-x\Big) + \frac{2-\beta}{2}\cos\Big(\frac{\beta}{2}(x-\pi)\Big).
$$

It is easy to prove that $g(\beta;x)$ decreases w.r.t. β, so $f(\beta;x)\le 0$. By Lemmas 4.7 and 4.8, $\mathrm{Re}(\lambda)\equiv 0$ for $\beta=1$ as $f(1;x)\equiv 0$, and $f(\beta;x)$ is not zero for $1<\beta\le 2$, we get $\mathrm{Re}(\lambda)<0$.

Case $(p,q)=(1,-1)$: the corresponding generating function $f(\beta;x)$ of $\frac{A+A^{\mathrm{T}}}{2}$ can be calculated in the following form with coefficients $w_k^{(\beta)}$ given by (4.127),

$$
\begin{aligned}
f(\beta;x) =& \frac{1}{2}\Big(\sum_{k=0}^{\infty} w_k^{(\beta)} e^{i(k-1)x} + \sum_{k=0}^{\infty} w_k^{(\beta)} e^{-i(k-1)x}\Big) \\
=& \frac{2+\beta}{8}\Big(e^{-ix}\sum_{k=0}^{\infty} g_k^{(\beta)} e^{ikx} + e^{ix}\sum_{k=0}^{\infty} g_k^{(\beta)} e^{-ikx}\Big) \\
&+ \frac{2-\beta}{8}\Big(e^{ix}\sum_{k=0}^{\infty} g_k^{(\beta)} e^{ikx} + e^{-ix}\sum_{k=0}^{\infty} g_k^{(\beta)} e^{-ikx}\Big) \\
=& \frac{2+\beta}{8}\Big(e^{-ix}(1-e^{ix})^{\beta} + e^{ix}(1-e^{-ix})^{\beta}\Big) \\
&+ \frac{2-\beta}{8}\Big(e^{ix}(1-e^{ix})^{\beta} + e^{-ix}(1-e^{-ix})^{\beta}\Big).
\end{aligned}
$$

Next we check $f(\beta;x) \leq 0$ for $1 < \beta \leq 2$. Since $f(\beta;x)$ is a real-valued and even function, we just consider its principal value on $[0,\pi]$. By simple calculation, we obtain

$$
f(\beta;x) = \Big(2\sin\Big(\frac{x}{2}\Big)\Big)^{\beta}\Big(\frac{\beta}{2}\sin\Big(\frac{\beta}{2}(x-\pi)\Big)\sin(x) + \cos\Big(\frac{\beta}{2}(x-\pi)\Big)\cos(x)\Big). \tag{4.130}
$$

Denoting

$$
g(\beta;x) = \frac{\beta}{2}\sin\Big(\frac{\beta}{2}(x-\pi)\Big)\sin(x) + \cos\Big(\frac{\beta}{2}(x-\pi)\Big)\cos(x),
$$

we can also check that $g(\beta;x)$ decreases w.r.t. β. Then

$$
f(\beta;x) \leq \Big(2\sin\Big(\frac{x}{2}\Big)\Big)^{\beta} g(1;x) = -\Big(2\sin\Big(\frac{x}{2}\Big)\Big)^{\beta}\sin^3\Big(\frac{x}{2}\Big) \leq 0.
$$

Therefore, by Lemmas 4.7 and 4.8, we get $\mathrm{Re}(\lambda) < 0$ for $1 < \beta \leq 2$.

From the above discussions and Lemma 4.8, we know, for $1 < \beta \leq 2$, the matrix $\frac{1}{2}(A + A^{\mathrm{T}})$ is negative definite, which implies matrix A is negative definite by Lemma 4.6. And the symmetric part of matrix $c_1 A + c_2 A^{\mathrm{T}}$ is $\frac{c_1+c_2}{2}(A + A^{\mathrm{T}})$. Thus we obtain $\mathrm{Re}(\lambda(c_1 A + c_2 A^{\mathrm{T}})) < 0$ for $1 < \beta \leq 2$. □

4.3.2 *Numerical schemes for one dimensional space fractional diffusion equation*

In this subsection, we develop second order finite difference schemes for the following one dimensional space fractional diffusion equation on $(x,t) \in \Omega \times (0,T]$:

$$
\frac{\partial u(x,t)}{\partial t} = K_1 \, {}_{-\infty}D_x^{\beta} u(x,t) + K_2 \, {}_xD_{\infty}^{\beta} u(x,t) + f(x,t), \tag{4.131}
$$

with the initial condition

$$u(x,0) = u_0(x), \quad x \in \Omega, \tag{4.132}$$

and the absorbing boundary condition

$$u(x,t) = 0, \quad (x,t) \in \mathbb{R}\backslash\Omega \times (0,T], \tag{4.133}$$

where $\Omega = (a,b)$, ${}_{-\infty}D_x^\beta$ and ${}_xD_\infty^\beta$ are the R-L fractional operators with $1 < \beta \le 2$. The diffusion coefficients K_1 and K_2 are nonnegative constants such that $K_1^2 + K_2^2 \neq 0$. In the analysis of the numerical method that follows, we assume that (4.131)–(4.133) has a unique and sufficiently smooth solution.

4.3.2.1 *CN-WSGD scheme*

We partition the interval (a,b) into a uniform mesh with the space step $h = (b-a)/N$ and the time step $\tau = T/M$, where N, M are two positive integers. And the set of grid points are denoted by $x_i = a + ih$ and $t_n = n\tau$ for $0 \le i \le N$ and $0 \le n \le M$. Let $t_{n+1/2} = (t_n + t_{n+1})/2$ for $0 \le n \le M-1$, and we use the following notations

$$u_i^n = u(x_i, t_n), \quad f_i^{n+1/2} = f(x_i, t_{n+1/2}).$$

Using the Crank-Nicolson (CN) technique for the time discretization and the WSGD operators for the space discretizations lead to

$$\begin{aligned}\frac{(u_i^{n+1} - u_i^n)}{\tau} &- \frac{1}{2}\Big(K_1\ {}_L\mathcal{D}_{h,p,q}^\beta u_i^{n+1} + K_2\ {}_R\mathcal{D}_{h,p,q}^\beta u_i^{n+1}\Big)\\ &= \frac{1}{2}\Big(K_1\ {}_L\mathcal{D}_{h,p,q}^\beta u_i^{n} + K_2\ {}_R\mathcal{D}_{h,p,q}^\beta u_i^{n}\Big) + f_i^{n+1/2} + \varepsilon_i^n,\end{aligned} \tag{4.134}$$

where

$$|\varepsilon_i^n| \le \tilde{c}(\tau^2 + h^2). \tag{4.135}$$

Substituting the approximations ${}_L\mathcal{D}_{h,p,q}^\beta u$, ${}_R\mathcal{D}_{h,p,q}^\beta u$ given by (4.126)into (4.134), we obtain that

$$\begin{aligned}&u_i^{n+1} - \frac{K_1\tau}{2h^\beta}\sum_{k=0}^{i+1} w_k^{(\beta)} u_{i-k+1}^{n+1} - \frac{K_2\tau}{2h^\beta}\sum_{k=0}^{N-i+1} w_k^{(\beta)} u_{i+k-1}^{n+1}\\ &= u_i^n + \frac{K_1\tau}{2h^\beta}\sum_{k=0}^{i+1} w_k^{(\beta)} u_{i-k+1}^{n} + \frac{K_2\tau}{2h^\beta}\sum_{k=0}^{N-i+1} w_k^{(\beta)} u_{i+k-1}^{n} + \tau f_i^{n+1/2} + \tau\varepsilon_i^n,\end{aligned} \tag{4.136}$$

where $w_k^{(\beta)}$ is defined in (4.127). Denoting U_i^n as the numerical approximation of u_i^n, for (4.131) we derive the CN-WSGD scheme

$$\begin{aligned} U_i^{n+1} - \frac{K_1\tau}{2h^\beta}\sum_{k=0}^{i+1} w_k^{(\beta)} U_{i-k+1}^{n+1} - \frac{K_2\tau}{2h^\beta}\sum_{k=0}^{N-i+1} w_k^{(\beta)} U_{i+k-1}^{n+1} \\ = U_i^n + \frac{K_1\tau}{2h^\beta}\sum_{k=0}^{i+1} w_k^{(\beta)} U_{i-k+1}^{n} + \frac{K_2\tau}{2h^\beta}\sum_{k=0}^{N-i+1} w_k^{(\beta)} U_{i+k-1}^{n} + \tau f_i^{n+1/2}. \end{aligned} \tag{4.137}$$

Introducing

$$U^n = \left(U_1^n, U_2^n, \cdots, U_{N-1}^n\right)^{\mathrm{T}},$$

$$F^n = \left(f_1^{n+1/2}, f_2^{n+1/2}, \cdots, f_{N-1}^{n+1/2}\right)^{\mathrm{T}},$$

we can rewrite (4.137) as the matrix vector form

$$\left(I - \frac{\tau}{2h^\beta}(K_1A + K_2A^{\mathrm{T}})\right)U^{n+1} = \left(I + \frac{\tau}{2h^\beta}(K_1A + K_2A^{\mathrm{T}})\right)U^n + \tau F^n, \tag{4.138}$$

where the matrix A is given by (4.128).

4.3.2.2 *Stability and convergence*

Now we consider the stability and convergence analysis for the CN-WSGD scheme (4.138).

Theorem 4.12. *The finite difference scheme (4.137) is unconditionally stable.*

Proof. Denoting $B = \frac{\tau}{2h^\beta}(K_1A + K_2A^{\mathrm{T}})$, the matrix form (4.138) can be rewritten as

$$(I - B)U^{n+1} = (I + B)U^n + \tau F^n. \tag{4.139}$$

Let $\widetilde{U}_i^n$ $(i = 1, \ldots, N-1;\ n = 1, \ldots, M)$ be the approximate solution of U_i^n, which is the exact solution of (4.138). Putting $\delta U_i^n = \widetilde{U}_i^n - U_i^n$, and denoting $\delta U^n = [\delta U_1^n, \delta U_2^n, \ldots, \delta U_{N-1}^n]$, then the perturbation equation of (4.139) is given by

$$(I - B)\delta U^{n+1} = (I + B)\delta U^n.$$

If denote λ as an eigenvalue of matrix B, then $\frac{1+\lambda}{1-\lambda}$ is the eigenvalue of matrix $(I-B)^{-1}(I+B)$. The result of Theorem 4.11 shows that the eigenvalues of matrix $\frac{B+B^{\mathrm{T}}}{2} = \frac{\tau(K_1+K_2)}{4h^\beta}(A+A^{\mathrm{T}})$ are negative. Thus $\mathrm{Re}(\lambda) < 0$, which implies that $|\frac{1+\lambda}{1-\lambda}| < 1$. Therefore, the spectral radius of matrix $(I-B)^{-1}(I+B)$ are less than one, which yields that $\left((I-B)^{-1}(I+B)\right)^n$ converges to zero matrix (see Theorem 1.5 in [Quarteroni *et al.* (2007)]). Then the difference scheme (4.137) is unconditionally stable. □

Remark 4.6. Considering the θ weighted scheme for the time discretization of (4.131), then the iteration matrix of the full discrete scheme is

$$(I-\theta B)^{-1}(I+(1-\theta)B). \tag{4.140}$$

If λ is an eigenvalue of matrix B, then the eigenvalue of (4.140) is $\frac{1+(1-\theta)\lambda}{1-\theta\lambda}$. As $\mathrm{Re}(\lambda) < 0$, it is easy to check that

$$\left|\frac{1+(1-\theta)\lambda}{1-\theta\lambda}\right| < 1 \tag{4.141}$$

for $\frac{1}{2} \le \theta \le 1$. Then the θ weighted WSGD scheme for (4.131) is unconditionally stable when $\frac{1}{2} \le \theta \le 1$.

Theorem 4.13. *Let u_i^n be the exact solution of problem (4.131), and U_i^n the solution of the finite difference scheme (4.137). Then for all $1 \le n \le M$, we have*

$$\|u^n - U^n\| \le c(\tau^2 + h^2), \tag{4.142}$$

where c denotes a positive constant and $\|\cdot\|$ stands for the discrete L^2 norm given in (4.91).

Proof. Let $e_i^n = u_i^n - U_i^n$. By (4.136) and (4.137), we have

$$(e^{n+1} - e^n) - \frac{K_1\tau}{2h^\beta}A(e^{n+1} + e^n) - \frac{K_2\tau}{2h^\beta}A^{\mathrm{T}}(e^{n+1} + e^n) = \tau\varepsilon^n, \tag{4.143}$$

where

$$e^n = \left(u_1^n - U_1^n, u_2^n - U_2^n, \cdots, u_{N-1}^n - U_{N-1}^n\right)^{\mathrm{T}},$$
$$\varepsilon^n = \left(\varepsilon_1^n, \varepsilon_2^n, \cdots, \varepsilon_{N-1}^n\right)^{\mathrm{T}}.$$

Multiplying (4.143) by h, and acting $(e^{n+1} + e^n)^{\mathrm{T}}$ on both sides, we obtain that

$$\begin{aligned} h(e^{n+1}+e^n)^{\mathrm{T}}I(e^{n+1}-e^n) &- \frac{K_1\tau}{2h^{\beta-1}}(e^{n+1}+e^n)^{\mathrm{T}}A(e^{n+1}+e^n) \\ -\frac{K_2\tau}{2h^{\beta-1}}(e^{n+1}+e^n)^{\mathrm{T}}A^{\mathrm{T}}(e^{n+1}+e^n) &= \tau h(e^{n+1}+e^n)^{\mathrm{T}}\varepsilon^n. \end{aligned} \tag{4.144}$$

By Theorem 4.11, it holds that

$$(e^{n+1}+e^n)^{\mathrm{T}}A(e^{n+1}+e^n)<0, \tag{4.145}$$

$$(e^{n+1}+e^n)^{\mathrm{T}}A^{\mathrm{T}}(e^{n+1}+e^n)<0. \tag{4.146}$$

It yields from (4.144)–(4.146) that

$$\|e^{n+1}\|^2-\|e^n\|^2\le \tau h(e^{n+1}+e^n)^{\mathrm{T}}\varepsilon^n\le \tau(\|e^{n+1}\|+\|e^n\|)\cdot\|\varepsilon^n\|. \tag{4.147}$$

Then we have

$$\|e^{n+1}\|-\|e^n\|\le \tau\|\varepsilon^n\|, \quad n\ge 0. \tag{4.148}$$

Consequently,

$$\|e^n\|\le \tau\sum_{k=1}^{n}\|\varepsilon^n\|\le c(\tau^2+h^2), \quad n\ge 1, \tag{4.149}$$

which is the result that we need. □

4.3.3 *Numerical schemes for two dimensional space fractional diffusion equation*

We next consider second order schemes for the following two dimensional space fractional diffusion equation in $\Omega\times(0,T]$:

$$\begin{aligned}\frac{\partial u(x,y,t)}{\partial t}&=\Big(K_1^+{}_{-\infty}D_x^{\beta}u(x,y,t)+K_2^+{}_xD_{\infty}^{\beta}u(x,y,t)\Big)\\&\quad+\Big(K_1^-{}_{-\infty}D_y^{\beta_1}u(x,y,t)+K_2^-{}_yD_{\infty}^{\beta_1}u(x,y,t)\Big)+f(x,y,t),\end{aligned} \tag{4.150}$$

with the initial-boundary conditions

$$u(x,y,0)=u_0(x,y), \quad (x,y)\in\Omega, \tag{4.151}$$

$$u(x,y,t)=0, \quad (x,y,t)\in\mathbb{R}^2\backslash\Omega\times[0,T], \tag{4.152}$$

where $\Omega=(a,b)\times(c,d)$, ${}_{-\infty}D_x^{\beta}$, ${}_xD_{\infty}^{\beta}$ and ${}_{-\infty}D_y^{\beta_1}$, ${}_yD_{\infty}^{\beta_1}$ are R-L fractional operators with $1<\beta,\beta_1\le 2$. The diffusion coefficients satisfy $K_i^+,\ K_i^-\ge 0,\ i=1,2,\ (K_1^+)^2+(K_2^+)^2\ne 0$ and $(K_1^-)^2+(K_2^-)^2\ne 0$. We assume that (4.150) has a unique and sufficiently smooth solution.

4.3.3.1 *CN-WSGD scheme*

Now we establish the CN difference scheme by using WSGD formulae (4.126) for problem (4.150). We partition the domain Ω into a uniform mesh with the space steps $h_x=(b-a)/N_x$, $h_y=(d-c)/N_y$ in each direction, and the time step is $\tau=T/M$, where N_x,N_y,M are positive integers.

The set of grid points is denoted by $x_i = a + ih_x$, $y_j = c + jh_y$ and $t_n = n\tau$ for $0 \le i \le N_x, 0 \le j \le N_y$ and $0 \le n \le M$. Let $t_{n+1/2} = (t_n + t_{n+1})/2$ for $0 \le n \le M - 1$, and we use the following notations

$$u_{i,j}^n = u(x_i, y_j, t_n), \quad f_{i,j}^{n+1/2} = f(x_i, y_j, t_{n+1/2}).$$

Similar to (4.134), we have

$$\begin{aligned}
&\Big(1 - \frac{K_1^+\tau}{2}{}_L\mathcal{D}^{\beta}_{h_x,p,q} - \frac{K_2^+\tau}{2}{}_R\mathcal{D}^{\beta}_{h_x,p,q} - \frac{K_1^-\tau}{2}{}_L\mathcal{D}^{\beta_1}_{h_y,p,q} - \frac{K_2^-\tau}{2}{}_R\mathcal{D}^{\beta_1}_{h_y,p,q}\Big)u_{i,j}^{n+1} \\
&= \Big(\frac{K_1^+\tau}{2}{}_L\mathcal{D}^{\beta}_{h_x,p,q} + \frac{K_2^+\tau}{2}{}_R\mathcal{D}^{\beta}_{h_x,p,q} + \frac{K_1^-\tau}{2}{}_L\mathcal{D}^{\beta_1}_{h_y,p,q} + \frac{K_2^-\tau}{2}{}_R\mathcal{D}^{\beta_1}_{h_y,p,q}\Big)u_{i,j}^{n} \\
&\quad + u_{i,j}^n + \tau f_{i,j}^{n+1/2} + \tau\hat{\varepsilon}_{i,j}^n,
\end{aligned} \tag{4.153}$$

where

$$|\hat{\varepsilon}_{i,j}^n| \le \tilde{c}(\tau^2 + h_x^2 + h_y^2) \tag{4.154}$$

denotes the local truncation error.

Unlike in Sec. 4.2, here we add a perturbation term to develop the ADI schemes for this two dimensional model. Denote

$$\delta_x^{\beta} = K_1^+\,{}_L\mathcal{D}^{\beta}_{h_x,p,q} + K_2^+\,{}_R\mathcal{D}^{\beta}_{h_x,p,q}, \quad \delta_y^{\beta_1} = K_1^-\,{}_L\mathcal{D}^{\beta_1}_{h_y,p,q} + K_2^-\,{}_R\mathcal{D}^{\beta_1}_{h_y,p,q}.$$

Using the Taylor expansion, we have

$$\begin{aligned}
&-\frac{\tau^2}{4}\delta_x^{\beta}\delta_y^{\beta_1}(u_{i,j}^{n+1} - u_{i,j}^n) \\
&\qquad = -\frac{\tau^3}{4}\Big((K_1^+\,{}_aD_x^{\beta} + K_2^+\,{}_xD_b^{\beta})(K_1^-\,{}_cD_y^{\beta_1} + K_2^-\,{}_yD_d^{\beta_1})u_t\Big)_{i,j}^{n+\frac{1}{2}} \\
&\qquad\quad + \tau^3 O(\tau^2 + h_x^2 + h_y^2).
\end{aligned} \tag{4.155}$$

Adding formula (4.155) to the right hand side of (4.153) and making the factorization lead to

$$\Big(1 - \frac{\tau}{2}\delta_x^{\beta}\Big)\Big(1 - \frac{\tau}{2}\delta_y^{\beta_1}\Big)u_{i,j}^{n+1} = \Big(1 + \frac{\tau}{2}\delta_x^{\beta}\Big)\Big(1 + \frac{\tau}{2}\delta_y^{\beta_1}\Big)u_{i,j}^{n} + \tau f_{i,j}^{n+1/2} + \tau\varepsilon_{i,j}^n, \tag{4.156}$$

where $\varepsilon_{i,j}^n = \hat{\varepsilon}_{i,j}^n + O(\tau^2)$. Denoting by $U_{i,j}^n$ the numerical approximation to $u_{i,j}^n$, for problem (4.150) we obtain the finite difference approximation

$$\Big(1 - \frac{\tau}{2}\delta_x^{\beta}\Big)\Big(1 - \frac{\tau}{2}\delta_y^{\beta_1}\Big)U_{i,j}^{n+1} = \Big(1 + \frac{\tau}{2}\delta_x^{\beta}\Big)\Big(1 + \frac{\tau}{2}\delta_y^{\beta_1}\Big)U_{i,j}^{n} + \tau f_{i,j}^{n+1/2}. \tag{4.157}$$

Then, we can use the following techniques to solve (4.157) efficiently:

(1) Peaceman-Rachford ADI scheme [Sun (2005)]:

$$\left(1-\frac{\tau}{2}\delta_x^{\beta}\right)V_{i,j}^{n} \ = \left(1+\frac{\tau}{2}\delta_y^{\beta_1}\right)U_{i,j}^{n}+\frac{\tau}{2}f_{i,j}^{n+1/2}, \tag{4.158a}$$

$$\left(1-\frac{\tau}{2}\delta_y^{\beta_1}\right)U_{i,j}^{n+1}=\left(1+\frac{\tau}{2}\delta_x^{\beta}\right)V_{i,j}^{n}+\frac{\tau}{2}f_{i,j}^{n+1/2}; \tag{4.158b}$$

(2) Douglas ADI scheme [Douglas and Kimy (2001)]:

$$\left(1-\frac{\tau}{2}\delta_x^{\beta}\right)V_{i,j}^{n} \ = \left(1+\frac{\tau}{2}\delta_x^{\beta}+\tau\delta_y^{\beta_1}\right)U_{i,j}^{n}+\tau f_{i,j}^{n+1/2}, \tag{4.159a}$$

$$\left(1-\frac{\tau}{2}\delta_y^{\beta_1}\right)U_{i,j}^{n+1}=V_{i,j}^{n}-\frac{\tau}{2}\delta_y^{\beta_1}U_{i,j}^{n}; \tag{4.159b}$$

(3) D'Yakonov ADI scheme [Sun (2005)]:

$$\left(1-\frac{\tau}{2}\delta_x^{\beta}\right)V_{i,j}^{n} \ = \left(1+\frac{\tau}{2}\delta_x^{\beta}\right)\left(1+\frac{\tau}{2}\delta_y^{\beta_1}\right)U_{i,j}^{n}+\tau f_{i,j}^{n+1/2}, \tag{4.160a}$$

$$\left(1-\frac{\tau}{2}\delta_y^{\beta_1}\right)U_{i,j}^{n+1}=V_{i,j}^{n}. \tag{4.160b}$$

In order to facilitate theoretical analysis, we introduce the vectors

$$\begin{aligned}
U^n =&(U_{1,1}^n,U_{2,1}^n,\cdots,U_{N_x-1,1}^n,U_{1,2}^n,U_{2,2}^n,\cdots,U_{N_x-1,2}^n,\\
&\cdots,U_{1,N_y-1}^n,U_{2,N_y-1}^n,\cdots,U_{N_x-1,N_y-1}^n)^{\mathrm{T}},\\
F^{n+1/2} =&(f_{1,1}^{n+1/2},f_{2,1}^{n+1/2},\cdots,f_{N_x-1,1}^{n+1/2},f_{1,2}^{n+1/2},f_{2,2}^{n+1/2},\cdots,f_{N_x-1,2}^{n+1/2},\\
&\cdots,f_{1,N_y-1}^{n+1/2},f_{2,N_y-1}^{n+1/2},\cdots,f_{N_x-1,N_y-1}^{n+1/2})^{\mathrm{T}},
\end{aligned}$$

and denote

$$\mathcal{D}_x=\frac{K_1^+\tau}{2h_x^{\beta}}I_y\otimes A_\beta+\frac{K_2^+\tau}{2h_x^{\beta}}I_y\otimes A_\beta^{\mathrm{T}},\ \ \mathcal{D}_y=\frac{K_1^-\tau}{2h_y^{\beta_1}}A_{\beta_1}\otimes I_x+\frac{K_2^-\tau}{2h_y^{\beta_1}}A_{\beta_1}^{\mathrm{T}}\otimes I_x, \tag{4.161}$$

where the symbol $\otimes$ denotes the Kronecker product, I_x and I_y are unit $(N_x-1)\times(N_x-1)$ and $(N_y-1)\times(N_y-1)$ matrices, respectively, and matrices A_β and A_{β_1} are defined in (4.128) corresponding to β,β_1, respectively. Then, the difference schemes (4.157) can be expressed as

$$(I-\mathcal{D}_x)(I-\mathcal{D}_y)U^{n+1}=(I+\mathcal{D}_x)(I+\mathcal{D}_y)U^n+\tau F^{n+1/2}. \tag{4.162}$$

4.3.3.2 *Stability and convergence*

Now we consider the stability and convergence analysis for the CN-WSGD scheme (4.157). For the convenient presentation, we choose $N=N_x=N_y$.

Theorem 4.14. *The difference scheme (4.157) is unconditionally stable for* $1<\beta,\beta_1\le 2$.

Proof. From the difference scheme (4.162), we have the relationship between the disturbance error δU^{n+1} in U^{n+1} and the disturbance error δU^n in U^n as follows

$$\delta U^{n+1} = (I - \mathcal{D}_y)^{-1}(I - \mathcal{D}_x)^{-1}(I + \mathcal{D}_x)(I + \mathcal{D}_y)\delta U^n, \tag{4.163}$$

where I is the $(N_x - 1) \times (N_y - 1)$ unit matrix. Using Lemma 4.10, we can check that $\mathcal{D}_x$ and $\mathcal{D}_y$ commute, i.e.,

$$\mathcal{D}_x\mathcal{D}_y = \mathcal{D}_y\mathcal{D}_x = \frac{\tau^2}{4h_x^{\beta}h_y^{\beta_1}}(K_1^- A_{\beta_1} + K_2^- A_{\beta_1}^{\mathrm{T}}) \otimes (K_1^+ A_\beta + K_2^+ A_\beta^{\mathrm{T}}). \tag{4.164}$$

Thus (4.163) can be rewritten as

$$\delta U^n = \left((I - \mathcal{D}_y)^{-1}(I + \mathcal{D}_y)\right)^n \left((I - \mathcal{D}_x)^{-1}(I + \mathcal{D}_x)\right)^n \delta U^0. \tag{4.165}$$

We can also calculate the symmetric parts of $\mathcal{D}_x$ and $\mathcal{D}_y$ by Lemma 4.11 as

$$\frac{\mathcal{D}_x + \mathcal{D}_x^{\mathrm{T}}}{2} = \frac{(K_1^+ + K_2^+)\tau}{2h_x^{\beta}} I_y \otimes \left(\frac{A_\beta + A_\beta^{\mathrm{T}}}{2}\right),$$
$$\frac{\mathcal{D}_y + \mathcal{D}_y^{\mathrm{T}}}{2} = \frac{(K_1^- + K_2^-)\tau}{2h_y^{\beta_1}} \left(\frac{A_{\beta_1} + A_{\beta_1}^{\mathrm{T}}}{2}\right) \otimes I_x.$$

From Theorem 4.11, the eigenvalues of $\frac{A_\beta + A_\beta^{\mathrm{T}}}{2}$ and $\frac{A_{\beta_1} + A_{\beta_1}^{\mathrm{T}}}{2}$ are all negative when $1 < \beta, \beta_1 \leq 2$. Letting λ_β and λ_{β_1} be arbitrary eigenvalues of matrices $\mathcal{D}_x$ and $\mathcal{D}_y$, respectively, then it yields from the consequences of Lemmas 4.7 and 4.9 that the real parts of λ_β and λ_{β_1} are both less than zero. Since $(1 + \lambda_\beta)/(1 - \lambda_\beta)$ and $(1 + \lambda_{\beta_1})/(1 - \lambda_{\beta_1})$ are eigenvalues of matrices $(I - \mathcal{D}_x)^{-1}(I + \mathcal{D}_x)$ and $(I - \mathcal{D}_y)^{-1}(I + \mathcal{D}_y)$, respectively, the spectral radius of each matrix is less than one, which follows that $\left((I - \mathcal{D}_x)^{-1}(I + \mathcal{D}_x)\right)^n$ and $\left((I - \mathcal{D}_y)^{-1}(I + \mathcal{D}_y)\right)^n$ converge to zero matrix (see Theorem 1.5 in [Quarteroni *et al.* (2007)]). Therefore the difference scheme (4.157) is unconditionally stable. □

Remark 4.7. For the similar reason described in Remark 4.6 and the proof of Theorem 4.14, we conclude that the WSGD scheme with θ weighted scheme for the time discretization for (4.150) is unconditionally stable when $\frac{1}{2} \leq \theta \leq 1$.

Lemma 4.14. *Let $\mathcal{D}_x$ and $\mathcal{D}_y$ be defined in (4.161). Then*

$$\|(I - \mathcal{D}_x)^{-1}(I - \mathcal{D}_y)^{-1}\| \leq 1,$$
$$\|(I - \mathcal{D}_\gamma)^{-1}(I + \mathcal{D}_\gamma)\| \leq 1, \quad \gamma = x, y,$$

where $\|\cdot\|$ denotes the discrete L^2 norm in (4.106).

Proof. From Theorem 4.11 and Lemma 4.9, we know that $\mathcal{D}_x + \mathcal{D}_x^{\mathrm{T}}$ and $\mathcal{D}_y + \mathcal{D}_y^{\mathrm{T}}$ are negative semi-definite and symmetric matrices. Then for any $v \in \mathbb{R}^{(N_x-1)\times(N_y-1)}$, we obtain that

$$v^{\mathrm{T}}v \leq v^{\mathrm{T}}(I - \mathcal{D}_\gamma^{\mathrm{T}})(I - \mathcal{D}_\gamma)v, \quad \gamma = x, y.$$

Replacing v and v^{T} by $(I - \mathcal{D}_\gamma)^{-1}v$ and $v^{\mathrm{T}}(I - \mathcal{D}_\gamma^{\mathrm{T}})^{-1}$, respectively, for any $v \in \mathbb{R}^{(N_x-1)\times(N_y-1)}$, we get

$$v^{\mathrm{T}}(I - \mathcal{D}_\gamma^{\mathrm{T}})^{-1}(I - \mathcal{D}_\gamma)^{-1}v \leq v^{\mathrm{T}}v, \quad \gamma = x, y.$$

Thus, it leads to

$$\|(I - \mathcal{D}_\gamma)^{-1}\| = \sup_{v\neq 0}\sqrt{\frac{v^{\mathrm{T}}(I - \mathcal{D}_\gamma^{\mathrm{T}})^{-1}(I - \mathcal{D}_\gamma)^{-1}v}{v^{\mathrm{T}}v}} \leq 1, \quad \gamma = x, y.$$

Consequently,

$$\|(I - \mathcal{D}_x)^{-1}(I - \mathcal{D}_y)^{-1}\| \leq \|(I - \mathcal{D}_x)^{-1}\|\|(I - \mathcal{D}_y)^{-1}\| \leq 1$$

holds.

Since $\mathcal{D}_x + \mathcal{D}_x^{\mathrm{T}}$ and $\mathcal{D}_y + \mathcal{D}_y^{\mathrm{T}}$ are negative semi-definite and symmetric, for any $v \in \mathbb{R}^{(N_x-1)\times(N_y-1)}$, we have

$$v^{\mathrm{T}}(I + \mathcal{D}_\gamma^{\mathrm{T}})(I + \mathcal{D}_\gamma)v \leq v^{\mathrm{T}}(I - \mathcal{D}_\gamma^{\mathrm{T}})(I - \mathcal{D}_\gamma)v, \quad \gamma = x, y.$$

By choosing vector $(I - \mathcal{D}_\gamma)^{-1}v$, we arrive at that for any $v \in \mathbb{R}^{(N_x-1)\times(N_y-1)}$,

$$v^{\mathrm{T}}(I - \mathcal{D}_\gamma^{\mathrm{T}})^{-1}(I + \mathcal{D}_\gamma^{\mathrm{T}})(I + \mathcal{D}_\gamma)(I - \mathcal{D}_\gamma)^{-1}v \leq v^{\mathrm{T}}v, \quad \gamma = x, y.$$

As $(I - \mathcal{D}_\gamma)^{-1}(I + \mathcal{D}_\gamma) = (I + \mathcal{D}_\gamma)(I - \mathcal{D}_\gamma)^{-1}$, then it yields that

$$\begin{aligned}
&\|(I - \mathcal{D}_\gamma)^{-1}(I + \mathcal{D}_\gamma)\| \\
&= \|(I + \mathcal{D}_\gamma)(I - \mathcal{D}_\gamma)^{-1}\| \\
&= \sup_{v\neq 0}\sqrt{\frac{v^{\mathrm{T}}(I - \mathcal{D}_\gamma^{\mathrm{T}})^{-1}(I + \mathcal{D}_\gamma^{\mathrm{T}})(I + \mathcal{D}_\gamma)(I - \mathcal{D}_\gamma)^{-1}v}{v^{\mathrm{T}}v}} \\
&\leq 1.
\end{aligned}$$

□

Theorem 4.15. *Let $u_{i,j}^n$ be the exact solution of (4.150) with $1 < \beta, \beta_1 \leq 2$, and $U_{i,j}^n$ the solution of the difference scheme (4.157). Then for all $1 \leq n \leq M$, we have*

$$\|u^n - U^n\| \leq c(\tau^2 + h_x^2 + h_y^2), \tag{4.166}$$

where c denotes a positive constant and $\|\cdot\|$ is the discrete L^2 norm in (4.106).

Proof. Let $e_{i,j}^n = u_{i,j}^n - U_{i,j}^n$. Subtracting (4.156) from (4.157) leads to

$$\big(I - \mathcal{D}_x\big)\big(I - \mathcal{D}_y\big)e^{n+1} = \big(I + \mathcal{D}_x\big)\big(I + \mathcal{D}_y\big)e^n + \tau\mathcal{E}^n, \tag{4.167}$$

where $\mathcal{D}_x$ and $\mathcal{D}_y$ are given in (4.161) and

$$\begin{aligned} e^n =&(e_{1,1}^n, e_{2,1}^n, \cdots, e_{N_x-1,1}^n, e_{1,2}^n, e_{2,2}^n, \cdots, e_{N_x-1,2}^n,\\ &\cdots, e_{1,N_y-1}^n, e_{2,N_y-1}^n, \cdots, e_{N_x-1,N_y-1}^n)^{\mathrm{T}},\\ \mathcal{E}^n =&(\varepsilon_{1,1}^n, \varepsilon_{2,1}^n, \cdots, \varepsilon_{N_x-1,1}^n, \varepsilon_{1,2}^n, \varepsilon_{2,2}^n, \cdots, \varepsilon_{N_x-1,2}^n,\\ &\cdots, \varepsilon_{1,N_y-1}^n, \varepsilon_{2,N_y-1}^n, \cdots, \varepsilon_{N_x-1,N_y-1}^n)^{\mathrm{T}}. \end{aligned}$$

Since $\mathcal{D}_x$ commutes with $\mathcal{D}_y$, denoting $P = \big(I - \mathcal{D}_x\big)^{-1}\big(I - \mathcal{D}_y\big)^{-1}\big(I + \mathcal{D}_x\big)\big(I + \mathcal{D}_y\big)$, it yields that

$$e^{n+1} = Pe^n + \tau\big(I - \mathcal{D}_x\big)^{-1}\big(I - \mathcal{D}_y\big)^{-1}\mathcal{E}^n. \tag{4.168}$$

Replacing n by k, iterating for all $0 \le k \le n-1$, and taking the discrete L^2 norm on both sides, we have that

$$\begin{aligned} \|e^n\| &\le \tau\|(I - \mathcal{D}_x)^{-1}(I - \mathcal{D}_y)^{-1}\|\sum_{k=0}^{n-1}\|P^k\|\cdot\|\mathcal{E}^{n-1-k}\|\\ &\le \tau\sum_{k=0}^{n-1}\|P^k\|\cdot\|\mathcal{E}^{n-1-k}\|, \end{aligned} \tag{4.169}$$

where Lemma 4.14 shows that $\|(I - \mathcal{D}_x)^{-1}(I - \mathcal{D}_y)^{-1}\| \le 1$.

Since $\mathcal{D}_x$ and $\mathcal{D}_y$ commute, matrix P can be rewritten as

$$P = (I - \mathcal{D}_x)^{-1}(I + \mathcal{D}_x)(I - \mathcal{D}_y)^{-1}(I + \mathcal{D}_y). \tag{4.170}$$

We then obtain from Lemma 4.14 that

$$\|P\| \le \|(I - \mathcal{D}_x)^{-1}(I + \mathcal{D}_x)\|\|(I - \mathcal{D}_y)^{-1}(I + \mathcal{D}_y)\| \le 1. \tag{4.171}$$

Then for any $1 \le k \le M$, $\|P^k\| \le \|P\|^k \le 1$ holds. We can get that

$$\|e^n\| \le \tau\sum_{k=0}^{n-1}\|\mathcal{E}^k\| \le c(\tau^2 + h_x^2 + h_y^2). \tag{4.172}$$

□

4.3.4 *Modifying the difference operators near the boundary of the domain*

Theoretically, when high order finite difference discretizations are used to solve the fractional PDEs, the potential analytical solution must have the global smoothness. More specifically, if the exact solution satisfies $u(x,t) = 0$ for $x \in \mathbb{R}\backslash\Omega$, in order to obtain the desired convergence order, its several derivatives should also be zero at $\partial\Omega$ (see Remark 4.2). These are different from the classical differential operators. In the following, we give a slight modification to the previous suggested high order difference operators, such that if the analytical solution is smooth enough on $\overline{\Omega}$, the desired high order accuracy can always be obtained.

In fact, let $\Omega = (a,b)$ and $u(\cdot,t) \in C^m(\overline{\Omega})$. We have

$$ {}_aD_x^\beta u(x,t) = {}_aD_x^\beta v_L(x,t) + \sum_{l=0}^{m} \frac{(x-a)^{l-\beta}}{\Gamma(l+1-\beta)} \partial_x^l u(a,t), \tag{4.173} $$

$$ {}_aD_x^\beta u(x,t) = {}_aD_x^\beta v_R(x,t) + \sum_{l=0}^{m} \frac{(b-x)^{l-\beta}}{\Gamma(l+1-\beta)} (-1)^l \partial_x^l u(b,t), \tag{4.174} $$

where

$$ v_L(x,t) = u(x,t) - \sum_{l=0}^{m} \frac{(x-a)^l}{\Gamma(l+1)} \partial_x^l u(a,t), \tag{4.175} $$

$$ v_R(x,t) = u(x,t) - \sum_{l=0}^{m} \frac{(b-x)^l}{\Gamma(l+1)} (-1)^l \partial_x^l u(b,t). \tag{4.176} $$

Obviously, we have made sure that $v_L(x,t)$ and its several derivatives w.r.t. x at $x = a$ are zero, and $v_R(x,t)$ and its several derivatives w.r.t. x at $x = b$ are zero. Then the high order discretizations for ${}_aD_x^\beta v_L(x,t)$ and ${}_xD_b^\beta v_R(x,t)$ keep their high accuracy for $x \in \Omega$.

In order to obtain the final discrete discretizations of ${}_aD_x^\beta u(x,t)$ and ${}_aD_x^\beta u(x,t)$, the appropriate approximations of $\partial_x^l u(a,t)$ and $(-1)^l \partial_x^l u(b,t)$ must be developed. Here we use the ν difference formula given in [Gustafsson (2008), p.83] and obtain

$$ \partial_x^l u(a,t) = \frac{1}{h^l} \sum_{q=0}^{l+\nu-1} d_q^l u(x_q,t) + \mathcal{O}(h^\nu), \tag{4.177} $$

$$ (-1)^l \partial_x^l u(b,t) = \frac{1}{h^l} \sum_{q=N-l-\nu+1}^{N} \overline{d}_q^l u(x_q,t) + \mathcal{O}(h^\nu), \tag{4.178} $$

Table 4.6: Coefficients d_q^l for approximations (4.177)

l	ν	x_0	x_1	x_2	x_3	x_4	x_5
	1	-1	1				
	2	$-\frac{3}{2}$	2	$-\frac{1}{2}$			
1	3	$-\frac{11}{6}$	3	$-\frac{3}{2}$	$\frac{1}{3}$		
	4	$-\frac{25}{12}$	4	-3	$\frac{4}{3}$	$-\frac{1}{4}$	
	1	1	-2	1			
	2	2	-5	4	-1		
2	3	$\frac{35}{12}$	$-\frac{26}{3}$	$\frac{19}{2}$	$-\frac{14}{3}$	$\frac{11}{12}$	
	4	$\frac{15}{4}$	$-\frac{77}{6}$	$\frac{107}{6}$	-13	$\frac{61}{12}$	$-\frac{5}{6}$

Table 4.7: Coefficients $\overline{d}_q^l$ for approximations (4.178)

l	ν	x_N	x_{N-1}	x_{N-2}	x_{N-3}	x_{N-4}	x_{N-5}
	1	-1	1				
	2	$-\frac{3}{2}$	2	$-\frac{1}{2}$			
1	3	$-\frac{11}{6}$	3	$-\frac{3}{2}$	$\frac{1}{3}$		
	4	$-\frac{25}{12}$	4	-3	$\frac{4}{3}$	$-\frac{1}{4}$	
	1	1	-2	1			
	2	2	-5	4	-1		
2	3	$\frac{35}{12}$	$-\frac{26}{3}$	$\frac{19}{2}$	$-\frac{14}{3}$	$\frac{11}{12}$	
	4	$\frac{15}{4}$	$-\frac{77}{6}$	$\frac{107}{6}$	-13	$\frac{61}{12}$	$-\frac{5}{6}$

where $x_0 = a, x_M = b$, and $x_q = a+qh$. The coefficients d_q^l and $\overline{d}_q^l$ $(l = 1, 2)$ are, respectively, given in Tables 4.6 and 4.7.

Similar to Remark 4.2, if $u(x,t) = 0$ for $x \in \mathbb{R}\backslash\Omega$ and $u(x,t)$ is smooth enough w.r.t. x on $\overline{\Omega}$, then for $\nu \geq 2$ we have

$$
\begin{aligned}
{}_{-\infty}D_x^{\beta}u(x,t)\,|_{x=x_i} &= {}_L\mathcal{D}_{h,p,q}^{\beta}v_L(x_i,t) + \sum_{l=1}^{m}\frac{(x_i-a)^{l-\beta}}{\Gamma(l+1-\beta)}\partial_x^l u(a,t) + \mathcal{O}(h^2) \\
&= \frac{1}{h^{\beta}}\sum_{k=0}^{i+1} w_k^{(\beta)}\left[u(x_{i-k+1},t) - \sum_{l=1}^{m}\frac{(x_{i-k+1}-a)^l}{\Gamma(l+1)}\left(\frac{1}{h^l}\sum_{q=1}^{l+\nu-1} d_q^l u(x_q,t)\right)\right] \\
&\quad + \sum_{l=1}^{m}\frac{(x_i-a)^{l-\beta}}{\Gamma(l+1-\beta)}\left(\frac{1}{h^l}\sum_{q=1}^{l+\nu-1} d_q^l u(x_q,t)\right) + O(h^2) \\
&= \frac{1}{h^{\beta}}\sum_{k=0}^{i+1} {}_L\omega_k^{\beta}u(x_{i-k+1},t) + \mathcal{O}(h^2), \qquad i = 1, 2, \cdots, N-1 \qquad (4.179)
\end{aligned}
$$

and

$$
\begin{aligned}
&{}_xD_\infty^\beta u(x,t)\,|_{x=x_i}\\
&= {}_R\mathcal{D}_{h,p,q}^\beta v_R(x_i,t) + \sum_{l=1}^{m} \frac{(b-x_i)^{l-\beta}}{\Gamma(l+1-\beta)}(-1)^l \partial_x^l u(b,t) + \mathcal{O}(h^2)\\
&= \sum_{k=0}^{N-i+1} \frac{w_k^{(\beta)}}{h^\beta}\left[u(x_{i+k-1},t) - \sum_{l=1}^{m}\frac{(b-x_{i+k-1})^l}{\Gamma(l+1)}\left(\sum_{q=N-l-\nu+1}^{N}\frac{\overline{d}_q^l}{h^l}u(x_q,t)\right)\right]\\
&\quad + \sum_{l=1}^{m}\frac{(b-x_i)^{l-\beta}}{\Gamma(l+1-\beta)}\left(\frac{1}{h^l}\sum_{q=N-l-\nu+1}^{N}\overline{d}_q^l u(x_q,t)\right) + O(h^2)\\
&= \frac{1}{h^\beta}\sum_{k=0}^{N-i+1} {}_R\omega_k^\beta u(x_{i+k-1},t) + \mathcal{O}(h^2), \quad i = 1,2,\cdots,N-1. \qquad (4.180)
\end{aligned}
$$

It is easy to check that

$$
\begin{aligned}
{}_L\omega_k^\beta &= \omega_k^{(\beta)} \quad \text{for } k \neq i+2-m-\nu, i+3-m-\mu,\cdots,i+1,\\
{}_R\omega_k^\beta &= \omega_k^{(\beta)} \quad \text{for } k \neq N+2-m-\nu-i,\cdots,N-i+1.
\end{aligned}
$$

Thus, the coefficients of $u(x_i)$ $(i = 1,2,\cdots,N-1)$ of the modified WSGD operators are different from the ones of the WSGD operators near the boundary of the domain. We can use them to solve the models given before without the limitations that the several derivatives of their exact solutions should be zero at the boundaries.

Remark 4.8. The techniques introduced here can be directly extended to the other high order finite difference schemes suggested in the following sections of this chapter, which will not be mentioned in that places. In addition, if the potential analytical solution has a very low regularity near the boundary of the domain [Jin *et al.* (2015a); Wang and Zhang (2015)], the modified WSGD operators can not improve the accuracy. In this case, the high order difference scheme with partition refinements near the boundary can be used [Zhao and Deng (2016)].

4.3.5 *Numerical results*

Example 4.3. Consider variable coefficients problem

$$
\frac{\partial u(x,t)}{\partial t} = x^\beta {}_{-\infty}D_x^\beta u(x,t) + (1-x)^\beta {}_xD_\infty^\beta u(x,t) + f(x,t), \qquad (4.181)
$$

in $(0,1)\times(0,1]$, with

$$u(x,t)=0,\quad (x,t)\in\mathbb{R}\backslash(0,1)\times[0,1],$$
$$u(x,0)=x^3(1-x)^3,\quad x\in(0,1),$$

and the source term

$$f(x,t)=-\mathrm{e}^{-t}\Big(\frac{\Gamma(4)}{\Gamma(4-\beta)}(x^3+(1-x)^3)-\frac{3\Gamma(5)}{\Gamma(5-\beta)}(x^4+(1-x)^4)$$
$$+x^3(1-x)^3+3\frac{\Gamma(6)}{\Gamma(6-\beta)}(x^5+(1-x)^5)-\frac{\Gamma(7)}{\Gamma(7-\beta)}(x^6+(1-x)^6)\Big).$$

Then for $(x,t)\in[0,1]\times[0,1]$, the exact solution is $u(x,t)=\mathrm{e}^{-t}x^3(1-x)^3$.

Table 4.8: The discrete L^2 or maximum norm errors and their convergence rates to Example 4.3 approximated by the CN-WSGD scheme at $t=1$ for different β with $\tau=h$.

		$(p,q)=(1,0)$				$(p,q)=(1,-1)$	
β	N	$\lVert u^M-U^M\rVert_\infty$	rate	$\lVert u^M-U^M\rVert$	rate	$\lVert u^M-U^M\rVert$	rate
1.1	16	1.77123e-04	-	7.32001e-05	-	1.92219e-04	-
	32	4.47870e-05	1.98	1.76184e-05	2.05	4.11452e-05	2.22
	64	1.08962e-05	2.04	4.36356e-06	2.01	1.00363e-05	2.04
	128	2.66784e-06	2.03	1.08906e-06	2.00	2.51523e-06	2.00
1.5	16	1.88510e-04	-	6.18902e-05	-	1.30433e-04	-
	32	4.48741e-05	2.07	1.46628e-05	2.08	2.80619e-05	2.22
	64	1.10524e-05	2.02	3.61334e-06	2.02	6.65178e-06	2.08
	128	2.74933e-06	2.01	8.99424e-07	2.01	1.63398e-06	2.03
1.9	16	1.61881e-04	-	4.02897e-05	-	5.75044e-05	-
	32	3.43080e-05	2.24	9.58213e-06	2.07	1.27751e-05	2.17
	64	7.72475e-06	2.15	2.35289e-06	2.03	3.02268e-06	2.08
	128	1.91676e-06	2.01	5.83977e-07	2.01	7.37420e-07	2.04

Example 4.4. Consider fractional diffusion equation

$$\frac{\partial u(x,y,t)}{\partial t}={}_{-\infty}D_x^{1.2}u(x,y,t)+{}_xD_{\infty}^{1.2}u(x,y,t)+{}_{-\infty}D_y^{1.8}u(x,y,t)$$
$$+{}_yD_{\infty}^{1.8}u(x,y,t)+f(x,y,t),$$

in $(0,1)^2\times(0,1]$, with the absorbing boundary conditions and the initial condition $u(x,y,0)=x^3(1-x)^3y^3(1-y)^3$, $(x,y)\in(0,1)^2$, where the source term

$$
\begin{aligned}
f(x,y,t) = -\mathrm{e}^{-t}\Big[&\Big(x^3(1-x)^3y^3(1-y)^3\Big)\\
&+\Big(\frac{\Gamma(4)}{\Gamma(2.8)}\big(x^{1.8}+(1-x)^{1.8}\big)-\frac{3\Gamma(5)}{\Gamma(3.8)}\big(x^{2.8}+(1-x)^{2.8}\big)\\
&+\frac{3\Gamma(6)}{\Gamma(4.8)}\big(x^{3.8}+(1-x)^{3.8}\big)-\frac{\Gamma(7)}{\Gamma(5.8)}\big(x^{4.8}+(1-x)^{4.8}\big)\Big)y^3(1-y)^3\\
&+\Big(\frac{\Gamma(4)}{\Gamma(2.2)}\big(y^{1.2}+(1-y)^{1.2}\big)-\frac{3\Gamma(5)}{\Gamma(3.2)}\big(y^{2.2}+(1-y)^{2.2}\big)\\
&+\frac{3\Gamma(6)}{\Gamma(4.2)}\big(y^{3.2}+(1-y)^{3.2}\big)-\frac{\Gamma(7)}{\Gamma(5.2)}\big(y^{4.2}+(1-y)^{4.2}\big)\Big)x^3(1-x)^3\Big].
\end{aligned}
$$

Then for $(x,y,t)\in[0,1]^2\times[0,1]$, the exact solution is $u(x,y,t)=\mathrm{e}^{-t}x^3(1-x)^3y^3(1-y)^3$. The numerical results are listed in Table 4.9, where three numerical schemes: PR-ADI (4.158), Douglas-ADI (4.159) and D'yakonov-ADI (4.160), are used to solve (4.157) effectively.

Table 4.9: The discrete L^2 or maximum norm errors and their convergence rates to Example 4.4 approximated by the CN-WSGD scheme at $t=1$ with $\tau=h_x=h_y$.

		$(p,q)=(1,0)$				$(p,q)=(1,-1)$	
Scheme	N	$\lVert u^M-U^M\rVert_\infty$	rate	$\lVert u^M-U^M\rVert$	rate	$\lVert u^M-U^M\rVert$	rate
PR-ADI	8	6.43195e-06	-	1.95007e-06	-	2.05016e-06	-
	16	1.54712e-06	2.06	4.84833e-07	2.01	6.06100e-07	1.76
	32	3.83522e-07	2.01	1.21460e-07	2.00	1.69028e-07	1.84
	64	9.57751e-08	2.00	3.04854e-08	1.99	4.50482e-08	1.91
	128	2.39462e-08	2.00	7.64237e-09	2.00	1.16567e-08	1.95
Douglas-ADI	8	6.43195e-06	-	1.95007e-06	-	2.05016e-06	-
	16	1.54712e-06	2.06	4.84833e-07	2.01	6.06100e-07	1.76
	32	3.83522e-07	2.01	1.21460e-07	2.00	1.69028e-07	1.84
	64	9.57751e-08	2.00	3.04854e-08	1.99	4.50482e-08	1.91
	128	2.39462e-08	2.00	7.64237e-09	2.00	1.16567e-08	1.95
D'yakonov-ADI	8	6.43195e-06	-	1.95007e-06	-	2.05016e-06	-
	16	1.54712e-06	2.06	4.84833e-07	2.01	6.06100e-07	1.76
	32	3.83522e-07	2.01	1.21460e-07	2.00	1.69028e-07	1.84
	64	9.57751e-08	2.00	3.04854e-08	1.99	4.50482e-08	1.91
	128	2.39462e-08	2.00	7.64237e-09	2.00	1.16567e-08	1.95

Example 4.5. Consider

$$\frac{\partial u(x,t)}{\partial t}={}_{-\infty}D_x^{\beta}u(x,t)+f(x,t), \tag{4.182}$$

in $(0,1)\times(0,1]$, with

$$u(x,t)=0,\quad (x,t)\in\mathbb{R}\backslash(0,1)\times[0,1],$$
$$u(x,0)=x^{\gamma}(1-x),\quad x\in(0,1),$$

and

$$f(x,t)=-\mathrm{e}^{-t}\Big(\frac{\Gamma(1+\gamma)}{\Gamma(1+\gamma-\beta)}x^{\gamma-\beta}-\frac{\Gamma(2+\gamma)}{\Gamma(2+\gamma-\beta)}x^{1+\gamma-\beta}+x^{\gamma}(1-x)\Big).$$

For $(x,t)\in[0,1]\times[0,1]$, its exact solution is $u(x,t)=\mathrm{e}^{-t}x^{\gamma}(1-x)$. The simulation results at $t=1$ are given in Tables 4.10 and 4.11.

Table 4.10: The discrete L^2 errors and convergence rates to Example 4.5 at $t=1$ for different γ and β with $\tau=h$. Here the '1st scheme' denotes the first order finite difference scheme given in Sec. 4.2.

		1st scheme		$(p,q)=(1,0)$		$(p,q)=(1,-1)$	
(γ,β)	N	$\|u^M-U^M\|$	rate	$\|u^M-U^M\|$	rate	$\|u^M-U^M\|$	rate
	2^7	1.2863e-03	-	3.3451e-03	-	1.5780e-03	-
(0.3, 1.3)	2^8	7.5479e-04	0.77	1.9167e-03	0.80	9.0000e-04	0.81
	2^9	4.4018e-04	0.78	1.0997e-03	0.80	5.1521e-04	0.80
	2^7	2.0978e-04	-	2.4894e-05	-	2.3006e-05	-
(0.8, 1.8)	2^8	1.0678e-04	0.97	1.0038e-05	1.31	9.2885e-06	1.31
	2^9	5.4037e-05	0.98	4.0612e-06	1.31	3.7646e-06	1.30
	2^7	1.3784e-04	-	6.8700e-05	-	6.7804e-05	-
(1, 1.5)	2^8	7.3150e-05	0.91	2.9854e-05	1.20	2.9560e-05	1.20
	2^9	3.8404e-05	0.93	1.2985e-05	1.20	1.2891e-05	1.20

4.4 Quasi-compact finite difference schemes for space fractional diffusion equations

In this section, based on the WSGD operators proposed in Sec. 4.3, we develop the third order quasi-compact finite difference schemes for the space fractional diffusion equations. When the order of fractional derivative β equals to 2, it becomes the compact difference operators for the second order space derivatives with fourth order accuracy, which have been widely used in numerically solving the linear and nonlinear, steady and evolution equations to achieve high order accuracy. The numerical stability and convergence w.r.t. the discrete L^2 norm are theoretically analyzed.

Table 4.11: The discrete L^2 errors and convergence rates to Example 4.5 approximated by the modified CN-WSGD scheme at $t = 1$ for different m and ν with $\gamma = 1, \beta = 1.8, \tau = h$.

		$(m,\nu) = (1,1)$		$(m,\nu) = (1,2)$		$(m,\nu) = (2,2)$	
(p, q)	N	$\|u^M - U^M\|$	rate	$\|u^M - U^M\|$	rate	$\|u^M - U^M\|$	rate
	2^7	2.2664e-06	-	1.7531e-06	-	5.6687e-07	-
(1, 0)	2^8	5.8009e-07	1.97	4.6762e-07	1.91	1.4173e-07	2.00
	2^9	1.4793e-07	1.97	1.2335e-07	1.92	3.5433e-08	2.00
	2^7	3.5257e-06	-	2.9988e-06	-	5.6686e-07	-
(1, -1)	2^8	8.9859e-07	1.97	7.8171e-07	1.94	1.4173e-07	2.00
	2^9	2.2801e-07	1.98	2.0229e-07	1.95	3.5433e-08	2.00

4.4.1 *Quasi-compact difference operators for the Riemann-Liouville fractional derivatives*

Let

$$ {}_LA_{h,p}^{\beta}u(x) := \frac{1}{h^{\beta}}\sum_{k=0}^{\infty} g_k^{\beta} u\left(x-(k-p)h\right), \tag{4.183}$$

$$ {}_RA_{h,p}^{\beta}u(x) := \frac{1}{h^{\beta}}\sum_{k=0}^{\infty} g_k^{\beta} u\left(x+(k-p)h\right). \tag{4.184}$$

Similar to Theorem 4.9 in Sec. 4.3 and Theorem 4.1 in Sec. 4.1, it is easy to prove that

Lemma 4.15 ([Zhou *et al.* (2013)]). *Let $u \in \mathscr{S}^{\beta+3}(\mathbb{R})$ and $u(x)$ have compact support. Define the WSGD operators by*

$$ {}_L\mathcal{D}_{h,p,q}^{\beta}u(x) = \frac{\beta-2q}{2(p-q)}\,{}_LA_{h,p}^{\beta}u(x) + \frac{2p-\beta}{2(p-q)}\,{}_LA_{h,q}^{\beta}u(x), \tag{4.185}$$

$$ {}_R\mathcal{D}_{h,p,q}^{\beta}u(x) = \frac{\beta-2q}{2(p-q)}\,{}_RA_{h,p}^{\beta}u(x) + \frac{2p-\beta}{2(p-q)}\,{}_RA_{h,q}^{\beta}u(x). \tag{4.186}$$

Then

$$ {}_L\mathcal{D}_{h,p,q}^{\beta}u(x) = {}_{-\infty}D_x^{\beta}u(x) + \left(c_{p,q,2}^{\beta}\,h^2\right)\,{}_{-\infty}D_x^{\beta+2}u(x) + \mathcal{O}(h^3), \tag{4.187}$$

$$ {}_R\mathcal{D}_{h,p,q}^{\beta}u(x) = {}_xD_{\infty}^{\beta}u(x) + \left(c_{p,q,2}^{\beta}\,h^2\right)\,{}_xD_{\infty}^{\beta+2}u(x) + O(h^3) \tag{4.188}$$

uniformly for $x \in \mathbb{R}$, where

$$ c_{p,q,2}^{\beta} = \frac{\beta-2q}{2(p-q)}a_{p,2}^{\beta} + \frac{2p-\beta}{2(p-q)}a_{q,2}^{\beta}, \tag{4.189}$$

$$ a_{r,2} = \frac{\beta+3\beta^2-12\beta r+12r^2}{24}, \quad r = p, q. \tag{4.190}$$

Denoting $\delta_x^2 u(x) = \frac{u(x-h)-2u(x)+u(x+h)}{h^2}$, then $\delta_x^2 u(x) = \frac{d^2}{dx^2}u(x) + \mathcal{O}(h^2)$. Combining

$$ {}_{-\infty}D_x^{\beta+2}u(x) = \frac{d^2}{dx^2}\left({}_{-\infty}D_x^{\beta}u(x)\right), \tag{4.191} $$

$$ {}_{x}D_{\infty}^{\beta+2}u(x) = \frac{d^2}{dx^2}\left({}_{x}D_{\infty}^{\beta}u(x)\right), \tag{4.192} $$

and Lemma 4.15, we obtain the quasi-compact WSGD operators

$$ {}_L\mathcal{D}_{h,p,q}^{\beta}u(x) = \mathcal{C}_x\, {}_{-\infty}D_x^{\beta}u(x) + \mathcal{O}(h^3), \tag{4.193} $$

$$ {}_R\mathcal{D}_{h,p,q}^{\beta}u(x) = \mathcal{C}_x\, {}_{x}D_{\infty}^{\beta}u(x) + O(h^3), \tag{4.194} $$

where $\mathcal{C}_x v(x) := \left(1 + c_{p,q,2}^{\beta} h^2 \delta_x^2\right) v(x)$.

For the special cases $(p,q) = (1,0), (1,-1)$ discussed in Sec. 4.3, the coefficients $C_{p,q,2}^{\beta}$ are

$$ C_{1,0,2}^{\beta} = \frac{1}{24}(7\beta - 3\beta^2), \quad C_{1,-1,2}^{\beta} = \frac{1}{24}(\beta - 3\beta^2 + 12), \tag{4.195} $$

respectively. It is easy to check that

(1) when $\beta = 2$, the quasi-compact WSGD operators (4.193) and (4.194) for $(p,q) = (1,0)$ and $(p,q) = (1,-1)$ reduce to the fourth order compact difference operator for the second order derivative.
(2) when $\beta = 1$ and $(p,q) = (1,0)$, the quasi-compact WSGD operators (4.193) and (4.194) all reduce to the fourth order compact difference operator for the first order derivative.

Remark 4.9. In fact, making use of the Taylor expansion and Corollary 4.1, we have

$$ {}_{-\infty}D_x^{\beta}u(x+ph) = {}_{-\infty}D_x^{\beta}u(x) + \sum_{l=1}^{m-1} \frac{(ph)^k}{k!} {}_{-\infty}D_x^{\beta+l}u(x) + \mathcal{O}(h^m), $$

$$ {}_LA_{h,p}^{\beta}u(x) = {}_{-\infty}D_x^{\beta}u(x) + \sum_{l=1}^{m-1} a_l^p\, {}_{-\infty}D_x^{\beta+l}u(x)h^l + \mathcal{O}(h^m). $$

Then if we choose b_i $(i = 1, 2, \cdots, 6)$ such that

$$ \begin{aligned} &b_1\, {}_{-\infty}D_x^{\beta}u(x-h) + b_2\, {}_{-\infty}D_x^{\beta}u(x) + b_3\, {}_{-\infty}D_x^{\beta}u(x+h), \\ &- b_4\, {}_LA_{h,-1}^{\beta}u(x) - b_5\, {}_LA_{h,0}^{\beta}u(x) - b_6\, {}_LA_{h,1}^{\beta}u(x) = \mathcal{O}(h^m), \end{aligned} $$

it yields that a group of quasi-compact difference operators with the form

$$ \begin{aligned} &b_1\, {}_{-\infty}D_x^{\beta}u(x-h) + b_2\, {}_{-\infty}D_x^{\beta}u(x) + b_3\, {}_{-\infty}D_x^{\beta}u(x+h) \\ &= b_4\, {}_LA_{h,-1}^{\beta}u(x) + b_5\, {}_LA_{h,0}^{\beta}u(x) + b_6\, {}_LA_{h,1}^{\beta}u(x) + \mathcal{O}(h^m). \end{aligned} $$

In particular, the above cases $(p,q)=(1,0)$ and $(p,q)=(1,-1)$ are included in the requirements

$$\begin{cases} b_1+b_2+b_3-(b_4+b_5+b_6)=0, \\ -b_1+b_3-b_4a_1^{-1}-b_5a_1^0-b_6a_1^1=0, \\ \frac{1}{2}(b_1+b_3)-b_4a_2^{-1}-b_5a_2^0-b_6a_2^1=0. \end{cases}$$

Moreover, if we further require

$$\begin{cases} b_1+b_2+b_3=1, \\ b_1=b_3, \\ \frac{1}{6}(-b_1+b_3)-b_4a_3^{-1}-b_5a_3^0-b_6a_3^1=0, \end{cases}$$

then a fourth order quasi-compact difference operator can be uniquely determined, which keeps of the properties of the third order quasi-compact difference scheme with $(p,q)=(1,0)$.

4.4.2 *Application of the quasi-compact difference operators to space fractional diffusion equations*

4.4.2.1 *Quasi-compact difference scheme of one dimensional problem*

In this subsection, we develop the third order quasi-compact finite difference scheme for the following one dimensional space fractional diffusion equation in $\Omega \times (0,T]$:

$$\frac{\partial u(x,t)}{\partial t} = K_1\ {}_{-\infty}D_x^{\beta}u(x,t) + K_2\ {}_xD_{\infty}^{\beta}u(x,t) + f(x,t), \tag{4.196}$$

with the initial condition

$$u(x,0)=u_0(x), \quad x\in\Omega, \tag{4.197}$$

and the absorbing boundary condition

$$u(x,t)=0, \quad (x,t)\in\mathbb{R}\backslash\Omega\times[0,T], \tag{4.198}$$

where $\Omega=(a,b)$, ${}_{-\infty}D_x^{\beta}$ and ${}_xD_{\infty}^{\beta}$ are R-L fractional operators with $1<\beta\le 2$. The diffusion coefficients K_1 and K_2 are nonnegative constants such that $K_1^2+K_2^2\neq 0$.

As previous section, we introduce a uniform mesh with the space step size $h=(b-a)/N$ and the time step size $\tau=T/M$,

$$\{(x_i,t_n)\mid x_i=a+ih,\ i=0,\cdots,N;\ t_n=n\tau,\ n=0,\cdots,M\},$$

where N,M are two positive integers. Denote by $t_{n+1/2}=(t_n+t_{n+1})/2$ for $0\le n\le M-1$, and introduce the following notations

$$u_i^n=u(x_i,t_n),\quad f_i^{n+1/2}=f(x_i,t_{n+1/2}).$$

Acting the invertible operator $\tau\mathcal{C}_x$ on both sides of (4.196), and using the CN techniques for time disretization and (4.193) and (4.194) for space discretization, we get

$$\begin{aligned}&\mathcal{C}_x u_i^{n+1} - \frac{K_1\tau}{2}{}_L\mathcal{D}_{h,p,q}^{\beta} u_i^{n+1} - \frac{K_2\tau}{2}{}_R\mathcal{D}_{h,p,q}^{\beta} u_i^{n+1}\\ &= \mathcal{C}_x u_i^{n} + \frac{K_1\tau}{2}{}_L\mathcal{D}_{h,p,q}^{\beta} u_i^{n} + \frac{K_2\tau}{2}{}_R\mathcal{D}_{h,p,q}^{\beta} u_i^{n} + \tau\mathcal{C}_x f_i^{n+1/2} + \tau\varepsilon_i^{n+1/2},\end{aligned} \tag{4.199}$$

where

$$\left|\varepsilon_i^{n+1/2}\right| \le \tilde{c}(\tau^2 + h^3). \tag{4.200}$$

Thus, the quasi-compact difference scheme for (4.196) is followed by replacing u_i^n with its numerical approximation U_i^n, i.e.,

$$\begin{aligned}&\mathcal{C}_x U_i^{n+1} - \frac{K_1\tau}{2h^\beta}\sum_{k=0}^{i+1} w_k^{(\beta)} U_{i-k+1}^{n+1} - \frac{K_2\tau}{2h^\beta}\sum_{k=0}^{N-i+1} w_k^{(\beta)} U_{i+k-1}^{n+1}\\ &= \mathcal{C}_x U_i^{n} + \frac{K_1\tau}{2h^\beta}\sum_{k=0}^{i+1} w_k^{(\beta)} U_{i-k+1}^{n} + \frac{K_2\tau}{2h^\beta}\sum_{k=0}^{N-i+1} w_k^{(\beta)} U_{i+k-1}^{n} + \tau\mathcal{C}_x f_i^{n+1/2},\end{aligned} \tag{4.201}$$

where the results in (4.126) have been used. The corresponding matrix vector form of (4.201) is given by

$$\left(C_\beta - \frac{\tau}{2h^\beta}(K_1A_\beta + K_2A_\beta^{\mathrm{T}})\right)U^{n+1} = \left(C_\beta + \frac{\tau}{2h^\beta}(K_1A_\beta + K_2A_\beta^{\mathrm{T}})\right)U^n + \tau C_\beta F^n, \tag{4.202}$$

where A_β is the matrix A given in (4.128), and

$$\begin{aligned} C_\beta &= I_{N-1} + c_{p,q,2}^{\beta}\,\mathrm{tridiag}(1,-2,1), \qquad (4.203)\\ U^n &= \left(U_1^n, U_2^n, \cdots, U_{N-1}^n\right)^{\mathrm{T}},\\ F^n &= \left(f_1^{n+1/2}, f_2^{n+1/2}, \cdots, f_{N-1}^{n+1/2}\right)^{\mathrm{T}}.\end{aligned}$$

Remark 4.10. Obviously, C_β is a symmetric $(N-1)\times(N-1)$ tri-diagonal matrix and its eigenvalues can be given by

$$\lambda_k(C_\beta) = 1 - 4\,c_{p,q,2}^{\beta}\sin^2\left(\frac{k\pi}{2N}\right), \quad k = 1, 2, \cdots, N-1. \tag{4.204}$$

For the case of $(p,q) = (1,0)$, we have

$$\lambda_k(C_\beta) > 1 - 4\,c_{1,0,2}^{\beta} \ge \frac{23}{72} > 0 \quad \forall 1 < \beta \le 2. \tag{4.205}$$

Then matrix C_β is symmetric positive definite. However, for the case of $(p,q)=(1,-1)$ and $1<\beta\le 2$, we have

$$\lambda_k(C_\beta) > 1-4C^\beta_{1,-1,2} > 0 \ \text{ iff } \ \frac{1+\sqrt{73}}{6} < \beta \le 2. \tag{4.206}$$

So if $\frac{1+\sqrt{73}}{6} < \beta \le 2$, the matrix C_β is symmetric positive definite.

4.4.2.2 *Quasi-compact difference scheme of two dimensional problem*

In the following, we discuss the quasi-compact difference scheme for the following two dimensional space fractional diffusion equation in $\Omega\times(0,T]$:

$$\begin{aligned}\frac{\partial u(x,y,t)}{\partial t} &= \Big(K_1^+ {}_{-\infty}D_x^\beta u(x,y,t) + K_2^+ {}_xD_\infty^\beta u(x,y,t)\Big)\\ &+\Big(K_1^- {}_{-\infty}D_y^{\beta_1} u(x,y,t) + K_2^- {}_yD_\infty^{\beta_1} u(x,y,t)\Big) + f(x,y,t),\end{aligned} \tag{4.207}$$

with the initial-boundary condition

$$u(x,y,0)=u_0(x,y),\quad (x,y)\in\Omega, \tag{4.208}$$

$$u(x,y,t)=0,\quad (x,y,t)\in\mathbb{R}^2\backslash\Omega\times[0,T], \tag{4.209}$$

where $\Omega=(a,b)\times(c,d)$, ${}_{-\infty}D_x^\beta$, ${}_xD_\infty^\beta$ and ${}_{-\infty}D_y^{\beta_1}$, ${}_yD_\infty^{\beta_1}$ are R-L fractional operators with $1<\beta,\beta_1\le 2$. The diffusion coefficients satisfy $K_i^+,\ K_i^-\ge 0,\ i=1,2,\ (K_1^+)^2+(K_2^+)^2\ne 0$ and $(K_1^-)^2+(K_2^-)^2\ne 0$.

We partition the domain Ω into a uniform mesh with the space step sizes $h_x=(b-a)/N_x, h_y=(d-c)/N_y$ in each direction, and the time step size is $\tau=T/M$, where N_x,N_y,M are positive integers. Then we can denote the grid points by $x_i=a+ih_x, y_j=c+jh_y$ and $t_n=n\tau$ for $0\le i\le N_x, 0\le j\le N_y$ and $0\le n\le M$. We use the following notations

$$u_{i,j}^n=u(x_i,y_j,t_n),\quad f_{i,j}^{n+1/2}=f(x_i,y_j,t_{n+1/2}).$$

Introducing the finite difference operators

$$\mathcal{C}_x u_{i,j}=(1+c^\beta_{p,q,2}h_x^2\delta_x^2)u_{i,j},\quad \mathcal{C}_y u_{i,j}=(1+c^{\beta_1}_{p,q,2}h_y^2\delta_y^2)u_{i,j}, \tag{4.210}$$

and acting $\tau\mathcal{C}_x\mathcal{C}_y$ on both sides of (4.207), similar to (4.199), we obtain

$$\begin{aligned}&\Big(\mathcal{C}_x\mathcal{C}_y-\frac{K_1^+\tau}{2}\mathcal{C}_{y\,L}\mathcal{D}^\beta_{h_x,p,q}-\frac{K_2^+\tau}{2}\mathcal{C}_{y\,R}\mathcal{D}^\beta_{h_x,p,q}\Big)u_{i,j}^{n+1}\\ &\quad-\Big(\frac{K_1^-\tau}{2}\mathcal{C}_{x\,L}\mathcal{D}^{\beta_1}_{h_y,p,q}-\frac{K_2^-\tau}{2}\mathcal{C}_{x\,R}\mathcal{D}^{\beta_1}_{h_y,p,q}\Big)u_{i,j}^{n+1}\\ &=\Big(\mathcal{C}_x\mathcal{C}_y+\frac{K_1^+\tau}{2}\mathcal{C}_{y\,L}\mathcal{D}^\beta_{h_x,p,q}+\frac{K_2^+\tau}{2}\mathcal{C}_{y\,R}\mathcal{D}^\beta_{h_x,p,q}+\frac{K_1^-\tau}{2}\mathcal{C}_{x\,L}\mathcal{D}^{\beta_1}_{h_y,p,q}\Big)u_{i,j}^n\\ &\quad+\Big(\frac{K_2^-\tau}{2}\mathcal{C}_{x\,R}\mathcal{D}^{\beta_1}_{h_y,p,q}\Big)u_{i,j}^n+\tau\mathcal{C}_x\mathcal{C}_y f_{i,j}^{n+1/2}+\tau\varepsilon_{i,j}^{n+1/2},\end{aligned} \tag{4.211}$$

where

$$\left|\varepsilon_{i,j}^{n+1/2}\right| \leq \tilde{c}(\tau^2 + h_x^3 + h_y^3) \tag{4.212}$$

denotes the local truncation error. Denote

$$\delta_x^\beta = K_1^+ {}_L\mathcal{D}_{h_x,p,q}^\beta + K_2^+ {}_R\mathcal{D}_{h_x,p,q}^\beta, \quad \delta_y^{\beta_1} = K_1^- {}_L\mathcal{D}_{h_y,p,q}^{\beta_1} + K_2^- {}_R\mathcal{D}_{h_y,p,q}^{\beta_1}.$$

Adding a small perturbation $-\frac{\tau^2}{4}\delta_x^\beta\delta_y^{\beta_1}(u_{i,j}^{n+1} - u_{i,j}^n)$ (see (4.155)) to the right hand side of (4.211) and making the factorization lead to

$$\begin{aligned}
&\left(\mathcal{C}_x - \frac{\tau}{2}\delta_x^\beta\right)\left(\mathcal{C}_y - \frac{\tau}{2}\delta_y^{\beta_1}\right)u_{i,j}^{n+1} \\
&= \left(\mathcal{C}_x + \frac{\tau}{2}\delta_x^\beta\right)\left(\mathcal{C}_y + \frac{\tau}{2}\delta_y^{\beta_1}\right)u_{i,j}^n + \tau\mathcal{C}_x\mathcal{C}_y f_{i,j}^{n+1/2} + \tau\hat{\varepsilon}_{i,j}^{n+1/2},
\end{aligned} \tag{4.213}$$

where

$$\hat{\varepsilon}_{i,j}^{n+1/2} = \varepsilon_{i,j}^{n+1/2} + O(\tau^2 + \tau^2 h_x^2 + \tau^2 h_y^2).$$

Thus the quasi-compact finite difference scheme for (4.207) can be given as

$$\begin{aligned}
&\left(\mathcal{C}_x - \frac{\tau}{2}\delta_x^\beta\right)\left(\mathcal{C}_y - \frac{\tau}{2}\delta_y^{\beta_1}\right)U_{i,j}^{n+1} \\
&= \left(\mathcal{C}_x + \frac{\tau}{2}\delta_x^\beta\right)\left(\mathcal{C}_y + \frac{\tau}{2}\delta_y^{\beta_1}\right)U_{i,j}^n + \tau\mathcal{C}_x\mathcal{C}_y f_{i,j}^{n+1/2},
\end{aligned} \tag{4.214}$$

where $U_{i,j}^n$ is the numerical approximation of $u_{i,j}^n$. Denote

$$\begin{aligned}
U^n &= (U_{1,1}^n, u_{2,1}^n, \cdots, U_{N_x-1,1}^n, u_{1,2}^n, U_{2,2}^n, \cdots, U_{N_x-1,2}^n, \cdots, \\
&\quad U_{1,N_y-1}^n, U_{2,N_y-1}^n, \cdots, U_{N_x-1,N_y-1}^n)^{\mathrm{T}} \\
F^n &= (f_{1,1}^{n+1/2}, f_{2,1}^{n+1/2}, \cdots, f_{N_x-1,1}^{n+1/2}, \cdots, \\
&\quad f_{1,N_y-1}^{n+1/2}, f_{2,N_y-1}^{n+1/2}, \cdots, f_{N_x-1,N_y-1}^{n+1/2})^{\mathrm{T}},
\end{aligned}$$

and

$$\mathcal{C}_x = I_y \otimes C_\beta, \quad \mathcal{C}_y = C_{\beta_1} \otimes I_x, \tag{4.215}$$

$$\mathcal{D}_x = \frac{K_1^+\tau}{2h_x^\beta} I_y \otimes A_\beta + \frac{K_2^+\tau}{2h_x^\beta} I_y \otimes A_\beta^{\mathrm{T}}, \tag{4.216}$$

$$\mathcal{D}_y = \frac{K_1^-\tau}{2h_y^{\beta_1}} A_{\beta_1} \otimes I_x + \frac{K_2^-\tau}{2h_y^{\beta_1}} A_{\beta_1}^{\mathrm{T}} \otimes I_x, \tag{4.217}$$

where the symbol $\otimes$ denotes the Kronecker product, I_x and I_y are $(N_x - 1) \times (N_x - 1)$ and $(N_y - 1) \times (N_y - 1)$ unit matrices, respectively, and the matrices A_β and A_{β_1} are defined in (4.128) corresponding to β and β_1, respectively, C_β and C_{β_1} are given in (4.203) corresponding to β and β_1, respectively. Then the corresponding matrix form of (4.214) is given as

$$(\mathcal{C}_x - \mathcal{D}_x)(\mathcal{C}_y - \mathcal{D}_y)U^{n+1} = (\mathcal{C}_x + \mathcal{D}_x)(\mathcal{C}_y + \mathcal{D}_y)U^n + \tau\mathcal{C}_x\mathcal{C}_y F^n. \tag{4.218}$$

In order to solve (4.214) effectively, similar to Sec. 4.3, we can introduce an intermediate variable $V_{i,j}^*$ and derive

(1) the quasi-compact Peaceman-Rachford ADI scheme:

$$\left(\mathcal{C}_x - \frac{\tau}{2}\delta_x^{\beta}\right)V_{i,j}^n = \left(\mathcal{C}_y + \frac{\tau}{2}\delta_y^{\beta_1}\right)U_{i,j}^n + \frac{\tau}{2}\mathcal{C}_y f_{i,j}^{n+1/2}, \tag{4.219a}$$

$$\left(\mathcal{C}_y - \frac{\tau}{2}\delta_y^{\beta_1}\right)U_{i,j}^{n+1} = \left(\mathcal{C}_x + \frac{\tau}{2}\delta_x^{\beta}\right)V_{i,j}^n + \frac{\tau}{2}\mathcal{C}_y f_{i,j}^{n+1/2}; \tag{4.219b}$$

(2) the quasi-compact Douglas ADI scheme:

$$\left(\mathcal{C}_x - \frac{\tau}{2}\delta_x^{\beta}\right)V_{i,j}^n = \left(\mathcal{C}_x\mathcal{C}_y + \frac{\tau}{2}\mathcal{C}_y\delta_x^{\beta} + \tau\mathcal{C}_x\delta_y^{\beta_1}\right)U_{i,j}^n + \tau\mathcal{C}_x\mathcal{C}_y f_{i,j}^{n+1/2}, \tag{4.220a}$$

$$\left(\mathcal{C}_y - \frac{\tau}{2}\delta_y^{\beta_1}\right)U_{i,j}^{n+1} = V_{i,j}^n - \frac{\tau}{2}\delta_y^{\beta_1}U_{i,j}^n; \tag{4.220b}$$

(3) the quasi-compact D'Yakonov ADI scheme:

$$\left(\mathcal{C}_x - \frac{\tau}{2}\delta_x^{\beta}\right)V_{i,j}^n = \left(\mathcal{C}_x + \frac{\tau}{2}\delta_x^{\beta}\right)\left(\mathcal{C}_y + \frac{\tau}{2}\delta_y^{\beta_1}\right)U_{i,j}^n + \tau\mathcal{C}_x\mathcal{C}_y f_{i,j}^{n+1/2}, \tag{4.221a}$$

$$\left(\mathcal{C}_y - \frac{\tau}{2}\delta_y^{\beta_1}\right)U_{i,j}^{n+1} = V_{i,j}^n. \tag{4.221b}$$

4.4.3 *Stability and convergence analyses*

We first give two auxiliary lemmas necessary to theoretical analysis.

Lemma 4.16 ([Marcus and Minc (1964); Laub (2005)]). *The matrix $A \in \mathbb{R}^{n\times n}$ is asymptotically stable if and only if there exists a symmetric and positive (or negative) definite solution $X \in \mathbb{R}^{n\times n}$ to the Lyapunov equation*

$$AX + XA^{\mathrm{T}} = C, \tag{4.222}$$

where $C = C^{\mathrm{T}} \in \mathbb{R}^{n\times n}$ is a negative (or positive) definite matrix. And a matrix A is called asymptotically stable if all its eigenvalues have real parts in the open left half plane, i.e., $Re(\lambda(A)) < 0$.

Lemma 4.17 ([Zhang *et al.* (2011)]). *Let A be an n-square symmetric and positive semi-definite matrix. Then there is a unique n-square symmetric and positive semi-definite matrix B such that $B^2 = A$. Such a matrix B is called the square root of A, denoted by $A^{1/2}$.*

4.4.3.1 *One dimensional case*

Next we consider the stability and convergence analysis for the scheme (4.202).

Theorem 4.16. *For the case of $(p,q) = (1,0)$, the difference scheme (4.202) is unconditionally stable for all $1 < \beta \leq 2$; and for the case of $(p,q) = (1,-1)$, the difference scheme (4.202) is also unconditionally stable for $\frac{1+\sqrt{73}}{6} < \beta \leq 2$.*

Proof. Denoting $D_\beta = \frac{\tau}{2h^\beta}(K_1 A_\beta + K_2 A_\beta^{\mathrm{T}})$, we rewrite (4.202) as

$$(C_\beta - D_\beta)U^{n+1} = (C_\beta + D_\beta)U^n + \tau C_\beta F^n. \tag{4.223}$$

From Remark 4.5, we know that C_β is a symmetric and positive definite matrix when $(p,q) = (1,0)$ with $1 < \beta \leq 2$ and $(p,q) = (1,-1)$ with $\frac{1+\sqrt{73}}{6} < \beta \leq 2$, which follows that C_β^{-1} is also symmetric and positive definite. On the other hand, Theorem 4.11 shows that the eigenvalues of the matrix $\frac{D_\beta + D_\beta^{\mathrm{T}}}{2} = \frac{\tau(K_1+K_2)}{4h^\beta}(A_\beta + A_\beta^{\mathrm{T}})$ are all negative for $1 < \beta \leq 2$. Thus $(D_\beta + D_\beta^{\mathrm{T}})$ is a symmetric and negative definite matrix. Then, for any on-zero vector $v = (v_1, v_2, \cdots, v_{N-1})^{\mathrm{T}} \in \mathbb{R}^{N-1}$, there exists

$$v^{\mathrm{T}}\Big((C_\beta^{-1}D_\beta)C_\beta^{-1} + C_\beta^{-1}(C_\beta^{-1}D_\beta)^{\mathrm{T}}\Big)v = v^{\mathrm{T}}C_\beta^{-1}(D_\beta + D_\beta^{\mathrm{T}})C_\beta^{-1}v < 0, \tag{4.224}$$

which means that the matrix $\big((C_\beta^{-1}D_\beta)C_\beta^{-1} + C_\beta^{-1}(C_\beta^{-1}D_\beta)^{\mathrm{T}}\big)$ is negative definite. Then it yields from Lemma 4.16 that all the eigenvalues of $(C_\beta^{-1}D_\beta)$ have negative real parts. In addition, λ is an eigenvalue of $(C_\beta^{-1}D_\beta)$ if and only if $\frac{1+\lambda}{1-\lambda}$ is an eigenvalue of matrix $(I - C_\beta^{-1}D_\beta)^{-1}(I + C_\beta^{-1}D_\beta)$, and $|\frac{1+\lambda}{1-\lambda}| < 1$ holds. Hence, the spectral radius of the matrix

$$(C_\beta - D_\beta)^{-1}(C_\beta + D_\beta) = (I - C_\beta^{-1}D_\beta)^{-1}(I + C_\beta^{-1}D_\beta)$$

is less than one, and the difference scheme (4.202) is stable. □

Theorem 4.17. *Let u_i^n be the exact solution of problem (4.196), and U_i^n be the solution of difference scheme (4.202) at grid point (x_i, t_n). Then the estimate*

$$\|u^n - U^n\| \leq c(\tau^2 + h^3), \quad 1 \leq n \leq M \tag{4.225}$$

holds for all $1 < \beta < 2$ with $(p,q) = (1,0)$ and $\frac{1+\sqrt{73}}{6} < \beta < 2$ with $(p,q) = (1,-1)$, where $\|\cdot\|$ denotes the discrete L^2 norm in (4.91).

Proof. Denoting $e_i^n = u_i^n - U_i^n$, from (4.199) and (4.201) we have

$$C_\beta(e^{n+1} - e^n) - \frac{K_1\tau}{2h^\beta}A_\beta(e^{n+1} + e^n) - \frac{K_2\tau}{2h^\beta}A_\beta^{\mathrm{T}}(e^{n+1} + e^n) = \tau\mathcal{E}^{n+1/2}, \tag{4.226}$$

where

$$e^n = (e_1^n, e_2^n, \cdots, e_{N-1}^n)^T,$$
$$\mathcal{E}^{n+1/2} = (\varepsilon_1^{n+1/2}, \varepsilon_2^{n+1/2}, \cdots, \varepsilon_{N-1}^{n+1/2})^T.$$

Multiplying (4.226) by $h(e^{n+1} + e^n)^{\mathrm{T}}$, we obtain that

$$\begin{aligned} &h(e^{n+1} + e^n)^{\mathrm{T}}C_\beta(e^{n+1} - e^n) - \frac{K_1\tau}{2h^{\beta-1}}(e^{n+1} + e^n)^{\mathrm{T}}A_\beta(e^{n+1} + e^n) \\ &- \frac{K_2\tau}{2h^{\beta-1}}(e^{n+1} + e^n)^{\mathrm{T}}A_\beta^{\mathrm{T}}(e^{n+1} + e^n) = \tau h(e^{n+1} + e^n)^{\mathrm{T}}\mathcal{E}^{n+1/2}. \end{aligned} \tag{4.227}$$

By Theorem 4.11, A_β and its transpose A_β^{T} are both negative semi-definite matrices for $1 \le \beta \le 2$. Thus

$$(e^{n+1}+e^n)^{\mathrm{T}}A_\beta(e^{n+1}+e^n) \le 0, \quad (e^{n+1}+e^n)^{\mathrm{T}}A_\beta^{\mathrm{T}}(e^{n+1}+e^n) \le 0. \tag{4.228}$$

Then (4.227) leads to

$$h(e^{n+1} + e^n)^{\mathrm{T}}C_\beta(e^{n+1} - e^n) \le \tau h(e^{n+1} + e^n)^{\mathrm{T}}\mathcal{E}^{n+1/2}. \tag{4.229}$$

As the matrix C_β is symmetric, we derive that

$$h(e^{n+1} + e^n)^{\mathrm{T}}C_\beta(e^{n+1} - e^n) = E^{n+1} - E^n, \tag{4.230}$$

where

$$E^n = h(e^n)^{\mathrm{T}}C_\beta(e^n) \ge (1 - 4\ c_{p,q,2}^\beta)_{\min}\|e^n\|^2. \tag{4.231}$$

Denote $\lambda = (1 - 4\ c_{p,q,2}^\beta)_{\min}$. From (4.206) and (4.205), it yields $\lambda = 1 - 4\left(c_{1,-1,2}^\beta\right)_{\max} > 0$ if $\frac{1+\sqrt{73}}{6} < \beta \le 2$ and $(p,q) = (1,-1)$; and $\lambda = \frac{23}{72}$ if $1 < \beta \le 2$ and $(p,q) = (1,0)$. Together with (4.229), it yields that

$$\begin{aligned} E^{k+1} - E^k &\le \tau h(e^{k+1} + e^k)^{\mathrm{T}}\mathcal{E}^{k+1/2} \\ &\le \frac{\tau\lambda}{2}(\|e^{k+1}\|^2 + \|e^k\|^2) + \frac{\tau}{\lambda}\|\mathcal{E}^{k+1/2}\|^2. \end{aligned} \tag{4.232}$$

Summing up for all $0 \le k \le n-1$, we have

$$\begin{aligned} \lambda\|e^n\|^2 &\le \tau h(e^n + e^{n-1})^{\mathrm{T}}\mathcal{E}^{n-1/2} + \frac{\tau\lambda}{2}\sum_{k=0}^{n-2}(\|e^{k+1}\|^2 + \|e^k\|^2) \\ &\quad + \frac{\tau}{\lambda}\sum_{k=1}^{n-2}\|\mathcal{E}^{k+1/2}\|^2 \\ &\le \frac{\lambda}{2}\|e^n\|^2 + \frac{\tau^2}{2\lambda}\|\mathcal{E}^{n-1/2}\|^2 + \tau\lambda\sum_{k=1}^{n-1}\|e^k\|^2 + \frac{\tau}{\lambda}\sum_{k=1}^{n-1}\|\mathcal{E}^{k+1/2}\|^2. \end{aligned} \tag{4.233}$$

Since $|\varepsilon_i^{k+1/2}| \le \tilde{c}(\tau^2 + h^3)$ for any $0 \le k \le n-1$, it leads to

$$\begin{aligned} \|e^n\|^2 &\le 2\tau \sum_{k=1}^{n-1} \|e^k\|^2 + \frac{2\tau}{\lambda^2} \sum_{k=1}^{n-1} \|\mathcal{E}^{k+1/2}\|^2 + \frac{\tau^2}{\lambda^2} \|\mathcal{E}^{n-1/2}\|^2 \\ &\le 2\tau \sum_{k=1}^{n-1} \|e^k\|^2 + c(\tau^2 + h^3)^2. \end{aligned} \tag{4.234}$$

Then, the desired results can be obtained by using the discrete Grönwall Lemma (see Appendix A). □

Remark 4.11. The truncation error in (4.200) becomes $\varepsilon_i^{n+1/2} = O(\tau^2 + h^4)$ when $\beta = 1, 2$ with $(p, q) = (1, 0)$ and $\beta = 2$ with $(p, q) = (1, -1)$, so when taking $\beta = 1, 2$, the compact finite difference schemes for the classical diffusion equations are recovered and the corresponding error estimate of the difference scheme (4.201) satisfies

$$\|u^n - U^n\| \le c(\tau^2 + h^4), \quad 1 \le n \le M. \tag{4.235}$$

4.4.3.2 *Two dimensional case*

In the theoretical analysis of the numerical method, we choose $N = N_x = N_y$ for simplification.

Theorem 4.18. *For the case of $(p, q) = (1, 0)$, the difference scheme (4.214) is unconditionally stable for $1 < \beta, \beta_1 \le 2$. For the case of $(p, q) = (1, -1)$, the difference scheme (4.214) is also unconditionally stable when $\frac{1+\sqrt{73}}{6} < \beta, \beta_1 \le 2$.*

Proof. Denoting the disturbances of U^{n+1} and U^n by δU^{n+1} and δU^n, respectively, we have from (4.218) that

$$\delta U^{n+1} = \left(\mathcal{C}_y - \mathcal{D}_y\right)^{-1} \left(\mathcal{C}_x - \mathcal{D}_x\right)^{-1} \left(\mathcal{C}_x + \mathcal{D}_x\right) \left(\mathcal{C}_y + \mathcal{D}_y\right) \delta U^n. \tag{4.236}$$

Using Lemma 4.10, we can check that $\mathcal{C}_x$ and $\mathcal{D}_x$ commute with $\mathcal{C}_y$ and $\mathcal{D}_y$. Hence, $(\mathcal{C}_y - \mathcal{D}_y)^{-1}$ and $(\mathcal{C}_y + \mathcal{D}_y)$ commute with $(\mathcal{C}_x - \mathcal{D}_x)^{-1}$ and $(\mathcal{C}_x + \mathcal{D}_x)$. Then from (4.236) we have

$$\delta U^n = \left((\mathcal{C}_y - \mathcal{D}_y)^{-1} (\mathcal{C}_y + \mathcal{D}_y) \right)^n \left((\mathcal{C}_x - \mathcal{D}_x)^{-1} (\mathcal{C}_x + \mathcal{D}_x) \right)^n \delta U^0. \tag{4.237}$$

From Remark 4.10 and Lemmas 4.9 and 4.10, $\mathcal{C}_x$ and $\mathcal{C}_y$ are symmetric and positive definite matrices in the cases of $(p, q) = (1, 0)$ with $1 < \beta, \beta_1 \le 2$ and $(p, q) = (1, -1)$ with $\frac{1+\sqrt{73}}{6} < \beta, \beta_1 \le 2$, so $\mathcal{C}_x^{-1}$ and $\mathcal{C}_y^{-1}$ are also symmetric and positive definite. On the other hand, Theorem 4.11 indicates that the eigenvalues of $\frac{A_\beta+A_\beta^{\mathrm{T}}}{2}$ and $\frac{A_{\beta_1}+A_{\beta_1}^{\mathrm{T}}}{2}$ are all negative. By Lemma 4.9, $(\mathcal{D}_x+\mathcal{D}_x^{\mathrm{T}})$ and $(\mathcal{D}_y+\mathcal{D}_y^{\mathrm{T}})$ are both symmetric and negative definite matrices. Then

$$\begin{aligned}&v^{\mathrm{T}}\Big((C_\gamma^{-1}\mathcal{D}_\gamma)C_\gamma^{-1}+C_\gamma^{-1}(C_\gamma^{-1}\mathcal{D}_\gamma)^{\mathrm{T}}\Big)v\\&=v^{\mathrm{T}}C_\gamma^{-1}(\mathcal{D}_\gamma+\mathcal{D}_\gamma^{\mathrm{T}})C_\gamma^{-1}v<0\end{aligned}\tag{4.238}$$

for $\gamma=x,y$, which means that the matrices $(C_\gamma^{-1}\mathcal{D}_\gamma)C_\gamma^{-1}+C_\gamma^{-1}(C_\gamma^{-1}\mathcal{D}_\gamma)^{\mathrm{T}}$ are symmetric and negative definite, from Lemma 4.16, the real parts of all the eigenvalues $\{\lambda_\gamma\}$ of $C_\gamma^{-1}\mathcal{D}_\gamma$ are negative. Thus the spectral radius of $(I-C_\gamma^{-1}\mathcal{D}_\gamma)^{-1}(I+C_\gamma^{-1}\mathcal{D}_\gamma)$ $(\gamma=x,y)$ are less than 1, which concludes that the difference scheme (4.214) is stable. □

Theorem 4.19. *Let $u_{i,j}^n$ be the exact solution of (4.207), and $U_{i,j}^n$ be the solution of the difference scheme (4.214). Then in the cases of $(p,q)=(1,0)$ with $1<\beta,\beta_1<2$ and $(p,q)=(1,-1)$ with $\frac{1+\sqrt{73}}{6}<\beta,\beta_1<2$, we have*

$$\|u^n-U^n\|\le c(\tau^2+h_x^3+h_y^3),\quad 1\le n\le M,\tag{4.239}$$

where c denotes a positive constant and $\|\cdot\|$ stands for the discrete L^2 norm in (4.106).

Proof. Let $e_{i,j}^n=u_{i,j}^n-U_{i,j}^n$. Subtracting (4.213) from (4.214) leads to

$$\big(C_x-\mathcal{D}_x\big)\big(C_y-\mathcal{D}_y\big)e^{n+1}=\big(C_x+\mathcal{D}_x\big)\big(C_y+\mathcal{D}_y\big)e^n+\tau\mathcal{E}^{n+1/2},\tag{4.240}$$

where

$$\begin{aligned}e^n=&\,(e_{1,1}^n,e_{2,1}^n,\cdots,e_{N_x-1,1}^n,e_{1,2}^n,e_{2,2}^n,\cdots,e_{N_x-1,2}^n,\cdots,\\&\,e_{1,N_y-1}^n,e_{2,N_y-1}^n,\cdots,e_{N_x-1,N_y-1}^n)^{\mathrm{T}},\\\mathcal{E}^{n+1/2}=&\,(\hat{\varepsilon}_{1,1}^{n+1/2},\hat{\varepsilon}_{2,1}^{n+1/2},\cdots,\hat{\varepsilon}_{N_x-1,1}^{n+1/2},\hat{\varepsilon}_{1,2}^{n+1/2},\hat{\varepsilon}_{2,2}^{n+1/2},\cdots,\hat{\varepsilon}_{N_x-1,2}^{n+1/2},\cdots,\\&\,\hat{\varepsilon}_{1,N_y-1}^{n+1/2},\hat{\varepsilon}_{2,N_y-1}^{n+1/2},\cdots,\hat{\varepsilon}_{N_x-1,N_y-1}^{n+1/2})^{\mathrm{T}},\end{aligned}$$

and the matrices C_x,C_y and $\mathcal{D}_x,\mathcal{D}_y$ are given by (4.215)- (4.217), respectively.

As stated in Theorem 4.18, under the cases of $(p,q)=(1,0)$ with $1<\beta,\beta_1\le 2$ and $(p,q)=(1,-1)$ with $\frac{1+\sqrt{73}}{6}<\beta,\beta_1\le 2$, the matrices C_β and C_{β_1} and their inverse are symmetric and positive definite. And from Lemmas 4.11 and 4.17, we know that $(C_x^{-1})^{1/2}=I_y\otimes(C_\beta^{-1})^{1/2}$ and $(C_y^{-1})^{1/2}=(C_{\beta_1}^{-1})^{1/2}\otimes I_x$ uniquely exist and are symmetric and positive semi-definite matrices. Then multiplying (4.240) by $(C_x^{-1})^{1/2}(C_y^{-1})^{1/2}$, and making the discrete L^2 norm on both sides, we have

$$\begin{aligned}
&\|(C_x^{-1})^{1/2}(C_y^{-1})^{1/2}(C_x - \mathcal{D}_x)(C_y - \mathcal{D}_y)e^{n+1}\| \\
&\le \|(C_x^{-1})^{1/2}(C_y^{-1})^{1/2}(C_x + \mathcal{D}_x)(C_y + \mathcal{D}_y)e^{n}\| \\
&\quad + \tau\|(C_x^{-1})^{1/2}(C_y^{-1})^{1/2}\mathcal{E}^{n+1/2}\|. \qquad (4.241)
\end{aligned}$$

Denote

$$\begin{aligned}
\tilde{e}_1^{n+1} &= (C_x^{-1})^{1/2}(C_y^{-1})^{1/2}(C_x - \mathcal{D}_x)(C_y - \mathcal{D}_y)e^{n+1}, \\
\tilde{e}_2^{n} &= (C_x^{-1})^{1/2}(C_y^{-1})^{1/2}(C_x + \mathcal{D}_x)(C_y + \mathcal{D}_y)e^{n}.
\end{aligned}$$

It can be noted that $(C_y - \mathcal{D}_y)$ commutes with $(C_x - \mathcal{D}_x)$, $(C_x^{-1})^{1/2}, (C_x - \mathcal{D}_x^{\mathrm{T}})$; $(C_y + \mathcal{D}_y)$ commutes with $(C_x + \mathcal{D}_x)$, $(C_x^{-1})^{1/2}, (C_x + \mathcal{D}_x^{\mathrm{T}})$; and $(C_y^{-1})^{1/2}$ commutes with $(C_x^{-1})^{1/2}$, $(C_x \pm \mathcal{D}_x^{\mathrm{T}})$. In addition, from Theorem 4.11 and Lemma 4.11, we have that $\mathcal{D}_\gamma + \mathcal{D}_\gamma^{\mathrm{T}}$ $(\gamma = x, y)$ are symmetric and negative definite matrices. Thus, using Lemma 4.10, we obtain

$$\begin{aligned}
&\left(\tilde{e}_1^{n+1}\right)^{\mathrm{T}} \tilde{e}_1^{n+1} \\
&= \left(e^{n+1}\right)^{\mathrm{T}} \left(C_y - \mathcal{D}_y^{\mathrm{T}} - \mathcal{D}_y + \mathcal{D}_y^{\mathrm{T}} C_y^{-1}\mathcal{D}_y\right)\left(C_x - \mathcal{D}_x^{\mathrm{T}} - \mathcal{D}_x + \mathcal{D}_x^{\mathrm{T}} C_x^{-1}\mathcal{D}_x\right)e^{n+1} \\
&\ge \left(e^{n+1}\right)^{\mathrm{T}} \left(C_y + \mathcal{D}_y^{\mathrm{T}} C_y^{-1}\mathcal{D}_y\right)\left(C_x + \mathcal{D}_x^{\mathrm{T}} C_x^{-1}\mathcal{D}_x\right)e^{n+1} \\
&\quad + \left(e^{n+1}\right)^{\mathrm{T}} \left(\mathcal{D}_y^{\mathrm{T}} + \mathcal{D}_y\right)\left(\mathcal{D}_x^{\mathrm{T}} + \mathcal{D}_x\right)e^{n+1},
\end{aligned}$$

and

$$\begin{aligned}
&(\tilde{e}_2^{n})^{\mathrm{T}} \tilde{e}_2^{n} \\
&= \left(e^{\mathrm{T}}\right) \left(C_y + \mathcal{D}_y^{\mathrm{T}} + \mathcal{D}_y + \mathcal{D}_y^{\mathrm{T}} C_y^{-1}\mathcal{D}_y\right)\left(C_x + \mathcal{D}_x^{\mathrm{T}} + \mathcal{D}_x + \mathcal{D}_x^{\mathrm{T}} C_x^{-1}\mathcal{D}_x\right)e^{n} \\
&\le \left(e^{n}\right)^{\mathrm{T}} \left(C_y + \mathcal{D}_y^{\mathrm{T}} C_y^{-1}\mathcal{D}_y\right)\left(C_x + \mathcal{D}_x^{\mathrm{T}} C_x^{-1}\mathcal{D}_x\right)e^{n} \\
&\quad + \left(e^{n}\right)^{\mathrm{T}} \left(\mathcal{D}_y^{\mathrm{T}} + \mathcal{D}_y\right)\left(\mathcal{D}_x^{\mathrm{T}} + \mathcal{D}_x\right)e^{n}.
\end{aligned}$$

Define

$$\begin{aligned}
(E^n)^2 = h_x h_y (e^n)^{\mathrm{T}} \Big(&\left(C_y + \mathcal{D}_y^{\mathrm{T}} C_y^{-1}\mathcal{D}_y\right)\left(C_x + \mathcal{D}_x^{\mathrm{T}} C_x^{-1}\mathcal{D}_x\right) \\
&+ \left(\mathcal{D}_y^{\mathrm{T}} + \mathcal{D}_y\right)\left(\mathcal{D}_x^{\mathrm{T}} + \mathcal{D}_x\right)\Big) e^n.
\end{aligned}$$

Then (4.241) becomes

$$\left|E^{k+1}\right| - \left|E^{k}\right| \le \tau\|(C_x^{-1})^{1/2}(C_y^{-1})^{1/2}\mathcal{E}^{k+1/2}\|. \qquad (4.242)$$

From (4.215)–(4.217), Lemmas 4.9–4.11, and Theorem 4.11, it is easy to check that the matrices $C_y\mathcal{D}_x^{\mathrm{T}} C_x^{-1}\mathcal{D}_x$, $C_x\mathcal{D}_y^{\mathrm{T}} C_y^{-1}\mathcal{D}_y$, $\mathcal{D}_y^{\mathrm{T}} C_y^{-1}\mathcal{D}_y\mathcal{D}_x^{\mathrm{T}} C_x^{-1}\mathcal{D}_x$

and $(\mathcal{D}_y^{\mathrm{T}}+\mathcal{D}_y)(\mathcal{D}_x^{\mathrm{T}}+\mathcal{D}_x)$ are all symmetric and positive definite. Then

$$\begin{aligned}|E^n| &\geq \sqrt{h_x h_y (e^n)^{\mathrm{T}} C_x C_y (e^n)} \\ &= \sqrt{h_x h_y (e^n)^{\mathrm{T}} (C_{\beta_1}\otimes C_\beta)(e^n)]} \\ &\geq \sqrt{\lambda_{\min}(C_\beta)\lambda_{\min}(C_{\beta_1})}\|e^n\|, \end{aligned} \tag{4.243}$$

where $\lambda_{\min}(C_\beta)$ and $\lambda_{\min}(C_{\beta_1})$ are the minimum eigenvalues of matrix C_β and C_{β_1}, respectively. As stated in Remark 4.5, we have $\lambda_{\min}(C_\beta) > 1 - 4c_{1,-1,2}^{\beta} > 0$, $\lambda_{\min}(C_{\beta_1}) > 1-4c_{1,-1,2}^{\beta_1} > 0$ if $\frac{1+\sqrt{73}}{6} < \beta, \beta_1 \leq 2$ and $(p,q) = (1,-1)$; and $\lambda_{\min}(C_\beta), \lambda_{\min}(C_{\beta_1}) > \frac{23}{72}$ if $1 \leq \beta, \beta_1 \leq 2$ and $(p,q) = (1,0)$. From the Rayleigh-Ritz Theorem (see Theorem 8.8 in [Zhang *et al.* (2011)]) and Lemma 4.9, we have for $k = 0, \cdots, n-1$ that

$$\begin{aligned}&\|(C_x^{-1})^{1/2}(C_y^{-1})^{1/2}\mathcal{E}^{k+1/2}\| \\ &= \sqrt{h_x h_y (\mathcal{E}^{k+1/2})^{\mathrm{T}} (C_x^{-1}C_y^{-1})\mathcal{E}^{k+1/2}} \\ &\leq \sqrt{\lambda_{\max}(C_x^{-1}C_y^{-1})}\|\mathcal{E}^{k+1/2}\| \\ &= \frac{1}{\sqrt{\lambda_{\min}(C_x C_y)}}\|\mathcal{E}^{k+1/2}\| \\ &= \frac{1}{\sqrt{\lambda_{\min}(C_\beta)\lambda_{\min}(C_{\beta_1})}}\|\mathcal{E}^{k+1/2}\|. \end{aligned} \tag{4.244}$$

Summing up (4.242) for all $0 \leq k \leq n-1$ shows that

$$\begin{aligned}|E^n| &\leq \tau \sum_{k=0}^{n-1} \|(C_x^{-1})^{1/2}(C_y^{-1})^{1/2}\mathcal{E}^{k+1/2}\| \\ &\leq \frac{\tau}{\sqrt{\lambda_{\min}(C_\beta)\lambda_{\min}(C_{\beta_1})}} \sum_{k=0}^{n-1} \|\mathcal{E}^{k+1/2}\|. \end{aligned} \tag{4.245}$$

Combining (4.245) and (4.243), and noticing $|\hat{\varepsilon}_{i,j}^{k+1/2}| \leq \tilde{c}(\tau^2 + h_x^3 + h_y^3)$ for $1 \leq i \leq N_x - 1$ and $1 \leq j \leq N_y - 1$, we obtain

$$\|e^n\| \leq \frac{c\,T}{\lambda_{\min}(C_\beta)\lambda_{\min}(C_{\beta_1})}(\tau^2 + h_x^3 + h_y^3). \tag{4.246}$$

□

Remark 4.12. If β, $\beta_1 = 1, 2$ with $(p,q) = (1,0)$ and β, $\beta_1 = 2$ with $(p,q) = (1,-1)$, then by the same reasoning of Theorem 4.19, it holds that the error estimate for the difference scheme (4.214), i.e.,

$$\|u^n - U^n\| \leq c(\tau^2 + h_x^4 + h_y^4), \quad 1 \leq n \leq M. \tag{4.247}$$

4.4.4 *Numerical results*

For saving computational time, we do one extrapolation to increase the accuracy to the third order in time (see Marchuk and Shaidurov, 1983). The detailed extrapolation algorithm is described as follows.

Step 1. Calculate ζ_1, ζ_2 from the following linear algebraic equations,

$$\begin{cases} \zeta_1 + \zeta_2 = 1, \\ \zeta_1 + \dfrac{\zeta_2}{4} = 0; \end{cases}$$

Step 2. Compute the solution U^n of the quasi-compact difference schemes with two time step sizes τ and $\tau/2$;

Step 3. Evaluate the extrapolation solution $W^n(\tau)$ by

$$W^n(\tau) = \zeta_1 \, U^n(\tau) + \zeta_2 \, U^n(\tau/2).$$

Example 4.6. Consider model (4.196) for $(x, t) \in (0, 1) \times (0, 1]$ with $K_1 = K_2 = 1$ and the exact solution

$$u(x, t) = \mathrm{e}^{-t} x^3 (1 - x)^3, \ (x, t) \in [0, 1] \times [0, 1].$$

The numerical results are given in Table 4.12, where

$$W_i^n(\tau) = -\frac{1}{3} U_i^n(\tau) + \frac{4}{3} U_i^n(\tau/2)$$

is the extrapolation solution, and U_i^n satisfies the quasi-compact scheme (4.201). It can be noted that for the case of $(p, q) = (1, -1)$, the numerical results are neither stable nor convergent when the order β is less than the critical value $\frac{1+\sqrt{73}}{6}$ (≈ 1.59), which coincides with the theoretical results.

Example 4.7. Consider model (4.207) in $(0, 1)^2 \times (0, 1]$ with $K_1^+ = K_2^+ = K_1^- = K_2^- = 1$, and the exact solution

$$u(x, t) = \mathrm{e}^{-t} x^3 (1 - x)^3 y^3 (1 - y)^3, \ (x, y, t) \in [0, 1]^2 \times [0, 1].$$

The numerical results are given in Table 4.13, where

$$W_{i,j}^n(\tau) = -\frac{1}{3} U_{i,j}^n(\tau) + \frac{4}{3} U_{i,j}^n(\tau/2)$$

is the numerical solution by extrapolation in time, and $U_{i,j}^n$ satisfies the quasi-compact Peaceman-Richardson (Q-CPR) scheme (4.219), quasi-compact Douglas (Q-C Douglas) scheme (4.220) and quasi-compact D'yakonov (Q-C D'yakonov) scheme (4.221), respectively. The third order accuracy both in time and space is verified, and in the computational process, the time costs are largely reduced.

Table 4.12: The discrete L^2 error and corresponding convergence for Example 4.6 approximated by the quasi-compact difference scheme at $t = 1$ for different β with $\tau = h$.

		$(p,q) = (1,0)$		$(p,q) = (1,-1)$	
β	N	$\|u^M - W^M\|$	rate	$\|u^M - W^M\|$	rate
1.2	8	4.12739e-05	-	1.17292e-01	-
	16	6.00551e-06	2.78	4.14981e-01	-1.82
	32	8.38665e-07	2.84	7.00900e+03	-14.04
	64	1.13698e-07	2.88	3.06749e+11	-25.38
	128	1.50548e-08	2.92	2.42775e+25	-46.17
	256	1.95986e-09	2.94	2.34722e+50	-83.00
$\frac{1+\sqrt{73}}{6}$	8	3.18317e-05	-	1.62093e-04	-
	16	4.15788e-06	2.94	2.50890e-05	2.69
	32	5.69588e-07	2.87	3.69218e-06	2.76
	64	7.81948e-08	2.86	5.16694e-07	2.84
	128	1.06194e-08	2.88	7.00251e-08	2.88
	256	1.42376e-09	2.90	9.30500e-09	2.91
1.8	8	2.87085e-05	-	8.54691e-05	-
	16	2.91599e-06	3.30	1.23638e-05	2.79
	32	3.47233e-07	3.07	1.76317e-06	2.81
	64	4.44726e-08	2.96	2.42232e-07	2.86
	128	5.85045e-09	2.93	3.24700e-08	2.90
	256	7.74722e-10	2.92	4.28918e-09	2.92

4.5 Effective fourth order discretizations for space fractional equations

In [Lubich (1986)], Lubich obtains the Kth order ($K \leq 6$) approximations of the βth derivative ($\beta > 0$) by the corresponding coefficients of the generating functions $\delta^{\beta,K}(\zeta)$, where

$$\delta^{\beta,K}(\zeta) = \left(\sum_{l=1}^{K} \frac{1}{l}(1-\zeta)^l\right)^{\beta}. \tag{4.248}$$

For $\beta = 1$, the scheme reduces to the classical $(K+1)$-point backward difference formula [Henrici (1962)]. For $K = 1$, the scheme (4.248) corresponds to the standard Grünwald discretization of βth derivative with first order accuracy, which is unconditionally unstable for time dependent problem. Taking $K = 2$, Cuesta et. al. discuss the convolution quadrature time discretization of fractional diffusion-wave equations [Cuesta *et al.* (2006)];

Table 4.13: The discrete L^2 errors and corresponding convergence for Example 4.7 approximated by the quasi-compact difference splitting schemes at $t = 1$ with $\tau = h_x = h_y$.

		$(p,q,\beta,\beta_1) = (1,0,1.1,1.7)$		$(p,q,\beta,\beta_1) = (1,-1,1.6,1.9)$	
Scheme	N	$\lVert u^M - W^M \rVert$	rate	$\lVert u^M - W^M \rVert$	rate
Q-CPR	8	2.63036e-007	-	1.33801e-006	-
	16	3.20134e-008	3.04	1.88287e-007	2.83
	32	4.19273e-009	2.93	2.48958e-008	2.92
	64	5.60865e-010	2.90	3.25063e-009	2.94
	128	7.51181e-011	2.90	4.22913e-010	2.94
	256	1.00029e-011	2.91	5.48530e-011	2.95
Q-C Douglas	8	2.54619e-007	-	1.15694e-006	-
	16	3.12227e-008	3.03	1.67892e-007	2.78
	32	4.12678e-009	2.92	2.30759e-008	2.86
	64	5.55710e-010	2.89	3.09798e-009	2.90
	128	7.47144e-011	2.89	4.09795e-010	2.92
	256	9.97005e-012	2.91	5.36563e-011	2.93
Q-C D'yakonov	8	2.54619e-007	-	1.15694e-006	-
	16	3.12227e-008	3.03	1.67892e-007	2.78
	32	4.12678e-009	2.92	2.30759e-008	2.86
	64	5.55710e-010	2.89	3.09798e-009	2.90
	128	7.47144e-011	2.89	4.09795e-010	2.92
	256	9.97005e-012	2.91	5.36563e-011	2.93

however, when applying the discretization scheme to space fractional operator with $\beta \in (1,2)$ for time dependent problem, the obtained scheme is also unstable, since the eigenvalues of the matrix corresponding to the discretized operator are greater than one. If using the shifted Lubich's formula, it reduces to the first order accuracy (see Theorem 4.47 in Sec. 4.1). This section weights and shifts Lubich's operators to obtain a class of fourth order discretization schemes, which are effective for time dependent problems. Then we use the fourth order schemes to solve the following two dimensional fractional diffusion equation with variable coefficients,

$$
\begin{aligned}
\frac{\partial u(x,y,t)}{\partial t} &= d_+(x,y)\,{}_{-\infty}D_x^{\beta}u(x,y,t) + d_-(x,y)\,{}_xD_{\infty}^{\beta}u(x,y,t) \\
&\quad + e_+(x,y)\,{}_{-\infty}D_y^{\beta_1}u(x,y,t) + e_-(x,y)\,{}_yD_{\infty}^{\beta_1}u(x,y,t) + f(x,y,t), \\
u(x,y,0) &= u_0(x,y), \qquad \text{for} \quad (x,y) \in \Omega, \\
u(x,y,t) &= 0, \qquad \text{for} \quad (x,y,t) \in \mathbb{R}^2 \backslash \Omega \times [0,T],
\end{aligned}
\tag{4.249}
$$

in $\Omega = (a,b)\times(c,d)$, $0 < t \le T$, where the orders of the fractional derivatives are $1 < \beta, \beta_1 < 2$, $f(x,y,t)$ is a source term, and all the variable coefficients are nonnegative.

4.5.1 *Derivation of a class of fourth order discretizations for space fractional operators*

In this subsection, we derive a class of fourth order approximations for R-L fractional derivatives, and prove that they are effective in solving space fractional PDEs, i.e., all the eigenvalues of the matrixes corresponding to the discretized operators have negative real parts.

4.5.1.1 *Fourth order Shifted Lubich's fractional difference approximations for the R-L fractional derivatives*

Let q_k^β satisfy $\sum_{k=0}^{\infty} q_k^\beta z^k = \left(\frac{3}{2} - 2z + \frac{1}{2}z^2\right)^\beta$ and define

$$
{}_L\widetilde{A}_p^\beta u(x) := \frac{1}{h^\beta} \sum_{k=0}^{\infty} q_k^\beta u\left(x - (k-p)h\right) \tag{4.250}
$$

and

$$
{}_R\widetilde{A}_p^\beta u(x) := \frac{1}{h^\beta} \sum_{k=0}^{\infty} q_k^\beta u(x + (k-p)h). \tag{4.251}
$$

By Theorem 4.2, if $u \in \mathscr{S}^{\beta+1}(\mathbb{R})$ and $u(x)$ has compact support, then

$$
{}_{-\infty}D_x^\beta u(x) = {}_L\widetilde{A}_p^\beta u(x) + \mathcal{O}(h) \tag{4.252}
$$

and

$$
{}_xD_\infty^\beta u(x) = {}_R\widetilde{A}_p^\beta u(x) + \mathcal{O}(h) \tag{4.253}
$$

uniformly for $x \in \mathbb{R}$ as $h \to 0$. The coefficients q_k^β can be calculated by recursive relationship

$$
q_0^\beta = \left(\frac{3}{2}\right)^\beta, \quad q_1^\beta = -\left(\frac{3}{2}\right)^\beta \frac{4\beta}{3}, \tag{4.254}
$$

$$
q_k^\beta = \frac{1}{3}\left(\frac{2\beta+2}{k} - 1\right) q_{k-2}^\beta + \frac{4}{3}\left(1 - \frac{\beta+1}{k}\right) q_{k-1}^\beta, \quad k \ge 2. \tag{4.255}
$$

In fact, under the reasonable regularity requirements of $u(x)$, similar to Theorem 4.9, the second and third order schemes can be easily derived. The obtained schemes are given as follows:

$$
\begin{aligned}
{}_{-\infty}D_x^\beta u(x) &= {}_{2L}\widetilde{A}_{p,q}^\beta u(x) + \mathcal{O}(h^2), \\
{}_{2L}\widetilde{A}_{p,q}^\beta u(x) &:= w_p\, {}_L\widetilde{A}_p^\beta u(x) + w_q\, {}_L\widetilde{A}_q^\beta u(x),
\end{aligned} \tag{4.256}
$$

where ${}_L\widetilde{A}_p^\beta$, ${}_L\widetilde{A}_q^\beta$ are defined in (4.250), $w_p = \frac{q}{q-p}$, $w_q = \frac{p}{p-q}$, $p \neq q$, and p, q are integers;

$$\begin{aligned} {}_{-\infty}D_x^\beta u(x) &= {}_{3L}\widetilde{A}_{p,q,r,s}^\beta u(x) + \mathcal{O}(h^3), \\ {}_{3L}\widetilde{A}_{p,q,r,s}^\beta u(x) &:= w_{p,q}\, {}_{2L}\widetilde{A}_{p,q}^\beta u(x) + w_{r,s}\, {}_{2L}\widetilde{A}_{r,s}^\beta u(x), \end{aligned} \tag{4.257}$$

where ${}_{2L}\widetilde{A}_{p,q}^\beta$ and ${}_{2L}\widetilde{A}_{r,s}^\beta$ are defined in (4.256), $w_{p,q} = \frac{3rs+2\beta}{3(rs-pq)}$, $w_{r,s} = \frac{3pq+2\beta}{3(pq-rs)}$, $rs \neq pq$, and p, q, r, s are integers.

Theorem 4.20 (Fourth order approximations). *Let $u(x) \in \mathscr{S}^{\beta+4}(\mathbb{R})$ and $u(x)$ have compact support. Denote*

$${}_{4L}\widetilde{A}_{p,q,r,s,\overline{p},\overline{q},\overline{r},\overline{s}}^\beta u(x) = w_{p,q,r,s}\, {}_{3L}\widetilde{A}_{p,q,r,s}^\beta u(x) + w_{\overline{p},\overline{q},\overline{r},\overline{s}}\, {}_{3L}\widetilde{A}_{\overline{p},\overline{q},\overline{r},\overline{s}}^\beta u(x), \tag{4.258}$$

where ${}_{3L}\widetilde{A}_{p,q,r,s}^\beta$ and ${}_{3L}\widetilde{A}_{\overline{p},\overline{q},\overline{r},\overline{s}}^\beta$ are defined in (4.257), and

$$w_{p,q,r,s} = \frac{a_{p,q,r,s}\,\overline{b}_{\overline{p},\overline{q},\overline{r},\overline{s}}}{a_{p,q,r,s}\,\overline{b}_{\overline{p},\overline{q},\overline{r},\overline{s}} - \overline{a}_{\overline{p},\overline{q},\overline{r},\overline{s}}\,b_{p,q,r,s}}, \tag{4.259}$$

$$w_{\overline{p},\overline{q},\overline{r},\overline{s}} = \frac{\overline{a}_{\overline{p},\overline{q},\overline{r},\overline{s}}\,b_{p,q,r,s}}{\overline{a}_{\overline{p},\overline{q},\overline{r},\overline{s}}\,b_{p,q,r,s} - a_{p,q,r,s}\,\overline{b}_{\overline{p},\overline{q},\overline{r},\overline{s}}}, \tag{4.260}$$

with

$$\begin{aligned} a_{p,q,r,s} &= rs - pq, \quad \overline{a}_{\overline{p},\overline{q},\overline{r},\overline{s}} = \overline{r}\,\overline{s} - \overline{p}\,\overline{q}, \\ b_{p,q,r,s} &= 6pqrs(r+s-p-q) + 4\beta\big[rs(r+s) - pq(p+q)\big] + 9\beta(rs-pq), \\ \overline{b}_{\overline{p},\overline{q},\overline{r},\overline{s}} &= 6\overline{p}\,\overline{q}\,\overline{r}\,\overline{s}(\overline{r}+\overline{s}-\overline{p}-\overline{q}) + 4\beta\big[\overline{r}\,\overline{s}(\overline{r}+\overline{s}) - \overline{p}\,\overline{q}(\overline{p}+\overline{q})\big] + 9\beta(\overline{r}\,\overline{s} - \overline{p}\,\overline{q}), \end{aligned}$$

and $a_{p,q,r,s}\,\overline{b}_{\overline{p},\overline{q},\overline{r},\overline{s}} \neq \overline{a}_{\overline{p},\overline{q},\overline{r},\overline{s}}\,b_{p,q,r,s}$ and p, q, r, s; $\overline{p}$, $\overline{q}$, $\overline{r}$, $\overline{s}$, are integers. Then

$${}_{-\infty}D_x^\beta u(x) = {}_{4L}\widetilde{A}_{p,q,r,s,\overline{p},\overline{q},\overline{r},\overline{s}}^\beta u(x) + \mathcal{O}(h^4). \tag{4.261}$$

Proof. Similar to the proof of Theorem 4.2, it is easy to check that

$$\begin{aligned} \mathcal{F}({}_{3L}\widetilde{A}_{p,q,r,s}^\beta u)(\omega) = (i\omega)^\beta\Big[1 + \\ \frac{6pqrs(r+s-p-q) + 4\beta\big[rs(r+s) - pq(p+q)\big] + 9\beta(rs-pq)}{36(rs-pq)} z^3 \\ + C_1 z^4 + \cdots\Big]\widehat{u}(\omega), \end{aligned}$$

and

$$\mathcal{F}({}_{3L}\widetilde{A}^{\beta}_{\overline{p},\overline{q},\overline{r},\overline{s}}u)(\omega) = (i\omega)^{\beta}\Big[1 + \frac{6\overline{p}\,\overline{q}\,\overline{r}\,\overline{s}(\overline{r}+\overline{s}-\overline{p}-\overline{q}) + 4\beta\big[\overline{r}\,\overline{s}(\overline{r}+\overline{s}) - \overline{p}\,\overline{q}(\overline{p}+\overline{q})\big] + 9\beta(\overline{r}\,\overline{s}-\overline{p}\,\overline{q})}{36(\overline{r}\,\overline{s}-\overline{p}\,\overline{q})} z^3 + C_2 z^4 + \cdots\Big]\widehat{u}(\omega),$$

with $z = i\omega h$, and C_1 and C_2 are constants independent of ω and z. Then there exists a constant C_3 such that

$$\mathcal{F}({}_{4L}\widetilde{A}^{\beta}_{p,q,r,s,\overline{p},\overline{q},\overline{r},\overline{s}}u)(\omega) = (-i\omega)^{\beta}\left(1 + C_3 z^4 + \cdots\right)\widehat{u}(\omega).$$

In the similar way to the proof of Lemma 4.2, we get

$${}_{-\infty}D^{\beta}_{x}u(x) = {}_{4L}\widetilde{A}^{\beta}_{p,q,r,s,\overline{p},\overline{q},\overline{r},\overline{s}}u(x) + \mathcal{O}(h^4).$$

□

When $u(x)$ satisfying $u(x) = 0$ for $x \in \mathbb{R}\backslash(a,b)$, letting $x_i = a + ih$, $i = 0,\ldots,0,1,\ldots,N_x-1,N_x$, and $h = (b-a)/N_x$ be the uniform space stepsize, then

$${}_{L}\widetilde{A}^{\beta}_{p}u(x_i) = \frac{1}{h^{\beta}}\sum_{k=0}^{i+p} q^{\beta}_{k}u(x_{i-k+p}), \tag{4.262}$$

where $u(x_{i-k+p}) = 0$ for $i-k+p \le 0$ and $i-k+p \ge N_x$. By (4.256), we have

$$\begin{aligned}{}_{-\infty}D^{\beta}_{x}u(x)\,|_{x=x_i} &= {}_{2L}\widetilde{A}^{\beta}_{p,q}u(x_i) + \mathcal{O}(h^2)\\ &= \frac{1}{h^{\beta}}\left(w_p\sum_{k=0}^{i+p} q^{\beta}_{k}u(x_{i-k+p}) + w_q\sum_{k=0}^{i+q} q^{\beta}_{k}u(x_{i-k+q})\right) + \mathcal{O}(h^2).\end{aligned} \tag{4.263}$$

Take $U = [u(x_1), u(x_2), \cdots, u(x_{N_x-1})]^{\mathrm{T}}$. Obviously, (4.263) can be rewritten as the matrix form (see Theorem 4.21 below)

$${}_{2L}\widetilde{A}^{\beta}_{p,q}U = \frac{1}{h^{\beta}}A^{\beta}_{p,q}U. \tag{4.264}$$

Similarly,

$${}_{3L}\widetilde{A}^{\beta}_{p,q,r,s}U = \frac{1}{h^{\beta}}A^{\beta}_{p,q,r,s}U, \quad \text{with } A^{\beta}_{p,q,r,s} = w_{p,q}A_{p,q} + w_{r,s}A_{r,s}, \tag{4.265}$$

and

$$\begin{aligned}&{}_{4L}\widetilde{A}^{\beta}_{p,q,r,s,\overline{p},\overline{q},\overline{r},\overline{s}}U = \frac{1}{h^{\beta}}A^{\beta}_{p,q,r,s,\overline{p},\overline{q},\overline{r},\overline{s}}U,\\ &\text{with } A^{\beta}_{p,q,r,s,\overline{p},\overline{q},\overline{r},\overline{s}} = w_{p,q,r,s}A^{\beta}_{p,q,r,s} + w_{\overline{p},\overline{q},\overline{r},\overline{s}}A^{\beta}_{\overline{p},\overline{q},\overline{r},\overline{s}}.\end{aligned} \tag{4.266}$$

The fourth order approximation for the right R-L fractional derivative

$$_{x}D_{\infty}^{\beta}u(x) = {}_{4R}\widetilde{A}_{p,q,r,s,\overline{p},\overline{q},\overline{r},\overline{s}}^{\beta}u(x) + \mathcal{O}(h^{4}) \tag{4.267}$$

can be derived by almost the same way as the derivation of the left R-L fractional derivative, just by replacing the $_{L}\widetilde{A}_{p}^{\beta}u(x)$ (see (4.256) and (4.257)) with the corresponding

$$_{R}\widetilde{A}_{p}^{\beta}u(x) = \frac{1}{h^{\beta_{1}}}\sum_{k=0}^{\infty} q_{k}^{\beta}u(x+(k-p)h), \tag{4.268}$$

for $p = p, q, r, s, \overline{p}, \overline{p}, \overline{r}, \overline{s}$. In particular, if $u(x) = 0$ for $x \in \mathbb{R}\backslash(a,b)$, we have

$$_{4R}\widetilde{A}_{p,q,r,s,\overline{p},\overline{q},\overline{r},\overline{s}}^{\beta}U = \frac{1}{h^{\beta}}\left(A_{p,q,r,s,\overline{p},\overline{q},\overline{r},\overline{s}}^{\beta}\right)^{\mathrm{T}}U. \tag{4.269}$$

4.5.1.2 *Effective fourth order discretization for space fractional derivatives*

Now we focus on how to choose the parameters $p, q, r, s, \overline{p}, \overline{q}, \overline{r}, \overline{s}$ such that all the eigenvalues of the matrix $A_{p,q}^{\beta}$ (or $A_{p,q,r,s}^{\beta}$ or $A_{p,q,r,s,\overline{p},\overline{q},\overline{r},\overline{s}}^{\beta}$) have negative real parts, which means that the corresponding schemes work for space fractional derivatives.

Theorem 4.21 (Effective second order schemes). *Let $A_{p,q}^{\beta}$ be given in (4.264) and $1 < \beta < 2$. Then any eigenvalue λ of $A_{p,q}^{\beta}$ satisfies*

$$\mathrm{Re}(\lambda(A_{p,q}^{\beta})) < 0, \quad \textit{for} \quad (p,q) = (1,q), \quad |q| \geq 2.$$

Moreover, the matrices $A_{p,q}^{\beta}$ and $(A_{p,q}^{\beta})^{\mathrm{T}}$ are negative definite.

Proof. **(1)** For $(p,q) = (1,q)$, $q \leq -2$, by (4.263), we have

$$A_{p,q}^{\beta} = \begin{bmatrix} \phi_{1}^{\beta} & \phi_{0}^{\beta} & & & \\ \phi_{2}^{\beta} & \phi_{1}^{\beta} & \phi_{0}^{\beta} & & \\ \vdots & \ddots & \ddots & \ddots & \\ \phi_{N_x-2}^{\beta} & \ddots & \ddots & \phi_{1}^{\beta} & \phi_{0}^{\beta} \\ \phi_{N_x-1}^{\beta} & \phi_{N_x-2}^{\beta} & \cdots & \phi_{2}^{\beta} & \phi_{1}^{\beta} \end{bmatrix}$$

with

$$\phi_{k}^{\beta} = \begin{cases} w_{p}q_{k}^{\beta}, & 0 \leq k \leq -q, \\ w_{p}q_{k}^{\beta} + w_{q}q_{k+q-1}^{\beta}, & k > -q. \end{cases}$$

Taking

$$H_{p,q} = \frac{A_{p,q}^{\beta} + \left(A_{p,q}^{\beta}\right)^{\mathrm{T}}}{2}, \tag{4.270}$$

then

$$f_{p,q}(\beta, x) = \frac{1}{2}\left(\sum_{k=0}^{\infty} \phi_k^{\beta} e^{i(k-1)x} + \sum_{k=0}^{\infty} \phi_k^{\beta} e^{-i(k-1)x}\right) \tag{4.271}$$

is the generating function of $H_{p,q}$, which is a 2π-periodic continuous real-valued even functions. Next, we will prove $f_{p,q}(\beta, x) \le 0$ for $x \in [0, \pi]$.

Note that

$$\begin{aligned}
f_{p,q}(\beta, x) &= \frac{1}{2}\left(w_p e^{-ix}\sum_{k=0}^{\infty} q_k^{\beta} e^{ikx} + w_q e^{-iqx}\sum_{k=0}^{\infty} q_k^{\beta} e^{ikx}\right) \\
&\quad + \frac{1}{2}\left(w_p e^{ix}\sum_{k=0}^{\infty} q_k^{\beta} e^{-ikx} + w_q e^{iqx}\sum_{k=0}^{\infty} q_k^{\beta} e^{-ikx}\right) \\
&= \frac{w_p}{2} e^{-ix}(1 - e^{ix})^{\beta}\left(1 + \frac{1}{2}(1 - e^{ix})\right)^{\beta} \\
&\quad + \frac{w_p}{2} e^{ix}(1 - e^{-ix})^{\beta}\left(1 + \frac{1}{2}(1 - e^{-ix})\right)^{\beta} \\
&\quad + \frac{w_q}{2} e^{-iqx}(1 - e^{ix})^{\beta}\left(1 + \frac{1}{2}(1 - e^{ix})\right)^{\beta} \\
&\quad + \frac{w_q}{2} e^{iqx}(1 - e^{-ix})^{\beta}\left(1 + \frac{1}{2}(1 - e^{-ix})\right)^{\beta}.
\end{aligned}$$

Because of

$$\begin{aligned}
(1 - e^{\pm ix})^{\beta} &= \left(2\sin\frac{x}{2}\right)^{\beta} e^{\pm i\beta(\frac{x}{2} - \frac{\pi}{2})}, \\
\left(1 + \frac{1}{2}(1 - e^{\pm ix})\right)^{\beta} &= \left(1 + 3\sin^2\frac{x}{2}\right)^{\frac{\beta}{2}} e^{\pm i\beta(\frac{x}{2} - \theta)},
\end{aligned}$$

where

$$\theta = 2\arctan\frac{2\sin\frac{x}{2}}{\cos\frac{x}{2} + \sqrt{1 + 3\sin^2\frac{x}{2}}} \in [0, \pi/2],$$

then, for $q \le -2$, there exists

$$\begin{aligned}
f_{p,q}(\beta, x) = &\left(2\sin\frac{x}{2}\right)^{\beta}\left(1 + 3\sin^2\frac{x}{2}\right)^{\frac{\beta}{2}} \times \\
&\left[w_p\cos\left(\beta(x - \frac{\pi}{2} - \theta) - x\right) + w_q\cos\left(\beta(x - \frac{\pi}{2} - \theta) - qx\right)\right].
\end{aligned}$$

(2) For $(p,q)=(1,q)$, $q\geq 2$, we have $A^{\beta}_{p,q}=\frac{1}{q-1}(qA^{\beta}_{1}-A^{\beta}_{q})$ and

$$A^{\beta}_{p,q}=\begin{bmatrix} \phi^{\beta}_{q} & \phi^{\beta}_{q-1} & \cdots & \phi^{\beta}_{0} & & & & & \\ \phi^{\beta}_{q+1} & \phi^{\beta}_{q} & \phi^{\beta}_{q-1} & \cdots & \phi^{\beta}_{0} & & & & \\ \phi^{\beta}_{q+2} & \phi^{\beta}_{q+1} & \phi^{\beta}_{q} & \phi^{\beta}_{q-1} & \cdots & \phi^{\beta}_{0} & & & \\ \vdots & \ddots & \ddots & \ddots & \ddots & \cdots & \ddots & & \\ \phi^{\beta}_{N_x-2} & \cdots & \ddots & \phi^{\beta}_{q+1} & \phi^{\beta}_{q} & \phi^{\beta}_{q-1} & \cdots & \phi^{\beta}_{0} \\ \vdots & \ddots & \cdots & \ddots & \ddots & \ddots & \ddots & \vdots \\ \phi^{\beta}_{q+N_x-3} & \cdots & \ddots & \cdots & \ddots & \phi^{\beta}_{q+1} & \phi^{\beta}_{q} & \phi^{\beta}_{q-1} \\ \phi^{\beta}_{q+N_x-2} & \phi^{\beta}_{q+N_x-3} & \cdots & \phi^{\beta}_{N_x-2} & \cdots & \phi^{\beta}_{q+2} & \phi^{\beta}_{q+1} & \phi^{\beta}_{q} \end{bmatrix},$$

with

$$\phi^{\beta}_{k}=\begin{cases} w_q q^{\beta}_{k}, & 0\leq k\leq q-2,\\ w_q q^{\beta}_{k}+w_p q^{\beta}_{k-q+1}, & k>q-2. \end{cases}$$

The generating functions of $A^{\beta}_{p,q}$ and $(A^{\beta}_{p,q})^{\mathrm{T}}$ are

$$f_{A^{\beta}_{p,q}}(x)=\sum_{k=0}^{\infty}\phi^{\beta}_{k}e^{i(k-q)x} \quad \text{and} \quad f_{(A^{\beta}_{p,q})^{\mathrm{T}}}(x)=\sum_{k=0}^{\infty}\phi^{\beta}_{k}e^{-i(k-q)x},$$

respectively. Denoting

$$H_{p,q}=\frac{A^{\beta}_{p,q}+\left(A^{\beta}_{p,q}\right)^{\mathrm{T}}}{2}, \tag{4.272}$$

then $f_{p,q}(\beta,x)=\frac{f_{A^{\beta}_{p,q}}(x)+f_{(A^{\beta}_{p,q})^{\mathrm{T}}}(x)}{2}$ is the generating function of $H_{p,q}$. In the similar way, for $q\geq 2$, there exists

$$\begin{aligned} f_{p,q}(\beta,x)=&\left(2\sin\frac{x}{2}\right)^{\beta}\left(1+3\sin^{2}\frac{x}{2}\right)^{\frac{\beta}{2}}\times\\ &\left[w_p\cos\left(\beta(x-\frac{\pi}{2}-\theta)-x\right)+w_q\cos\left(\beta(x-\frac{\pi}{2}-\theta)-qx\right)\right]. \end{aligned}$$

It can be noted that $f_{p,q}(\beta,x)$ has the same form when $q\leq -2$ and $q\geq 2$, $p=1$. And we can check that, for $(p,q)=(1,q)$, $|q|\geq 2$, there exists (see Figs. 4.1-4.4)

$$\begin{aligned} f_{p,q}(\beta,x)=&\left(2\sin\frac{x}{2}\right)^{\beta}\left(1+3\sin^{2}\frac{x}{2}\right)^{\frac{\beta}{2}}\times\\ &\left[w_p\cos\left(\beta(x-\frac{\pi}{2}-\theta)-x\right)+w_q\cos\left(\beta(x-\frac{\pi}{2}-\theta)-qx\right)\right]\leq 0. \end{aligned} \tag{4.273}$$

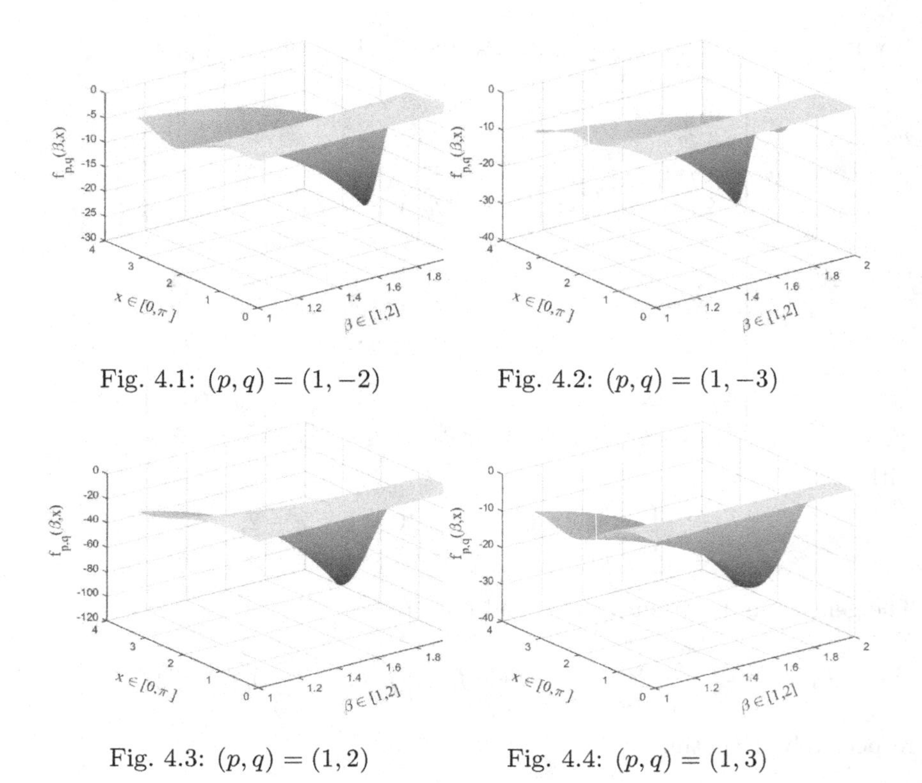

Fig. 4.1: $(p, q) = (1, -2)$

Fig. 4.2: $(p, q) = (1, -3)$

Fig. 4.3: $(p, q) = (1, 2)$

Fig. 4.4: $(p, q) = (1, 3)$

Since $f_{p,q}(\beta, x)$ is not zero for any given $\beta \in (1, 2)$, from Lemma 4.8, it implies that $\lambda(H_{p,q}) < 0$ and $H_{p,q}$ is negative definite. Then we get $\mathrm{Re}(\lambda(A^{\beta}_{p,q})) < 0$ from Lemma 4.7, and the matrices $A^{\beta}_{p,q}$ and $(A^{\beta}_{p,q})^{\mathrm{T}}$ are negative definite by Lemma 4.6. $\square$

Theorem 4.22 (Effective third order schemes). *Let $A^{\beta}_{p,q,r,s}$ with $1 < \beta < 2$ be given in (4.265). Then any eigenvalue λ of $A^{\beta}_{p,q,r,s}$ satisfies*

$$\mathrm{Re}(\lambda(A^{\beta}_{p,q,r,s})) < 0,$$

for $(p, q, r, s) = (1, q, 1, s), \quad |q| \geq 2, \quad |s| \geq 2, \quad$ and $\quad qs < 0$; moreover, the matrices $A^{\beta}_{p,q,r,s}$ and $(A^{\beta}_{p,q,r,s})^{\mathrm{T}}$ are negative definite.

Proof. Taking

$$H_{p,q,r,s} = \frac{A^{\beta}_{p,q,r,s} + \left(A^{\beta}_{p,q,r,s}\right)^{\mathrm{T}}}{2} = w_{p,q}H_{p,q} + w_{r,s}H_{r,s},$$

where $H_{p,q}$ and $H_{r,s}$ are defined in (4.272), then

$$f_{p,q,r,s}(\beta,x) = w_{p,q}f_{p,q}(\beta,x) + w_{r,s}f_{r,s}(\beta,x) \tag{4.274}$$

is the generating function of $H_{p,q,r,s}$, where $f_{p,q}(\beta,x)$ and $f_{r,s}(\beta,x)$ are given by (4.273). Since $|q| \geq 2$, $|s| \geq 2$, and $qs < 0$, we can check that $w_{p,q} = w_{1,q} = \frac{3s+2\beta}{3(s-q)} > 0$, $w_{r,s} = w_{1,s} = \frac{3q+2\beta}{3(q-s)} > 0$. Then from (4.273) and (4.274), we get $f_{p,q,r,s}(\beta,x) \leq 0$.

Again, from Lemmas 4.6-4.8, the desired results are obtained. □

Theorem 4.23 (Effective fourth order schemes). *Let $A^{\beta}_{p,q,r,s,\overline{p},\overline{q},\overline{r},\overline{s}}$ with $1<\beta<2$ be given in (4.266), where $(p,q,r,s,\overline{p},\overline{q},\overline{r},\overline{s}) = (1,2,1,-2,1,\overline{q},1,\overline{s})$ $|\overline{q}| \geq 2$, $|\overline{s}| \geq 2$, $(\overline{q},\overline{s}) \neq (2,-2)$ and $\overline{q}\,\overline{s}<0$. Then any eigenvalue λ of $A^{\beta}_{p,q,r,s,\overline{p},\overline{q},\overline{r},\overline{s}}$ satisfies*

$$\mathrm{Re}(\lambda(A^{\beta}_{p,q,r,s,\overline{p},\overline{q},\overline{r},\overline{s}})) < 0,$$

and the matrices $A^{\beta}_{p,q,r,s,\overline{p},\overline{q},\overline{r},\overline{s}}$ and $(A^{\beta}_{p,q,r,s,\overline{p},\overline{q},\overline{r},\overline{s}})^{\mathrm{T}}$ are negative definite.

Moreover, if $(p,q,r,s,\overline{p},\overline{q},\overline{r},\overline{s})$ takes the following values

$$\begin{aligned}
(p,q,r,s,\overline{p},\overline{q},\overline{r},\overline{s}) &= (1,2,1,0,1,2,1,-2),\\
(p,q,r,s,\overline{p},\overline{q},\overline{r},\overline{s}) &= (1,2,1,0,1,-1,1,-2),\\
(p,q,r,s,\overline{p},\overline{q},\overline{r},\overline{s}) &= (1,2,1,-1,1,2,1,-2),\\
(p,q,r,s,\overline{p},\overline{q},\overline{r},\overline{s}) &= (1,2,1,-1,1,-1,1,-2),\\
(p,q,r,s,\overline{p},\overline{q},\overline{r},\overline{s}) &= (1,0,1,-1,1,2,1,-2),\\
(p,q,r,s,\overline{p},\overline{q},\overline{r},\overline{s}) &= (1,0,1,-2,1,2,1,-2),\\
(p,q,r,s,\overline{p},\overline{q},\overline{r},\overline{s}) &= (1,-1,1,-2,1,2,1,-2),
\end{aligned}$$

then $\mathrm{Re}(\lambda(A^{\beta}_{p,q,r,s,\overline{p},\overline{q},\overline{r},\overline{s}})) < 0$, the matrices $A^{\beta}_{p,q,r,s,\overline{p},\overline{q},\overline{r},\overline{s}}$ and $(A^{\beta}_{p,q,r,s,\overline{p},\overline{q},\overline{r},\overline{s}})^{\mathrm{T}}$ are negative definite.

Proof. In the similar way to the proofs of Theorems 4.21 and 4.22, we obtain the desired results. □

4.5.2 *Applying the fourth order approximations to the space fractional diffusion equations*

We use two subsections to derive the full discretization of (4.249). For completeness, we also present the scheme for the one dimensional case of (4.249).

4.5.2.1 *Numerical scheme for 1D problem*

We first consider the one dimensional case of (4.249) with variable coefficients, namely,

$$\frac{\partial u(x,t)}{\partial t} = d_+(x)_{-\infty}D_x^\beta u(x,t) + d_-(x)_x D_\infty^\beta u(x,t) + f(x,t), \qquad (4.275)$$

with $u(x,0) = u_0(x)$ for $x \in (a,b)$ and $u(x,t) = 0$ for $(x,t) \in \mathbb{R}\backslash(a,b) \times [0,T]$. In the time direction, we use the C-N scheme. The fourth order left fractional approximation operator (4.261), and right fractional approximation operator (4.267) are respectively used to discretize the left R-L fractional derivative, and right R-L fractional derivative.

Let the mesh points $x_i = a + ih$, $i = 0, 1, \ldots, N_x$, and $t_n = n\tau$, $0 \le n \le M$, where $h = (b-a)/N_x$, $\tau = T/M$, i.e., h is the uniform space stepsize and τ the time steplength. Taking $u_i^n = u(x_i, t_n)$, and $d_{+,i} = d_+(x_i)$, $d_{-,i} = d_-(x_i)$, $f_i^{n+1/2} = f(x_i, t_{n+1/2})$, where $t_{n+1/2} = (t_n + t_{n+1})/2$, then for $1 \le i \le N_x - 1$ and $0 \le n \le M-1$, (4.275) can be rewritten as

$$\begin{aligned}&\left[1 - \frac{\tau}{2}\left(d_{+,i}\,{}_{4L}\widetilde{A}^\beta_{p,q,r,s,\overline{p},\overline{q},\overline{r},\overline{s}} + d_{-,i}\,{}_{4R}\widetilde{A}^\beta_{p,q,r,s,\overline{p},\overline{q},\overline{r},\overline{s}}\right)\right] u_i^{n+1}\\ &\quad= \left[1 + \frac{\tau}{2}\left(d_{+,i}\,{}_{4L}\widetilde{A}^\beta_{p,q,r,s,\overline{p},\overline{q},\overline{r},\overline{s}} + d_{-,i}\,{}_{4R}\widetilde{A}^\beta_{p,q,r,s,\overline{p},\overline{q},\overline{r},\overline{s}}\right)\right] u_i^n \qquad (4.276)\\ &\quad\quad+ \tau f(x_i, t_{n+1/2}) + \varepsilon_i^n\end{aligned}$$

with

$$|\varepsilon_i^n| \le \widetilde{c}\tau\left(\tau^2 + h^4\right). \qquad (4.277)$$

Denoting U_i^n as the numerical approximation of u_i^n, we derive the difference scheme for (4.275), i.e.,

$$\begin{aligned}&\left[1 - \frac{\tau}{2}\left(d_{+,i}\,{}_{4L}\widetilde{A}^\beta_{p,q,r,s,\overline{p},\overline{q},\overline{r},\overline{s}} + d_{-,i}\,{}_{4R}\widetilde{A}^\beta_{p,q,r,s,\overline{p},\overline{q},\overline{r},\overline{s}}\right)\right] U_i^{n+1}\\ &\quad= \left[1 + \frac{\tau}{2}\left(d_{+,i}\,{}_{4L}\widetilde{A}^\beta_{p,q,r,s,\overline{p},\overline{q},\overline{r},\overline{s}} + d_{-,i}\,{}_{4R}\widetilde{A}^\beta_{p,q,r,s,\overline{p},\overline{q},\overline{r},\overline{s}}\right)\right] U_i^n + \tau f_i^{n+1/2}.\end{aligned} \qquad (4.278)$$

For convenience, we denote

$$U^n = \left(U_1^n, U_2^n, \ldots, U_{N_x-1}^n\right)^{\mathrm{T}}, \quad F^{n+1/2} = \left(f_1^{n+1/2}, f_2^{n+1/2}, \ldots, f_{N_x-1}^{n+1/2}\right)^{\mathrm{T}}.$$

Then, the matrix vector form of the finite difference scheme (4.278) is

$$\begin{aligned}&\left[I - \frac{\tau}{2h^\beta}\left(D_+A_\beta + D_-A_\beta^{\mathrm{T}}\right)\right] U^{n+1}\\ &\quad= \left[I + \frac{\tau}{2h^\beta}\left(D_+A_\beta + D_-A_\beta^{\mathrm{T}}\right)\right] U^n + \tau F^{n+1/2},\end{aligned} \qquad (4.279)$$

where $A_\beta = A^\beta_{p,q,r,s,\overline{p},\overline{q},\overline{r},\overline{s}}$ is defined by (4.266), and

$$D_+ = \begin{bmatrix} d_{+,1} & & & \\ & d_{+,2} & & \\ & & \ddots & \\ & & & d_{+,N_x-1} \end{bmatrix}, \quad D_- = \begin{bmatrix} d_{-,1} & & & \\ & d_{-,2} & & \\ & & \ddots & \\ & & & d_{-,N_x-1} \end{bmatrix}. \tag{4.280}$$

4.5.2.2 *Numerical scheme for 2D problem*

We now examine the full discretization scheme of (4.249). For effectively performing the theoretical analysis, we suppose $d_+(x) = d_+(x,y)$, $d_-(x) = d_-(x,y)$, and $e_+(y) = e_+(x,y)$, $e_-(y) = e_-(x,y)$.

Analogously we still use the C-N scheme to do the discretization in time direction. Let the mesh points $x_i = a + ih_x$, $i = 0,1,\ldots,N_x$, and $y_j = c + jh_y$, $j = 0,1,\ldots,N_y$, $t_n = n\tau$, $0 \le n \le M$, and $h_x = (b-a)/N_x$, $h_y = (d-c)/N_y$, $\tau = T/M$; and $d_{+,i} = d_+(x_i,y_j)$, $d_{-,i} = d_-(x_i,y_j)$, and $e_{+,j} = e_+(x_i,y_j)$, $e_{-,j} = e_-(x_i,y_j)$. Taking $u^n_{i,j} = u(x_i,y_j,t_n)$ and $f^{n+1/2}_{i,j} = f(x_i,y_j,t_{n+1/2})$, where $t_{n+1/2} = (t_n + t_{n+1})/2$, then (4.249) can be rewritten as

$$\begin{aligned}
&\Big[1 - \frac{\tau}{2}\Big(d_{+,i}\, {}_{4L}\widetilde{A}^\beta_{p,q,r,s,\overline{p},\overline{q},\overline{r},\overline{s}} + d_{-,i}\, {}_{4R}\widetilde{A}^\beta_{p,q,r,s,\overline{p},\overline{q},\overline{r},\overline{s}} \\
&\qquad + e_{+,j}\, {}_{4L}\widetilde{A}^{\beta_1}_{p,q,r,s,\overline{p},\overline{q},\overline{r},\overline{s}} + e_{-,j}\, {}_{4R}\widetilde{A}^{\beta_1}_{p,q,r,s,\overline{p},\overline{q},\overline{r},\overline{s}}\Big)\Big] u^{n+1}_{i,j} \\
=&\Big[1 + \frac{\tau}{2}\Big(d_{+,i}\, {}_{4L}\widetilde{A}^\beta_{p,q,r,s,\overline{p},\overline{q},\overline{r},\overline{s}} + d_{-,i}\, {}_{4R}\widetilde{A}^\beta_{p,q,r,s,\overline{p},\overline{q},\overline{r},\overline{s}} \\
&\qquad + e_{+,j}\, {}_{4L}\widetilde{A}^{\beta_1}_{p,q,r,s,\overline{p},\overline{q},\overline{r},\overline{s}} + e_{-,j}\, {}_{4R}\widetilde{A}^{\beta_1}_{p,q,r,s,\overline{p},\overline{q},\overline{r},\overline{s}}\Big)\Big] u^{n}_{i,j} \\
&\qquad + \tau f(x_i,y_j,t_{n+1/2}) + \varepsilon^n_{i,j}
\end{aligned} \tag{4.281}$$

with

$$|\varepsilon^n_{i,j}| \le \widetilde{c}\tau\left(\tau^2 + (h_x)^4 + (h_y)^4\right). \tag{4.282}$$

Adding a perturbation

$$-\frac{\tau^2}{4}\delta_{\beta,x}\delta_{\beta_1,y}(u^{n+1}_{i,j} - u^n_{i,j}) = \mathcal{O}(\tau^3)$$

to the right hand side of (4.281) and defining

$$\begin{aligned}
\delta_{\beta,x} &:= d_{+,i}\, {}_{4L}\widetilde{A}^\beta_{p,q,r,s,\overline{p},\overline{q},\overline{r},\overline{s}} + d_{-,i}\, {}_{4R}\widetilde{A}^\beta_{p,q,r,s,\overline{p},\overline{q},\overline{r},\overline{s}}, \\
\delta_{\beta_1,y} &:= e_{+,j}\, {}_{4L}\widetilde{A}^{\beta_1}_{p,q,r,s,\overline{p},\overline{q},\overline{r},\overline{s}} + e_{-,j}\, {}_{4R}\widetilde{A}^{\beta_1}_{p,q,r,s,\overline{p},\overline{q},\overline{r},\overline{s}},
\end{aligned}$$

we obtain the finite difference approximation for problem (4.249), i.e.,

$$\begin{aligned}&\left(1-\frac{\tau}{2}\delta_{\beta,x}\right)\left(1-\frac{\tau}{2}\delta_{\beta_1,y}\right)U_{i,j}^{n+1}\\&=\left(1+\frac{\tau}{2}\delta_{\beta,x}\right)\left(1+\frac{\tau}{2}\delta_{\beta_1,y}\right)U_{i,j}^{n}+\tau f_{i,j}^{n+1/2},\end{aligned}\tag{4.283}$$

where $U_{i,j}^n$ is the numerical approximation of $u_{i,j}^n$, and the local truncation error is $\mathcal{O}(\tau(\tau^2+(h_x)^4+(h_y)^4)$. Similar to Sec. 4.3, the system (4.283) can be solved by

(1) PR-ADI scheme [Peaceman and Rachford (1955)]:

$$\left(1-\frac{\tau}{2}\delta_{\beta,x}\right)U_{i,j}^{*}=\left(1+\frac{\tau}{2}\delta_{\beta_1,y}\right)U_{i,j}^{n}+\frac{\tau}{2}f_{i,j}^{n+1/2},\tag{4.284}$$

$$\left(1-\frac{\tau}{2}\delta_{\beta_1,y}\right)U_{i,j}^{n+1}=\left(1+\frac{\tau}{2}\delta_{\beta,x}\right)U_{i,j}^{*}+\frac{\tau}{2}f_{i,j}^{n+1/2};\tag{4.285}$$

(2) D-ADI scheme [Douglas (1955)]:

$$\left(1-\frac{\tau}{2}\delta_{\beta,x}\right)U_{i,j}^{*}=\left(1+\frac{\tau}{2}\delta_{\beta,x}+\tau\delta_{\beta_1,y}\right)U_{i,j}^{n}+\tau f_{i,j}^{n+1/2},\tag{4.286}$$

$$\left(1-\frac{\tau}{2}\delta_{\beta_1,y}\right)U_{i,j}^{n+1}=u_{i,j}^{*}-\frac{\tau}{2}\delta_{\beta_1,y}U_{i,j}^{n}.\tag{4.287}$$

For simplification, in the following of this section, we assume that $N_x = N_y$. Take

$$U^n=\left(U_{1,1}^n,U_{2,1}^n,\dots,U_{N_x-1,1}^n,\dots,U_{1,N_y-1}^n,U_{2,N_y-1}^n,\dots,U_{N_x-1,N_y-1}^n\right)^{\mathrm{T}},$$

$$F^{n+1/2}=\left(f_{1,1}^{n+1/2},\dots,f_{N_x-1,1}^{n+1/2},\dots,f_{1,N_y-1}^{n+1/2},f_{2,N_y-1}^{n+1/2},\dots,f_{N_x-1,N_y-1}^{n+1/2}\right)^{\mathrm{T}},$$

and denote

$$\begin{aligned}\mathcal{A}_x&=\frac{\tau}{2(h_x)^\beta}\left[(I\otimes D_+)(I\otimes A_\beta)+(I\otimes D_-)(I\otimes A_\beta^{\mathrm{T}})\right]\\&=\frac{\tau}{2(h_x)^\beta}I\otimes\left(D_+A_\beta+D_-A_\beta^{\mathrm{T}}\right),\\\mathcal{A}_y&=\frac{\tau}{2(h_y)^{\beta_1}}\left[(E_+\otimes I)(A_{\beta_1}\otimes I)+(E_-\otimes I)(A_{\beta_1}^{\mathrm{T}}\otimes I)\right]\\&=\frac{\tau}{2(h_y)^{\beta_1}}\left(E_+A_{\beta_1}+E_-A_{\beta_1}^{\mathrm{T}}\right)\otimes I,\end{aligned}\tag{4.288}$$

where I denotes the unit matrix and the symbol $\otimes$ the Kronecker product [Laub (2005)], and $A_\beta = A^\beta_{p,q,r,s,\overline{p},\overline{q},\overline{r},\overline{s}}$, $A_{\beta_1} = A^{\beta_1}_{p,q,r,s,\overline{p},\overline{q},\overline{r},\overline{s}}$ are defined by (4.266). The matrices D_+ and D_- are defined by (4.280), and

$$E_+=\begin{bmatrix}e_{+,1}&&&\\&e_{+,2}&&\\&&\ddots&\\&&&e_{+,N_y-1}\end{bmatrix},\quad E_-=\begin{bmatrix}e_{-,1}&&&\\&e_{-,2}&&\\&&\ddots&\\&&&e_{-,N_y-1}\end{bmatrix}.\tag{4.289}$$

Therefore, the finite difference scheme (4.283) has the following form

$$(I - \mathcal{A}_x)(I - \mathcal{A}_y)U^{n+1} = (I + \mathcal{A}_x)(I + \mathcal{A}_y)U^n + \tau F^{n+1/2}. \qquad (4.290)$$

4.5.3 *Stability and convergence analyses*

In this subsection, we theoretically prove that the difference scheme is unconditionally stable and fourth order of convergence in space directions and second order of convergence in time direction. In the following, the matrices D_+, D_- and E_+, E_- are defined by (4.280) and (4.289), respectively.

Theorem 4.24. *Let* $A_\beta = A^\beta_{p,q,r,s,\overline{p},\overline{q},\overline{r},\overline{s}}$ *be defined by (4.266) and* $D_- = \kappa_\beta D_+$, *where* κ_β *is any given nonnegative constant. Then* $\mathrm{Re}\left(\lambda\left(D_+(A_\beta + \kappa_\beta A_\beta^{\mathrm{T}})\right)\right) < 0$.

Proof. Since $D_+^{-\frac{1}{2}}\left[D_+(A_\beta + \kappa_\beta A_\beta^{\mathrm{T}})\right] D_+^{\frac{1}{2}} = D_+^{\frac{1}{2}}(A_\beta + \kappa_\beta A_\beta^{\mathrm{T}})D_+^{\frac{1}{2}}$, it means that $D_+(A_\beta + \kappa_\beta A_\beta^{\mathrm{T}})$ and $D_+^{\frac{1}{2}}(A_\beta + \kappa_\beta A_\beta^{\mathrm{T}})D_+^{\frac{1}{2}}$ are similar. From Theorem 4.23, we know A_β and A_β^{T} are negative definite, and thanks to Definition 4.1, it implies that

$$\left(D_+^{\frac{1}{2}}(A_\beta + \kappa_\beta A_\beta^{\mathrm{T}})D_+^{\frac{1}{2}}x, x\right) = \left((A_\beta + \kappa_\beta A_\beta^{\mathrm{T}})D_+^{\frac{1}{2}}x, D_+^{\frac{1}{2}}x\right) < 0$$

for all $x \in \mathbb{R}^n$ and $x \neq 0$, i.e., the matrix $\widetilde{A} := D_+^{\frac{1}{2}}(A_\beta + \kappa_\beta A_\beta^{\mathrm{T}})D_+^{\frac{1}{2}}$ is negative definite. From Lemma 4.6, $\widetilde{H} = \frac{\widetilde{A}+\widetilde{A}^{\mathrm{T}}}{2}$ is negative definite and $\lambda_{\max}(\widetilde{H}) < 0$; and according to Lemma 4.7, we obtain $\mathrm{Re}(\lambda(\widetilde{A})) \leq \lambda_{\max}(\widetilde{H}) < 0$. Therefore, $\mathrm{Re}\left(\lambda\left(D_+(A_\beta + \kappa_\beta A_\beta^{\mathrm{T}})\right)\right) = \mathrm{Re}(\lambda(\widetilde{A})) < 0$. □

Theorem 4.25. *Let* $\mathcal{A}_x$ *and* $\mathcal{A}_y$ *be defined by (4.288) and* $D_- = \kappa_\beta D_+$, $E_- = \kappa_{\beta_1} E_+$, *where* κ_β *and* κ_{β_1} *are any given nonnegative constant. Then* $\mathrm{Re}\left(\lambda(\mathcal{A}_x)\right) < 0$ *and* $\mathrm{Re}\left(\lambda(\mathcal{A}_y)\right) < 0$.

Proof. From (4.288), there exists

$$\mathcal{A}_x = \frac{\tau}{2(h_x)^\beta} I \otimes \left(D_+(A_\beta + \kappa_\beta A_\beta^{\mathrm{T}})\right),$$

$$\mathcal{A}_y = \frac{\tau}{2(h_y)^{\beta_1}} \left(E_+(A_{\beta_1} + \kappa_{\beta_1} A_{\beta_1}^{\mathrm{T}})\right) \otimes I.$$

By Theorem 4.24, we get $\mathrm{Re}\left(\lambda\left(D_+(A_\beta + \kappa_\beta A_\beta^{\mathrm{T}})\right)\right) < 0$ and $\mathrm{Re}\left(\lambda\left(E_+(A_{\beta_1} + \kappa_{\beta_1} A_{\beta_1}^{\mathrm{T}})\right)\right) < 0$. Then, according to Lemma 4.9, it implies that $\mathrm{Re}\left(\lambda(\mathcal{A}_x)\right) < 0$ and $\mathrm{Re}\left(\lambda(\mathcal{A}_y)\right) < 0$. □

4.5.3.1 *Stability and convergence for 1D*

Theorem 4.26. *Let $D_- = \kappa_\beta D_+$. Then the difference scheme (4.279) with $\beta \in (1,2)$ is unconditionally stable.*

Proof. The perturbation equation of (4.279) is

$$(I - A)\delta U^{n+1} = (I + A)\delta U^n,$$

i.e., $\delta U^{n+1} = (I - A)^{-1}(I + A)\delta U^n$, with

$$A = \frac{\tau}{2h^\beta} D_+(A_\beta + \kappa_\beta A_\beta^{\mathrm{T}}). \tag{4.291}$$

Denoting λ as an eigenvalue of the matrix A, then from Theorem 4.24, we get $\mathrm{Re}(\lambda(A)) < 0$. Thus, the spectral radius of the matrix $(I-A)^{-1}(I+A)$ is less than one, which means that the scheme (4.279) is unconditionally stable. □

Theorem 4.27. *Let u_i^n be the exact solution of (4.275) with constant coefficients and $\beta \in (1,2)$, and U_i^n the solution of the finite difference scheme (4.279), and $D_- = \kappa_\beta D_+$. Then there exists a positive constant c such that*

$$\|u^n - U^n\| \le c(\tau^2 + h^4).$$

Proof. Denoting $e_i^n = u_i^n - U_i^n$, then

$$(I - A)e^{n+1} = (I + A)e^n + \mathcal{E}^n,$$

where A is defined by (4.291), and $\mathcal{E}^n = \left(\varepsilon_1^n, \varepsilon_2^n, \ldots, \varepsilon_{N_x-1}^n\right)^{\mathrm{T}}$ with $|\varepsilon_i^n| \le \widetilde{c}\tau(\tau^2 + h^4)$. The above equation can be rewritten as

$$e^{n+1} = (I - A)^{-1}(I + A)e^n + (I - A)^{-1}\mathcal{E}^n.$$

Similar to the proof of Lemma 4.14 in Sec. 4.3, we have that $\left\|(I - A)^{-1}(I + A)\right\|$ and $\left\|(I - A)^{-1}\right\|$ are less than one Then, using $e^0 = \mathbf{0}$, we obtain

$$\begin{aligned}\|e^n\| &\le \left\|(I - A)^{-1}(I + A)\right\| \cdot \left\|e^{n-1}\right\| + \left\|(I - A)^{-1}\right\| \cdot |\mathcal{E}^{n-1}| \\ &\le \left\|e^{n-1}\right\| + |\mathcal{E}^{n-1}| \le \sum_{k=0}^{n-1} |\mathcal{E}^k| \le c(\tau^2 + h^4).\end{aligned}$$

Thus, we complete the proof. □

4.5.3.2 *Stability and convergence for 2D*

Theorem 4.28. *Let $D_- = \kappa_\beta D_+$ and $E_- = \kappa_{\beta_1} E_+$. Then the difference scheme (4.290) with $1 < \beta, \beta_1 < 2$ is unconditionally stable.*

Proof. The perturbation equation of (4.290) is

$$(I - \mathcal{A}_x)(I - \mathcal{A}_y)\delta U^{n+1} = (I + \mathcal{A}_x)(I + \mathcal{A}_y)\delta U^n, \tag{4.292}$$

where $\mathcal{A}_x$ and $\mathcal{A}_y$ are given in (4.288). According to Lemma 4.10 and (4.288), it is easy to check that $\mathcal{A}_x$ and $\mathcal{A}_y$ commute, i.e.,

$$\mathcal{A}_x\mathcal{A}_y = \mathcal{A}_y\mathcal{A}_x = \frac{\tau^2}{4(h_x)^\beta(h_y)^{\beta_1}}\left(E_+A_{\beta_1} + E_-A_{\beta_1}^{\mathrm{T}}\right) \otimes \left(D_+A_\beta + D_-A_\beta^{\mathrm{T}}\right). \tag{4.293}$$

Therefore,

$$\delta U^{n+1} = (I - \mathcal{A}_x)^{-1}(I + \mathcal{A}_x)(I - \mathcal{A}_y)^{-1}(I + \mathcal{A}_y)\delta U^n. \tag{4.294}$$

From Theorem 4.25, we have $\mathrm{Re}\,(\lambda(\mathcal{A}_x)) < 0$ and $\mathrm{Re}\,(\lambda(\mathcal{A}_y)) < 0$. Thus, the spectral radius of the matrix $(I-\mathcal{A}_x)^{-1}(I+\mathcal{A}_x)$ and $(I-\mathcal{A}_y)^{-1}(I+\mathcal{A}_y)$ are less than one. Then the difference scheme (4.290) is unconditionally stable. □

Theorem 4.29. *Let $u_{i,j}^n$ be the exact solution of (4.249) with constant coefficients and $1< \beta, \beta_1 < 2$, $U_{i,j}^n$ the solution of the finite difference scheme (4.290), and $D_-= \kappa_\beta D_+$ and $E_- = \kappa_{\beta_1} E_+$. Then there exists a positive constant c such that*

$$\|u^n - U^n\| \le c(\tau^2 + (h_x)^4 + (h_y)^4).$$

Proof. Taking $e_{i,j}^n = u_{i,j}^n - U_{i,j}^n$. Then

$$(I - \mathcal{A}_x)(I - \mathcal{A}_y)e^{n+1} = (I + \mathcal{A}_x)(I + \mathcal{A}_y)e^n + \mathcal{E}^n, \tag{4.295}$$

where $\mathcal{A}_x$ and $\mathcal{A}_y$ are given in (4.288), and

$$e^n = \left(e_{1,1}^n, e_{2,1}^n, \ldots, e_{N_x-1,1}^n, \ldots, e_{1,N_y-1}^n, e_{2,N_y-1}^n, \ldots, e_{N_x-1,N_y-1}^n\right)^{\mathrm{T}},$$

$$\mathcal{E}^n = \left(\hat{\varepsilon}_{1,1}^n, \hat{\varepsilon}_{2,1}^n, \ldots, \hat{\varepsilon}_{N_x-1,1}^n, \ldots, \hat{\varepsilon}_{1,N_y-1}^n, \hat{\varepsilon}_{2,N_y-1}^n, \ldots, \hat{\varepsilon}_{N_x-1,N_y-1}^n\right)^{\mathrm{T}},$$

with $|\hat{\varepsilon}_{i,j}^{n+1}| \le \widetilde{c}\tau(\tau^2 + (h_x)^4 + (h_y)^4)$.

From (4.293), $\mathcal{A}_x$ and $\mathcal{A}_y$ commute, then (4.295) can be rewritten as

$$\begin{aligned} e^{n+1} = {} & (I - \mathcal{A}_x)^{-1}(I + \mathcal{A}_x)(I - \mathcal{A}_y)^{-1}(I + \mathcal{A}_y)e^n \\ & + (I - \mathcal{A}_x)^{-1}(I - \mathcal{A}_y)^{-1}\mathcal{E}^n. \end{aligned}$$

Again, similar to the proof of Lemma 4.14 in Sec. 4.3, we know that $\|(I-A_\nu)^{-1}(I+A_\nu)\|$ and $\|(I-A_\nu)^{-1}\|$ are less than one, where $\nu = x,\ y$. Then there exists

$$\|e^n\| \le \sum_{k=0}^{n-1} |\mathcal{E}^k| \le c(\tau^2 + (h_x)^4 + (h_y)^4).$$

Thus, we complete the proof. □

4.5.4 *Numerical results*

Example 4.8. Consider the one-dimensional fractional diffusion equation (4.275) in $(0,2)\times(0,1]$, with the variable coefficients $d_+(x) = x^\beta$, $d_-(x) = 2x^\beta$, the forcing function

$$\begin{aligned} f(x,t) &= \cos(t+1)x^4(2-x)^4 - x^\beta \sin(t+1)\Big[\frac{\Gamma(9)}{\Gamma(9-\beta)}(x^{8-\beta} + 2(2-x)^{8-\beta}) \\ &- 8\frac{\Gamma(8)}{\Gamma(8-\beta)}(x^{7-\beta} + 2(2-x)^{7-\beta}) + 24\frac{\Gamma(7)}{\Gamma(7-\beta)}(x^{6-\beta} + 2(2-x)^{6-\beta}) \\ &- 32\frac{\Gamma(6)}{\Gamma(6-\beta)}(x^{5-\beta} + 2(2-x)^{5-\beta}) + 16\frac{\Gamma(5)}{\Gamma(5-\beta)}(x^{4-\beta} + 2(2-x)^{4-\beta})\Big], \end{aligned}$$

and the initial condition $u(x,0) = \sin(1)x^4(2-x)^4, x \in (0,2)$.

Then for $(x,t) \in [0,2]\times[0,1]$, the exact solution of the equation is $u(x,t) = \sin(t+1)x^4(2-x)^4$. The numerical results are listed in Table 4.14, which confirms the desired convergence with error $\mathcal{O}(\tau^2 + h^4)$.

Table 4.14: The discrete maximum errors and convergence orders for the scheme (4.279) of the one-dimensional fractional diffusion equation (4.275) at $t = 1$ and $\tau = h^2$.

$(p,q,r,s,\overline{p},\overline{q},\overline{r},\overline{s})$	h	$\beta = 1.1$	rate	$\beta = 1.9$	rate
	1/10	4.7842e-03	-	5.8264e-03	-
(1,2,1,0,1,2,1,-2)	1/20	2.5436e-04	4.23	5.9999e-04	3.28
	1/40	1.9662e-05	3.69	4.6242e-05	3.70
	1/60	4.1748e-06	3.82	9.7725e-06	3.83
	1/10	8.5475e-03	-	5.5003e-03	-
(1,2,1,-3,1,2,1,-2)	1/20	4.9722e-04	4.10	5.7476e-04	3.26
	1/40	3.9559e-05	3.65	4.4490e-05	3.69
	1/60	8.6604e-06	3.74	9.4148e-06	3.83

Example 4.9. Consider the two dimensional fractional diffusion equation (4.249) in $(0,2)^2 \times (0,1]$ with the variable coefficients $d_+(x,y) = x^\beta$,

$d_-(x,y) = 2x^\beta$, and $e_+(x,y) = y^{\beta_1}$, $e_-(x,y) = 2y^{\beta_1}$, and the initial condition $u(x,y,0) = \sin(1)x^4(2-x)^4y^4(2-y)^4$, $(x,y) \in (0,2)^2$. The forcing function can be derived from the exact solution $u(x,y,t) = \sin(t+1)x^4(2-x)^4y^4(2-y)^4, (x,y,t) \in [0,2]^2 \times [0,1]$.

Table 4.15 displays the maximum errors of the scheme (4.290), and confirms the desired convergence with the error $\mathcal{O}(\tau^2 + (h_x)^4 + (h_y)^4)$.

Table 4.15: The discrete maximum errors and convergence orders for the scheme (4.290) of the two dimensional fractional diffusion equation (4.249) at $t = 1$ and $\tau = (h_x)^2 = (h_y)^2$.

$(p,q,r,s,\overline{p},\overline{q},\overline{r},\overline{s})$	h_x	$\beta = \beta_1 = 1.1$	rate	$\beta = 1.8, \beta_1 = 1.9$	rate
	1/10	8.6154e-03	-	6.6647e-03	-
(1,2,1,0,1,2,1,-2)	1/20	5.4115e-04	3.99	4.5632e-04	3.87
	1/30	1.2626e-04	3.59	9.0006e-05	4.00
	1/40	4.3328e-05	3.72	2.8287e-05	4.02

4.6 Second order difference scheme for the fractional substantial differential equation with truncated Lévy flights

The CTRW with the power law waiting time and/or jump length can microscopically describe anomalous diffusion; its macroscopic limit in unbounded domain leads to the time and/or space fractional diffusion equation [Barkai *et al.* (2000)]. In fact, based on the fractional Fourier law and conservation law, the fractional diffusion equation can also be easily derived. The space fractional diffusion equation defined in unbounded domain characterizes the probability distribution of the positions of the particles of the Lévy flights, which has the divergent second order moment. The more practical transport problems often take place in bounded domain and the involved observables have finite second moments; exponentially tempering the Lévy measure of the Lévy flights, i.e., making the Lévy density decay as $e^{-\lambda|x|}|x|^{-1-\beta}$ with $\lambda > 0$ [del Castillo-Negrete (2009); Hanert and Piret (1995)] leads to the space tempered fractional diffusion equation [Baeumera and Meerschaert (2010); Cartea and del Castillo-Negrete (2007); del Castillo-Negrete (2009); Hanert and Piret (2014); Sabzikar *et al.* (2015)];

in other words, the space tempered fractional diffusion equation describes the probability density function (PDF) of the truncated Lévy flights.

On the other hand, the functionals of the trajectories of the particles often attract practical interest, e.g., the time spent by a particle in a given domain, the macroscopic measured signal in the nuclear magnetic resonance experiments; and the functionals are usually defined as $A = \int_0^t U[x(\tau)]d\tau$. When $x(t)$ is the trajectory of a Brownian particle, Kac derived a Schrödinger-like equation for the distribution of the functionals of diffusive motion; with the deep understanding of the anomalous diffusion, the distribution of the functionals of the paths of anomalous diffusion naturally attracts the interests of scientists, and the corresponding fractional Feynman-Kac equation is derived [Carmi *et al.* (2010); Carmi and Barkai (2011); Turgeman *et al.* (2009)], which involves the fractional substantial derivative [Chen and Deng (2015); Friedrich *et al.* (2006); Zhang and Deng (2017)].

Hence, in this section, we focus on providing effective and high accurate numerical algorithms for the model

$$\begin{cases} {}^s_cD_t^\alpha G(x,\rho,t) = K\left({}_aD_x^{\beta,\lambda} + {}_xD_b^{\beta,\lambda}\right)G(x,\rho,t), & (x,t)\in\Omega\times(0,T], \\ G(x,\rho,0) = G_0(x,\rho), \quad x\in\Omega, \\ G(x,\rho,t) = 0, \quad (x,t)\in\mathbb{R}\backslash\Omega\times[0,T]. \end{cases} \tag{4.296}$$

Here, $\Omega = (a,b)$, $K = -\frac{1}{2\cos(\beta\pi/2)}$, $1<\beta<2$, ρ is the Fourier pair of A, and $G(x,A,t)$ is the joint PDF of finding the particle at time t with the functional value A and the initial position of the particle at x. The fractional substantial derivative ${}^s_cD_t^\alpha G(x,\rho,t)$ $(0<\alpha<1)$ is defined by

$${}^s_cD_t^\alpha G(x,\rho,t) = e^{J\rho U(x)t}\,{}^C_0D_t^\alpha\left(e^{-J\rho U(x)t}G(x,\rho,t)\right), \tag{4.297}$$

where $J=\sqrt{-1}$ and $U(x)$ is a prescribed function; and the tempered Riesz R-L fractional derivative is

$$\begin{aligned} &\left({}_{-\infty}D_x^{\beta,\lambda} + {}_xD_\infty^{\beta,\lambda}\right)G(x,\rho,t) \\ &= \left({}_{-\infty}\mathbb{D}_x^{\beta,\lambda}G(x,\rho,t) - \lambda^\beta G(x,\rho,t)\right) \\ &\quad + \left({}_x\mathbb{D}_\infty^{\beta,\lambda}G(x,\rho,t) - \lambda^\beta G(x,\rho,t)\right), \end{aligned} \tag{4.298}$$

where $\lambda\geq 0$, and ${}_{-\infty}D_x^{\beta,\lambda}$, ${}_xD_\infty^{\beta,\lambda}$, ${}_{-\infty}\mathbb{D}_x^{\beta,\lambda}$, and ${}_x\mathbb{D}_\infty^{\beta,\lambda}$ are given in (4.59)–(4.62), respectively.

4.6.1 Second order approximations for the tempered fractional derivatives

For convenience, we use the notations

$$\begin{aligned} {}_{-\infty}\nabla_x^{\beta,\lambda}G(x) &= \left({}_{-\infty}\mathbb{D}_x^{\beta,\lambda}G(x,\rho,t) - \lambda^\beta G(x,\rho,t)\right), \\ {}_x\nabla_\infty^{\beta,\lambda}G(x) &= \left({}_x\mathbb{D}_\infty^{\beta,\lambda}G(x,\rho,t) - \lambda^\beta G(x,\rho,t)\right). \end{aligned} \tag{4.299}$$

Let

$$\widetilde{A}_p^{\beta,\lambda}G(x) = \frac{1}{h^\beta}\sum_{j=0}^{\infty} g_j^\beta e^{-(j-p)\lambda h}G(x-(j-p)h) - e^{ph\lambda}\frac{(1-e^{-h\lambda})^\beta}{h^\beta}G(x). \tag{4.300}$$

From Theorem 4.3, we know that if $G(x) \in \mathscr{S}^{\beta+1}(\mathbb{R})$ and $G(x)$ has compact support, it holds that

$${}_{-\infty}\nabla_x^{\beta,\lambda}G(x) = \widetilde{A}_p^{\beta,\lambda}G(x) + \mathcal{O}(h),$$

where

$$\left(1-e^{-\lambda h}\zeta\right)^\beta = \sum_{j=0}^{\infty} g_j^\beta e^{-j\lambda h}\zeta^j \tag{4.301}$$

with g_j^β satisfying (see Proposition 4.1)

$$g_0^\beta = 1, \quad g_j^\beta = \left(1 - \frac{\beta+1}{j}\right) g_{j-1}^\beta, \quad j \geq 1. \tag{4.302}$$

Lemma 4.18. *Let $G(x) \in \mathscr{S}^{\beta+2}(\mathbb{R})$ and $G(x)$ have compact support. Define*

$$\widetilde{A}_{r_1,r_2,r_3}^{\beta,\lambda}G(x) = \left(r_1\widetilde{A}_1^{\beta,\lambda} + r_2\widetilde{A}_0^{\beta,\lambda} + r_3\tilde{A}_{-1}^{\beta,\lambda}\right)G(x), \tag{4.303}$$

where $\widetilde{A}_1^{\beta,\lambda}$, $\widetilde{A}_0^{\beta,\lambda}$ and $\widetilde{A}_{-1}^{\beta,\lambda}$ are defined by (4.300). Then

$${}_{-\infty}\nabla_x^{\beta,\lambda}G(x) = \widetilde{A}_{r_1,r_2,r_3}^{\beta,\lambda}G(x) + \mathcal{O}(h^2),$$

where r_1, r_2, r_3 satisfy

$$r_1 + r_2 + r_3 = 1 \text{ and } r_1 - r_3 = \frac{\beta}{2}. \tag{4.304}$$

Proof. Let $W_{\beta,p}(z) = e^{pz}\left(\frac{1-e^{-z}}{z}\right)^\beta$. Note that

$$\begin{aligned} &\mathscr{F}\left[A_{r_1,r_2,r_3}^{\beta,\lambda}G\right](\omega) \\ &= -\lambda^\beta\Big[r_1W_{\beta,1}(\lambda h) + r_2W_{\beta,0}(\lambda h) + r_3W_{\beta,-1}(\lambda h)\Big]\widehat{G}(\omega) \\ &\quad + (\lambda+i\omega)^\beta\Big[r_1W_{\beta,1}\left((\lambda+i\omega)h\right) + r_2W_{\beta,0}\left((\lambda+i\omega)h\right) \\ &\quad + r_3W_{\beta,-1}\left((\lambda+i\omega)h\right)\Big]\widehat{G}(\omega), \end{aligned}$$

and

$$\left| W_{\beta,p}(z) - \left(1 + \left(p - \frac{\beta}{2}\right) z\right) \right| \leq C|z|^2 \text{ for bounded } \mathrm{Re}(z).$$

Then the proof is similar to Theorem 4.3. □

Assume that $G(x) = 0$ for $x \in \mathbb{R} \backslash \Omega$. Take the mesh points $x_i = a + ih, i = 0, 1, \cdots, N$ with $h = \frac{b-a}{N}$. Then for $p \in \mathbb{Z}$, we have

$$\widetilde{A}_p^{\beta,\lambda} G(x_i) = \frac{1}{h^\beta} \sum_{j=0}^{i+p} e^{-(j-p)\lambda h} g_j^\beta G(x_{i-j+p}) - e^{ph\lambda} \frac{(1-e^{-h\lambda})^\beta}{h^\beta} G(x_i), \tag{4.305}$$

and

$$\begin{aligned} {}_a\nabla_x^{\beta,\lambda} G(x) \mid_{x=x_i} &= \widetilde{A}_{r_1,r_2,r_3}^{\beta,\lambda} G(x_i) + \mathcal{O}(h^2) \\ &= \frac{1}{h^\beta} \sum_{j=0}^{i+1} e^{-(j-1)\lambda h} \omega_j^\beta G(x_{i-j+1}) + \mathcal{O}(h^2). \end{aligned} \tag{4.306}$$

Here

$$\widetilde{A}_{r_1,r_2,r_3}^{\beta,\lambda} G(x_i) = \left(r_1 \widetilde{A}_1^{\beta,\lambda} + r_2 \widetilde{A}_0^{\beta,\lambda} + r_3 \widetilde{A}_{-1}^{\beta,\lambda}\right) G(x_i), \tag{4.307}$$

and

$$\begin{cases} \omega_0^\beta = r_1 g_0^\beta, \\ \omega_1^\beta = r_1 g_1^\beta + r_2 g_0^\beta - \left(r_1 e^{\lambda h} + r_2 + r_3 e^{-\lambda h}\right)\left(1 - e^{-\lambda h}\right)^\beta, \\ \omega_j^\beta = r_1 g_j^\beta + r_2 g_{j-1}^\beta + r_3 g_{j-2}^\beta, \quad 2 \leq j \leq N-1. \end{cases} \tag{4.308}$$

Let $\widetilde{G} = (G(x_1), G(x_2), \cdots, G(x_{N-1}))^{\mathrm{T}}$. Then we can rewrite (4.307) as the matrix form

$$\widetilde{A}_{r_1,r_2,r_3}^{\beta,\lambda} \widetilde{G} = \frac{1}{h^\beta} A^{\beta,\lambda} \widetilde{G} \tag{4.309}$$

with

$$A^{\beta,\lambda} = \begin{bmatrix} e^{0\lambda h}\omega_1^\beta & e^{\lambda h}\omega_0^\beta & & & & \\ e^{-\lambda h}\omega_2^\beta & e^{0\lambda h}\omega_1^\beta & e^{\lambda h}\omega_0^\beta & & & \\ e^{-2\lambda h}\omega_3^\beta & e^{-\lambda h}\omega_2^\beta & e^{0\lambda h}\omega_1^\beta & \ddots & & \\ \vdots & \cdots & \ddots & \ddots & \ddots & \\ e^{-(N-3)\lambda h}\omega_{N-2}^\beta & \cdots & \cdots & \ddots & e^{0\lambda h}\omega_1^\beta & e^{\lambda h}\omega_0^\beta \\ e^{-(N-2)\lambda h}\omega_{N-1}^\beta & e^{-(N-3)\lambda h}\omega_{N-2}^\beta & \cdots & \cdots & e^{-\lambda h}\omega_2^\beta & e^{0\lambda h}\omega_1^\beta \end{bmatrix}. \tag{4.310}$$

Similarly, the second order approximation for the right tempered R-L fractional derivative is given as

$$\begin{aligned} {}_x\nabla_b^{\beta,\lambda} G(x_i) &= \widetilde{B}_{r_1,r_2,r_3}^{\beta,\lambda} G(x_i) + \mathcal{O}(h^2) \\ &= \frac{1}{h^\beta} \sum_{j=0}^{N-i+1} e^{-(j-1)\lambda h} \omega_j^\beta G(x_{i+j-1}) + \mathcal{O}(h^2), \end{aligned} \tag{4.311}$$

where ω_j^β is defined by (4.308), and the matrix form is

$$\widetilde{B}_{r_1,r_2,r_3}^{\beta,\lambda} U = \frac{1}{h^\beta} B^{\beta,\lambda} U \quad \text{with} \quad B^{\beta,\lambda} = (A^{\beta,\lambda})^{\mathrm{T}}. \tag{4.312}$$

The system (4.304) has infinitely many solutions. With the help of the knowledge of linear algebra, the solutions of (4.304) can be collected by the following three sets

$$\mathcal{S}_1^\beta(r_1, r_2, r_3) = \left\{ r_1 \text{ is given}, r_2 = \frac{2+\beta}{2} - 2r_1, \ r_3 = r_1 - \frac{\beta}{2} \right\}, \tag{4.313}$$

or

$$\mathcal{S}_2^\beta(r_1, r_2, r_3) = \left\{ r_1 = \frac{2+\beta}{4} - \frac{r_2}{2}, \ r_2 \text{ is given}, \ r_3 = \frac{2-\beta}{4} - \frac{r_2}{2} \right\}, \tag{4.314}$$

or

$$\mathcal{S}_3^\beta(r_1, r_2, r_3) = \left\{ r_1 = \frac{\beta}{2} + r_3, \ r_2 = \frac{2-\beta}{2} - 2r_3, \ r_3 \text{ is given} \right\}. \tag{4.315}$$

The parameter values presented in the sets $\mathcal{S}_j^\beta, j = 1, 2, 3$ produce infinite number of second order approximations for the tempered R-L fractional derivative. Particularly, if taking $\lambda = 0$ and $\alpha_j = 0$ in $\mathcal{S}_j^\beta$, $j = 1, 2, 3$, they recover the second order approximations presented in Sec. 4.3 for the R-L fractional derivative.

Denote $H = A^{\beta,\lambda} + (A^{\beta,\lambda})^{\mathrm{T}}$. Using (4.310), we obtain

$$H = \begin{bmatrix} 2\phi_1^\beta & \phi_0^\beta + \phi_2^\beta & \phi_3^\beta & \cdots & \phi_{N-2}^\beta & \phi_{N-1}^\beta \\ \phi_0^\beta + \phi_2^\beta & 2\phi_1^\beta & \phi_0^\beta + \phi_2^\beta & \phi_3^\beta & \cdots & \phi_{N-2}^\beta \\ \phi_3^\beta & \phi_0^\beta + \phi_2^\beta & 2\phi_1^\beta & \phi_0^\beta + \phi_2^\beta & \ddots & \vdots \\ \vdots & \ddots & \ddots & \ddots & \ddots & \phi_3^\beta \\ \phi_{N-2}^\beta & \ddots & \ddots & \ddots & 2\phi_1^\beta & \phi_0^\beta + \phi_2^\beta \\ \phi_{N-1}^\beta & \phi_{N-2}^\beta & \cdots & \cdots & \phi_0^\beta + \phi_2^\beta & 2\phi_1^\beta \end{bmatrix}, \tag{4.316}$$

where

$$\begin{aligned} &\phi_0^\beta = e^{\lambda h} \omega_0^\beta; \quad \phi_1^\beta = e^{0\lambda h} \omega_1^\beta; \\ &\phi_2^\beta = e^{-\lambda h} \omega_2^\beta; \quad \phi_j^\beta = e^{-(j-1)\lambda h} \omega_j^\beta, \quad j \geq 3, \end{aligned} \tag{4.317}$$

with ω_j^β $(j \geq 0)$ being given in (4.308).

Theorem 4.30. *Let $A^{\beta,\lambda}$ be given in (4.310) with $\lambda \geq 0$ and $1 < \beta < 2$. If r_1, r_2 and r_3 are chosen in set $\mathcal{S}_1^\beta(r_1,r_2,r_3)$ with $\max\left\{\frac{2(\beta^2+3\beta-4)}{\beta^2+3\beta+2}, \frac{\beta^2+3\beta}{\beta^2+3\beta+4}\right\} \leq r_1 \leq \frac{3(\beta^2+3\beta-2)}{2(\beta^2+3\beta+2)}$, or set $\mathcal{S}_2^\beta(r_1,r_2,r_3)$ with $\frac{(\beta-4)(\beta^2+3\beta+2)+24}{2(\beta^2+3\beta+2)} \leq r_2 \leq \min\left\{\frac{(\beta-2)(\beta^2+3\beta+4)+16}{2(\beta^2+3\beta+4)}, \frac{(\beta-6)(\beta^2+3\beta+2)+48}{2(\beta^2+3\beta+2)}\right\}$, or set $\mathcal{S}_3^\beta(r_1,r_2,r_3)$ with $\max\left\{\frac{(2-\beta)(\beta^2+\beta-8)}{\beta^2+3\beta+2}, \frac{(1-\beta)(\beta^2+2\beta)}{2(\beta^2+3\beta+4)}\right\} \leq r_3 \leq \frac{(2-\beta)(\beta^2+2\beta-3)}{2(\beta^2+3\beta+2)}$, then the elements, denoted by ϕ_j^β, of $H = A^{\beta,\lambda} + (A^{\beta,\lambda})^{\mathrm{T}}$ defined by (4.316) satisfy*

$$\phi_1^\beta < 0,\ \phi_0^\beta + \phi_2^\beta \geq 0,\ \phi_3^\beta > 0, \ \text{ and } \ \phi_{\mathrm{j}}^\beta > 0\ (\mathrm{j} \geq 4). \tag{4.318}$$

Proof. We only prove the conclusion for r_1, r_2 and r_3 selected in set $\mathcal{S}_3^\beta(r_1,r_2,r_3)$. The conclusions for r_1, r_2 and r_3 selected in sets $\mathcal{S}_1^\beta(r_1,r_2,r_3)$ and $\mathcal{S}_2^\beta(r_1,r_2,r_3)$ can be proved in a similar manner.

Case $\phi_1^\beta < 0$: From (4.302) and (4.304), we can check that $\phi_1^\beta < 0$ when $r_3 > -\frac{\beta-1}{2}$. And it can also be noted that in this case $r_1 > 0$.

Case $\phi_0^\beta + \phi_2^\beta \geq 0$: Using $r_1 \geq 0$, $\lambda \geq 0$, and $h > 0$, we obtain

$$\begin{aligned}
\phi_0^\beta + \phi_2^\beta &= e^{\lambda h}\omega_0^\beta + e^{-\lambda h}\omega_2^\beta = e^{\lambda h} r_1 g_0^\beta + e^{-\lambda h}\left(r_1 g_2^\beta + r_2 g_1^\beta + r_3 g_0^\beta\right) \\
&\geq e^{-\lambda h}\left(r_1 g_0^\beta + r_1 g_2^\beta + r_2 g_1^\beta + r_3 g_0^\beta\right) \\
&= e^{-\lambda h}\left[r_3 \frac{\beta^2+3\beta+4}{2} + \frac{\beta(\beta-1)(\beta+2)}{4}\right] \geq 0,
\end{aligned}$$

which leads to $r_3 \geq -\frac{\beta(\beta-1)(\beta+2)}{2(\beta^2+3\beta+4)}$.

Case $\phi_3^\beta > 0$: According to (4.302), (4.304), and (4.316), we get

$$\begin{aligned}
\phi_3^\beta &= e^{-2\lambda h}\omega_3^\beta = e^{-2\lambda h}\left(r_1 g_3^\beta + r_2 g_2^\beta + r_3 g_1^\beta\right) \\
&= e^{-2\lambda h}\left[r_3 \frac{-\beta(\beta+1)(\beta+2)}{6} + \frac{\beta(\beta-1)(2-\beta)(\beta+3)}{12}\right] \geq 0,
\end{aligned}$$

if and only if $r_3 \leq \frac{(\beta-1)(2-\beta)(\beta+3)}{2(\beta+1)(\beta+2)}$.

Case $\phi_j^\beta > 0$, $j \geq 4$: From (4.302), (4.304) and (4.317), there exists

$$\begin{aligned}
\phi_j^\beta &= e^{-(j-1)\lambda h}\omega_j^\beta = e^{-(j-1)\lambda h}\left(r_1 g_j^\beta + r_2 g_{j-1}^\beta + r_3 g_{j-2}^\beta\right) \\
&= e^{-(j-1)\lambda h}\left[r_3 \frac{(\beta+1)(\beta+2)}{j(j-1)} + \frac{(j-\beta-2)(2j-\beta^2-\beta)}{2j(j-1)}\right] g_{j-2}^\beta \geq 0,
\end{aligned}$$

which results in $r_3 \geq -\frac{(2-\beta)(8-\beta^2-\beta)}{2(\beta+1)(\beta+2)}$. Then, the proof is completed.

□

Corollary 4.2. *Denote the elements of H defined in (4.316) by $h_{i,j}$ and let r_1, r_2 and r_3 satisfy Theorem 4.30. Then there exist*

(1) $h_{i,i} = 2\phi_1^\beta < 0, \quad h_{i,j} > 0, \quad (j \neq i);$

(2) $\sum_{j=0}^{\infty} h_{i,j} = 0 \quad and \quad -h_{i,i} > \sum_{j=0, j\neq i}^{N} h_{i,j};$

(3) Matrix H is negative definite.

Proof. We only need to prove (2) and (3). From (4.316), (4.317), and (4.308), we have

$$\begin{aligned}
\frac{1}{2}\sum_{j=-\infty}^{\infty} h_{i,j} &= \sum_{j=0}^{\infty} e^{-(j-1)\lambda h}\omega_j^\beta \\
&= \left(r_1 e^{\lambda h} + r_2 + r_3 e^{-\lambda h}\right)\sum_{j=0}^{\infty} e^{-j\lambda h} g_j^\beta \\
&\quad - \left(r_1 e^{\lambda h} + r_2 + r_3 e^{-\lambda h}\right)\left(1 - e^{-\lambda h}\right)^\beta \\
&= 0,
\end{aligned}$$

where $\sum_{j=0}^{\infty} e^{-j\lambda h} g_j^\beta = \left(1 - e^{-\lambda h}\right)^\beta$ has been used. Combining with $h_{i,j} > 0\,(j \neq i)$, we arrive at (2).

By the Gerschgorin theorem [Isaacson and Keller (1996)], it is easy to see that all the eigenvalues of matrix H are negative. Then (3) follows from the fact that H is symmetric. □

4.6.2 *Derivation of the numerical schemes*

Take the mesh points $x_i = a + ih, i = 0, 1, \ldots, N$, and $t_n = n\tau, n = 0, 1, \ldots, M$, where $h = (b-a)/N$, $\tau = T/M$ are the uniform space stepsize and time steplength, respectively. Denote $G_{i,\rho}^n$ as the numerical approximation to $G(x_i, \rho, t_n)$.

For the time substantial fractional derivative, by (4.297) and Proposition B.15, we have

$$\begin{aligned}
&{}^s_c D_t^\alpha G(x, \rho, t) \\
&= e^{J\rho U(x)t}{}_0 D_t^\alpha \left[e^{-J\rho U(x)t}\left(G(x,\rho,t) - e^{J\rho U(x)t} G(x,\rho,0)\right)\right].
\end{aligned} \tag{4.319}$$

Regarding $e^{-J\rho U(x)t}\left(G(x,\rho,t) - e^{J\rho U(x)t}G(x,\rho,0)\right)$ as a whole and using

Lemma 4.2 (see also [Deng and Chen (2014)]) we have

$$
\begin{aligned}
&{}_c^sD_t^\alpha G(x,\rho,t)|_{(x=x_i,t=t_n)}\\
&= e^{J\rho U(x_i)t_n}{}_0D_t^\alpha\left[e^{-J\rho U(x_i)t}\left(G(x_i,\rho,t)-e^{J\rho U(x_i)t}G(x,\rho,0)\right)\right]\Bigg|_{t=t_n}\\
&= \frac{1}{\tau^\alpha}\sum_{k=0}^{n}d_{i,k}^{1,\alpha}\left(G(x_i,\rho,t_{n-k})-e^{J\rho U_i(n-k)\tau}G(x_i,\rho,0)\right)+\mathcal{O}(\tau), \quad (4.320)
\end{aligned}
$$

where

$$d_{i,k}^{1,\alpha}=e^{J\rho U_i k\tau}g_k^\alpha \text{ with } g_k^\alpha \text{ satisfing (4.302)}, \qquad U_i=U(x_i). \tag{4.321}$$

Combining (4.306), (4.311), (4.298), and (4.299), for $i=1,2,\cdots,N-1$, we obtain the approximation operator of the Riesz tempered fractional derivative

$$
\begin{aligned}
&\left({}_{-\infty}D_x^{\beta,\lambda}+{}_xD_\infty^{\beta,\lambda}\right)G(x,\rho,t)\Big|_{(x=x_i,t=t_n)}\\
&=-\kappa_\beta\left({}_a\nabla_x^{\beta,\lambda}+{}_x\nabla_b^{\beta,\lambda}\right)G(x,\rho,t)\Big|_{(x=x_i,t=t_n)}\\
&=-\tfrac{\kappa_\beta}{h^\beta}\textstyle\sum_{j=0}^{N}\omega_{i,j}^\beta G(x_j,\rho,t_n)+\mathcal{O}(h^2).
\end{aligned}
\tag{4.322}
$$

Here

$$
\omega_{i,j}^\beta=\begin{cases}
e^{-(i-j)\lambda h}\omega_{i-j+1}^\beta, & j<i-1,\\
e^{\lambda h}\omega_0^\beta+e^{-\lambda h}\omega_2^\beta, & j=i-1,\\
2\omega_1^\beta, & j=i,\\
e^{\lambda h}\omega_0^\beta+e^{-\lambda h}\omega_2^\beta, & j=i+1,\\
e^{-(j-i)\lambda h}\omega_{j-i+1}^\beta, & j>i+1,
\end{cases}
\tag{4.323}
$$

where ω_j^β is given in (4.308) with the parameters r_1, r_2 and r_3 satisfying Theorem 4.30. Then by Corollary 4.2, we have

$$
\begin{aligned}
&(1)\ \ \omega_{i,i}^\beta<0, \quad \omega_{i,j}^\beta>0\ (j\neq i);\\
&(2)\ \ \sum_{j=0}^{N}\omega_{i,j}^\beta<0 \ \text{ and } \ -\omega_{i,i}^\beta>\sum_{j=0,j\neq i}^{N}\omega_{i,j}^\beta.
\end{aligned}
\tag{4.324}
$$

From (4.320) and (4.322), we rewrite (4.296) as

$$
\begin{aligned}
&\sum_{k=0}^{n}d_{i,k}^{1,\alpha}G(x_i,\rho,t_{n-k})-\sum_{k=0}^{n}d_{i,k}^{1,\alpha}e^{J\rho U_i(n-k)\tau}G(x_i,\rho,0)\\
&=\kappa\sum_{j=0}^{N}\omega_{i,j}^\beta G(x_j,\rho,t_n)+\varepsilon_i^n,
\end{aligned}
\tag{4.325}
$$

where

$$\kappa = -\frac{K\tau^{\alpha}}{h^{\beta}} > 0, \quad \beta \in (1,2), \tag{4.326}$$

and

$$|\varepsilon_i^n| = |\tau^{\alpha} r_i^n| \le C_G \tau^{\alpha}(\tau + h^2), \tag{4.327}$$

with C_G being a constant independent of τ and h.

From (4.325), the resulting finite difference scheme of (4.296) is

$$G_{i,\rho}^n - \kappa \sum_{j=0}^{N} \omega_{i,j}^{\beta} G_{j,\rho}^n = \sum_{k=0}^{n-1} g_k^{\alpha} e^{J\rho U_i n\tau} G_{i,\rho}^0 - \sum_{k=1}^{n-1} g_k^{\alpha} e^{J\rho U_i k\tau} G_{i,\rho}^{n-k}, \quad n \ge 1. \tag{4.328}$$

Introducing the grid function

$$\widetilde{G}^n = \left(G_{1,\rho}^n, G_{2,\rho}^n, \dots, G_{N-1,\rho}^n\right)^T,$$

we can also rewrite (4.328) as the matrix form

$$(I - \kappa H)\,\widetilde{G}^n = \sum_{k=0}^{n-1} g_k^{\alpha} B^n \widetilde{G}^0 - \sum_{k=1}^{n-1} g_k^{\alpha} B^k \widetilde{G}^{n-k}, \tag{4.329}$$

where H is defined by (4.316), I is the identity matrix, and

$$B^k = \operatorname{diag}\left(e^{J\rho U_1 k\tau}, e^{J\rho U_2 k\tau}, \cdots, e^{J\rho U_{N-1} k\tau}\right). \tag{4.330}$$

4.6.3 *Stability and convergence analyses*

Denote the grid function space $\mathcal{V} = \{v_i \,|\, 0 \le i \le N, v_0 = v_N = 0\}$. For any $u, v \in \mathcal{V}$, we define the discrete L^2 inner product and norms

$$(u,v) = h\sum_{i=1}^{N-1} u_i \bar{v}_i, \quad \|v\| = \sqrt{(v,v)}; \quad \|v\|_{\infty} = \max_{1\le i\le N-1} |v_i|.$$

Lemma 4.19. *The coefficients g_k^{α} $(0 < \alpha < 1)$ in (4.321) satisfy*

$$g_0^{\alpha} = 1, \quad g_k^{\alpha} < 0\,(k \ge 1), \quad \sum_{k=0}^{n-1} g_k^{\alpha} > 0, \quad \sum_{k=0}^{\infty} g_k^{\alpha} = 0; \tag{4.331}$$

$$\frac{1}{(n+1)^{\alpha}\Gamma(1-\alpha)} \le \sum_{k=0}^{n} g_k^{\alpha} \le \frac{1}{n^{\alpha}\Gamma(1-\alpha)}\ (n \ge 1). \tag{4.332}$$

Proof. One can verify (4.331) easily from the recursive relation (4.302). We only give the proof of (4.332). Let $Q_n^\alpha = \sum_{k=0}^n g_k^\alpha$, from (4.336), the sequences $\{Q_k^\alpha\}$ is nonnegative and decreasing with k. Considering the coefficients of the power series of $(1-z)^{\alpha-1} = (1-z)^{-1}(1-z)^\alpha$, it is easy to obtain $\sum_{n=0}^\infty Q_n^\alpha z^n = (1-z)^{\alpha-1}$. Using the Stirling formula (see [Samko *et al.* (1993), p. 16]), we have

$$Q_{n-1}^\alpha = (-1)^{n-1}\binom{\alpha-1}{n-1} = \frac{n^{-\alpha}}{\Gamma(1-\alpha)} + \mathcal{O}(n^{-1-\alpha}), \quad n \to \infty. \tag{4.333}$$

In addition,

$$\frac{Q_{n-1}^\alpha}{Q_n^\alpha} = \frac{\Gamma(n-\alpha)}{\Gamma(1-\alpha)\Gamma(n)} = \frac{n}{n-\alpha} > 1 + \frac{\alpha}{n} \geq \left(\frac{n+1}{n}\right)^\alpha. \tag{4.334}$$

Therefore, the sequence $\{\Gamma(1-\alpha)(n+1)^\alpha Q_n^\alpha\}$ is decreasing and $\Gamma(1-\alpha)(n+1)^\alpha Q_n^\alpha \to 1$, which means that $n^{-\alpha}\Gamma^{-1}(1-\alpha) \leq Q_{n-1}^\alpha$.

Using the log-convexity of Gamma function on the positive reals (i.e., the Bohr-Mollerup Theorem), it holds that

$$\begin{aligned}\Gamma(n+1-\alpha) &= \Gamma(\alpha n + (1-\alpha)(n+1)) \\ &\leq \Gamma(n)^\alpha \Gamma(n+1)^{1-\alpha} \leq \Gamma(n) n^{1-\alpha}.\end{aligned} \tag{4.335}$$

Combining with [Samko *et al.* (1993), p. 15]

$$(-1)^n \binom{\alpha-1}{n} = \binom{n-\alpha}{n},$$

we actually have $Q_n^\alpha \leq \Gamma^{-1}(1-\alpha)n^{-\alpha}$.

□

Lemma 4.20. *Let $R \geq 0$; $\varepsilon^k \geq 0$ $(k = 0, 1, \ldots, M)$ and satisfy*

$$\varepsilon^n \leq -\sum_{k=1}^{n-1} g_k^\alpha \varepsilon^{n-k} + R, \quad n \geq 1. \tag{4.336}$$

Then we have the following estimates:

(1) when $0 < \alpha < 1$,

$$\varepsilon^n \leq \left(\sum_{k=0}^{n-1} g_k^\alpha\right)^{-1} R \leq n^\alpha \Gamma(1-\alpha) R; \tag{4.337}$$

(2) when $\alpha \to 1$,

$$\varepsilon^n \leq nR. \tag{4.338}$$

Proof. It is worth noting that the first term on the right hand side of (4.336) automatically vanishes when $n = 1$.

(a) Case $0 < \beta < 1$: By the mathematical induction, we prove the estimate

$$\epsilon^n \leq \left(\sum_{k=0}^{n-1} g_k^\alpha\right)^{-1} R.$$

Equation (4.336) holds obviously for $n = 1$. Supposing that

$$\epsilon^s \leq \left(\sum_{i=0}^{s-1} g_i^\alpha\right)^{-1} R, \quad s = 1, 2, \cdots, n-1,$$

then from (4.336) we have

$$\begin{aligned}
\epsilon^n &\leq -\sum_{k=1}^{n-1} g_k \epsilon^{n-k} + R \leq -\sum_{k=1}^{n-1} g_k^\alpha \left(\sum_{i=0}^{n-k-1} g_i^\alpha\right)^{-1} R + R \\
&\leq \left(1 - \sum_{k=0}^{n-1}\right)\left(\sum_{i=0}^{n-1} g_i^\alpha\right)^{-1} R + R \leq \left(\sum_{i=0}^{n-1} g_i^\alpha\right)^{-1} R.
\end{aligned}$$

According to (4.332) and the above inequality, it leads to

$$\epsilon^n \leq \left(\sum_{i=0}^{n-1} g_i^\alpha\right)^{-1} R \leq n^\alpha \Gamma(1-\alpha) R.$$

(b) Case $\beta \to 1$: Since $\Gamma(1-\beta) \to \infty$ as $\beta \to 1$ in the estimate (4.337), we need to look for an estimate of other form. By the mathematical induction, we prove that the estimate

$$\epsilon^n \leq nR.$$

Equation (4.336) holds obviously for $n = 1$. Supposing that

$$\epsilon^s \leq sR, \quad s = 1, 2, \cdots, n-1,$$

thus, from (4.336) we get

$$\begin{aligned}
\epsilon^n &\leq -\sum_{k=1}^{n-1} g_k^\alpha \epsilon^{n-k} + R \leq -\sum_{k=1}^{n-1} g_k^\alpha (n-k) R + R \\
&\leq -\sum_{k=1}^{n-1} g_k^\alpha (n-1) R + R \leq (n-1)R + R \leq nR.
\end{aligned}$$

□

Theorem 4.31. *The difference scheme (4.328) is unconditionally stable.*

Proof. Let $\widetilde{G}^n_{i,\rho}$ be the approximate solution of $G^n_{i,\rho}$, which is the exact solution of (4.296). Taking $\varepsilon^n_i = \widetilde{G}^n_{i,\rho} - G^n_{i,\rho}$, then from (4.328) we get the following perturbation equation

$$(1-\kappa\omega^{\beta}_{i,i})\varepsilon^1_i - \kappa\sum_{j=0,j\neq i}^{N}\omega^{\beta}_{i,j}\varepsilon^1_j = e^{J\rho U_i\tau}\varepsilon^0_i,\quad n=1,$$

$$(1-\kappa\omega^{\beta}_{i,i})\varepsilon^n_i - \kappa\sum_{j=0,j\neq i}^{N}\omega^{\beta}_{i,j}\varepsilon^n_j \tag{4.339}$$

$$=\sum_{k=0}^{n-1}g^{\alpha}_k e^{J\rho U_i n\tau}\varepsilon^0_i - \sum_{k=1}^{n-1}g^{\alpha}_k e^{J\rho U_i k\tau}\varepsilon^{n-k}_i,\quad n>1.$$

Denote $\varepsilon^n = (\varepsilon^n_0, \varepsilon^n_1, \dots, \varepsilon^n_N)^{\mathrm{T}}$ and $\|\varepsilon^n\|_\infty = \max_{0\le i\le N}|\varepsilon^n_i|$. Next we prove that $\|\varepsilon^n\|_\infty \le \|\varepsilon^0\|_\infty$ by the mathematical induction.

For $n=1$, suppose $|\varepsilon^1_{i_0}| = ||\varepsilon^1||_\infty = \max_{0\le i\le N}|\varepsilon^1_i|$. From (4.339), we obtain

$$(1-\kappa\omega^{\beta}_{i_0,i_0})\varepsilon^1_{i_0} - \kappa\sum_{j=0,j\neq i_0}^{N}\omega^{\beta}_{i_0,j}\varepsilon^1_j = e^{J\rho U_{i_0}\tau}\varepsilon^0_{i_0}. \tag{4.340}$$

Then

$$\begin{aligned}
||\varepsilon^1||_\infty = |\varepsilon^1_{i_0}| &\le |\varepsilon^1_{i_0}| - \kappa\sum_{j=0}^{N}\omega^{\beta}_{i_0,j}|\varepsilon^1_{i_0}| \\
&= (1-\kappa\omega^{\beta}_{i_0,i_0})|\varepsilon^1_{i_0}| - \kappa\sum_{j=0,j\neq i_0}^{N}\omega^{\beta}_{i_0,j}|\varepsilon^1_{i_0}| \\
&\le (1-\kappa\omega^{\beta}_{i_0,i_0})|\varepsilon^1_{i_0}| - \kappa\sum_{j=0,j\neq i_0}^{N}\omega^{\beta}_{i_0,j}|\varepsilon^1_{j}| \\
&\le \left|(1-\kappa\omega^{\beta}_{i_0,i_0})\varepsilon^1_{i_0} - \kappa\sum_{j=0,j\neq i_0}^{N}\omega^{\beta}_{i_0,j}\varepsilon^1_{j}\right| \\
&= |e^{J\rho U_{i_0}\tau}\varepsilon^0_{i_0}| = |\varepsilon^0_{i_0}| \le ||\varepsilon^0||_\infty.
\end{aligned}$$

Supposing $|\varepsilon_{i_0}^n| = \|\varepsilon^n\|_\infty = \max_{0\le i\le N} |\varepsilon_i^n|$, from (4.324) and (4.339), we obtain

$$
\begin{aligned}
\|\varepsilon^n\|_\infty &\le \left|(1-\kappa\omega_{i_0,i_0}^\beta)\varepsilon_{i_0}^n - \kappa\sum_{j=0,j\ne i_0}^{N}\omega_{i_0,j}^\beta\varepsilon_j^n\right| \\
&= \left|\sum_{k=0}^{n-1} g_k^\alpha e^{J\rho U_{i_0}n\tau}\varepsilon_{i_0}^0 - \sum_{k=1}^{n-1} g_k^\alpha e^{J\rho U_{i_0}k\tau}\varepsilon_{i_0}^{n-k}\right| \qquad (4.341)\\
&\le \left|\sum_{k=0}^{n-1} g_k^\alpha e^{J\rho U_{i_0}n\tau}\varepsilon_{i_0}^0\right| + \left|\sum_{k=1}^{n-1} g_k^\alpha e^{J\rho U_{i_0}k\tau}\varepsilon_{i_0}^{n-k}\right|.
\end{aligned}
$$

Using Lemma 4.19, we get

$$
\left|\sum_{k=0}^{n-1} g_k^\alpha e^{J\rho U_{i_0}n\tau}\varepsilon_{i_0}^0\right| = \left|\varepsilon_{i_0}^0\right|\cdot\left|e^{J\rho U_{i_0}n\tau}\right|\cdot\left|\sum_{k=0}^{n-1} g_k^\alpha\right| = \sum_{k=0}^{n-1} g_k^\alpha\|\varepsilon^0\|_\infty, \qquad (4.342)
$$

and

$$
\left|\sum_{k=1}^{n-1} g_k^\alpha e^{J\rho U_{i_0}k\tau}\varepsilon_{i_0}^{n-k}\right| \le \sum_{k=1}^{n-1}|g_k^\alpha|\cdot\|\varepsilon^{n-k}\|_\infty = -\sum_{k=1}^{n-1} g_k^\alpha\|\varepsilon^{n-k}\|_\infty. \qquad (4.343)
$$

Thus, according to (4.341)–(4.343), there exists

$$
\|\varepsilon^n\|_\infty \le \sum_{k=0}^{n-1} g_k^\alpha\left\|\varepsilon^0\right\|_\infty - \sum_{k=1}^{n-1} g_k^\alpha\left\|\varepsilon^{n-k}\right\|_\infty. \qquad (4.344)
$$

Next by the mathematical induction, we prove that the following inequality

$$
\|\varepsilon^n\|_\infty \le \left\|\varepsilon^0\right\|_\infty \quad \forall n\ge 1.
$$

In fact, for $n=1$, (4.344) holds obviously. Suppose that

$$
\|\varepsilon^s\|_\infty \le \left\|\varepsilon^0\right\|_\infty, \quad s=1,2,\ldots,n-1.
$$

Then from (4.344), it implies that

$$
\begin{aligned}
\|\varepsilon^n\|_\infty &\le \sum_{k=0}^{n-1} g_k^\alpha\left\|\varepsilon^0\right\|_\infty - \sum_{k=1}^{n-1} g_k^\alpha\left\|\varepsilon^{n-k}\right\|_\infty \\
&\le \sum_{k=0}^{n-1} g_k^\alpha\left\|\varepsilon^0\right\|_\infty - \sum_{k=1}^{n-1} g_k^\alpha\left|\varepsilon^0\right\|_\infty \\
&= \left\|\varepsilon^0\right\|_\infty.
\end{aligned}
$$

Thus, the proof is completed. □

Theorem 4.32. *Let $G(x_i, \rho, t_n)$ be the exact solution of (4.296), and $G^n_{i,\rho}$ the solution of the finite difference scheme (4.328). Then the maximum norm error estimates are*

$$\left\| G^n - \widetilde{G}^n \right\|_\infty \leq C_G \Gamma(1-\alpha) T^\alpha (\tau + h^2), \quad \text{for } \ 0 < \alpha < 1,$$

and

$$\left\| G^n - \widetilde{G}^n \right\|_\infty \leq C_G T \tau^{\alpha - 1} (\tau + h^2), \quad \text{for } \ \alpha \to 1,$$

where C_G is given by (4.327), and

$$G^n = \left(G(x_1, \rho, t_n), G(x_2, \rho, t_n), \cdots, G(x_{N-1}, \rho, t_n)\right)^{\mathrm{T}}.$$

Proof. Let $e^n_i = G(x_i, \rho, t_n) - G^n_{i,\rho}$ and $e^n = (e^n_0, e^n_1, \ldots, e^n_N)^{\mathrm{T}}$. Subtracting (4.325) from (4.328) and using $e^0_i = 0$, we obtain

$$\begin{aligned}
&(1 - \kappa\omega^\beta_{i,i})e^1_i - \kappa \sum_{j=0, j\neq i}^{N} \omega^\beta_{i,j} e^1_j = \varepsilon^1_i, \quad n = 1, \\
&(1 - \kappa\omega^\beta_{i,i})e^n_i - \kappa \sum_{j=0, j\neq i}^{N} \omega^\beta_{i,j} e^n_j = -\sum_{k=1}^{n-1} g^\alpha_k e^{J\rho U_i k\tau} e^{n-k}_i + \varepsilon^n_i, \quad n > 1,
\end{aligned} \tag{4.345}$$

where ε^n_i is given in (4.327).

Denoting that $||e^n||_\infty = \max_{0\leq i\leq N} |e^n_i|$ and $\varepsilon_{\max} = \max_{0\leq i\leq N, 0\leq n\leq M} |\varepsilon^n_i|$, the desired result can be proved by using the mathematical induction.

For $n = 1$, supposing $|e^1_{i_0}| = \left\|e^1\right\|_\infty$ and using (4.345), we get

$$(1 - \kappa\omega^\beta_{i_0,i_0})e^1_{i_0} - \kappa \sum_{j=0, j\neq i_0}^{N} \omega^\beta_{i_0,j} e^1_j = \varepsilon^1_{i_0}.$$

According to (4.324) and the above equation, we obtain

$$||e^1||_\infty = |e^1_{i_0}| \leq \left|(1 - \kappa\omega^\beta_{i_0,i_0})e^1_{i_0} - \kappa \sum_{j=0, j\neq i_0}^{N} \omega^\beta_{i_0,j} e^1_j\right| = |R^1_{i_0}| \leq \varepsilon_{\max}.$$

Supposing $|e_{i_0}^n| = \|e^n\|_\infty$, and from (4.343), (4.345), (4.324), there exists

$$
\begin{aligned}
\|e^n\|_\infty &\leq \left|(1-\kappa\omega_{i_0,i_0}^{\beta})e_{i_0}^n - \kappa\sum_{j=0,j\neq i_0}^{N}\omega_{i_0,j}^{\beta}e_j^n\right| \\
&= \left|-\sum_{k=1}^{n-1} g_k^{\alpha} e^{J\rho U_{i_0}k\tau} e_{i_0}^{n-k} + \varepsilon_{i_0}^n\right| \\
&\leq \left|\sum_{k=1}^{n-1} g_k^{\alpha} e^{J\rho U_{i_0}k\tau} e_{i_0}^{n-k}\right| + \varepsilon_{\max} \\
&\leq -\sum_{k=1}^{n-1} g_k^{\alpha}\|e^{n-k}\|_\infty + \varepsilon_{\max}.
\end{aligned}
$$

Hence, from Lemma 4.20, we have the following estimates

(1) when $0<\alpha<1$,

$$
\begin{aligned}
\|e^n\|_\infty &\leq \left(\sum_{k=0}^{n-1} g_k^{\alpha}\right)^{-1}\varepsilon_{\max} \leq n^{\alpha}\Gamma(1-\alpha)\varepsilon_{\max} \\
&\leq C_G\Gamma(1-\alpha)T^{\alpha}(\tau+h^2);
\end{aligned}
$$

(2) when $\alpha\to 1$,

$$
\|e^n\|_\infty \leq n\varepsilon_{\max} \leq C_G T\tau^{\alpha-1}(\tau+h^2).
$$

□

In the following theorem, we present the convergence result under the L^2 norm; because of the similarity, we omit the proof of unconditional stability.

Theorem 4.33. *Let $G(x_i,\rho,t_n)$ be the exact solution of (4.296), and $G_{i,\rho}^n$ the solution of the finite difference scheme (4.328). Then the L^2 error estimates are*

$$
\left\|G^n - \widetilde{G}^n\right\| \leq (b-a)^{\frac{1}{2}}C_G\Gamma(1-\alpha)T^{\beta}(\tau+h^2) \quad \text{for } 0<\alpha<1,
$$

and

$$
\left\|G^n - \widetilde{G}^n\right\| \leq (b-a)^{\frac{1}{2}}C_G T\tau^{\alpha-1}(\tau+h^2) \quad \text{for } \alpha\to 1,
$$

where C_G is given by (4.327), and

$$
G^n = \left(G(x_1,\rho,t_n), G(x_2,\rho,t_n),\cdots,G(x_{N-1},\rho,t_n)\right)^{\mathrm{T}}.
$$

Proof. Denote $e_i^n = G(x_i, \rho, t_n) - G_{i,\rho}^n$ and $e^n = \left(e_1^n, e_2^n, \ldots, e_{N-1}^n\right)^{\mathrm{T}}$. Subtracting (4.325) from (4.328) and using $e_i^0 = 0$, we obtain

$$(I - \kappa H)\, e^n = -\sum_{k=1}^{n-1} g_k^\alpha B^k e^{n-k} + \varepsilon^n,$$

where H is defined by (4.316), and $\varepsilon^n = \left(\varepsilon_1^n, \varepsilon_2^n, \ldots, \varepsilon_{N-1}^n\right)^{\mathrm{T}}$.

Performing the inner product in both sides of the above equation by e^n leads to

$$((I - \kappa H)\, e^n, e^n) = \left(-\sum_{k=1}^{n-1} g_k^\alpha B^k e^{n-k}, e^n \right) + (\varepsilon^n, e^n). \tag{4.346}$$

By Corollary 4.2, we have $(-\kappa H e^n, e^n) \geq 0$. Then

$$\|e^n\|^2 \leq \left(-\sum_{k=1}^{n-1} g_k^\alpha B^k e^{n-k}, e^n \right) + (\varepsilon^n, e^n),$$

by the Cauchy-Schwarz inequality, which results in

$$\|e^n\| \leq \left\| -\sum_{k=1}^{n-1} g_k^\alpha B^k e^{n-k} \right\| + \|\varepsilon^n\|. \tag{4.347}$$

Thus

$$\|e^n\| \leq -\sum_{k=1}^{n-1} g_k^\alpha \left\|e^{n-k}\right\| + \varepsilon_{\max},$$

where $\varepsilon_{\max} \leq (b-a)^{\frac{1}{2}} C_G \tau^\alpha (\tau + h^2)$.

Hence, from Lemma 4.20, we have the estimates

(1) when $0 < \alpha < 1$,

$$\begin{aligned} \|e^n\| &\leq \left(\sum_{k=0}^{n-1} g_k^\alpha \right)^{-1} \varepsilon_{\max} \leq n^\alpha \Gamma(1-\alpha) \varepsilon_{\max} \\ &\leq (b-a)^{\frac{1}{2}} C_G \Gamma(1-\alpha) T^\alpha (\tau + h^2); \end{aligned}$$

(2) when $\alpha \to 1$,

$$\|e^n\| \leq n \varepsilon_{\max} \leq (b-a)^{\frac{1}{2}} C_G T \tau^{\alpha-1} (\tau + h^2).$$

$\square$

Remark 4.13. In fact, if the matrix H in (4.346) satisfies

$$\mathrm{Re}\,(He^n, e^n) = (e^n)^* \frac{H + H^*}{2} e^n \le 0,$$

by

$$\begin{aligned}
&(e^n, e^n) - \mathrm{Re}\Big(He^n, e^n\Big) \\
&\le \left| \Big(-\sum_{k=1}^{n-1} g_k^{\alpha} B^k e^{n-k}, e^n \Big) + (\varepsilon^n, e^n) \right| \\
&\le \left(\Big\| -\sum_{k=1}^{n-1} g_k^{\alpha} B^k e^{n-k} \Big\| + \|\varepsilon^n\| \right) \|e^n\|,
\end{aligned}$$

the above L^2 error estimate still holds. Therefore, the techniques of the L^2 theoretical analysis here also apply to the more general non-symmetric tempered operators [Li and Deng (2016)].

4.6.4 *Numerical results*

Example 4.10. Consider (4.296) in a finite domain with $\Omega = (a, b)$ $(a = 0, b = 1), 0 < t \le 1$, $U(x) = x$, $\rho = 1$, $J = \sqrt{-1}$; the forcing function

$$\begin{aligned}
f(x, \rho, t) =& \frac{\Gamma(4+\alpha)}{\Gamma(4)} e^{J\rho x t} t^3 \sin(x^\gamma) \sin((1-x)^\gamma) \\
&+ \frac{1}{2\cos(\beta\pi/2)} (t^{3+\alpha} + 1) e^{-\lambda x} {}_aD_x^{\beta} [e^{(\lambda + J\rho t)x} \sin(x^\gamma) \sin((1-x)^\gamma)] \\
&+ \frac{1}{2\cos(\beta\pi/2)} (t^{3+\alpha} + 1) e^{\lambda x} {}_xD_b^{\beta} [e^{(-\lambda + J\rho t)x} \sin(x^\gamma) \sin((1-x)^\gamma)] \\
&- \frac{\lambda^\beta}{\cos(\beta\pi/2)} e^{J\rho x t} (t^{3+\alpha} + 1) \sin(x^\gamma) \sin((1-x)^\gamma);
\end{aligned} \tag{4.348}$$

the initial condition $G(x, \rho, 0) = \sin(x^\gamma) \sin((1-x)^\gamma)$, $x \in \Omega$, and the boundary condition $G(x, \rho, t) = 0$, $(x, t) \in \mathbb{R} \backslash \Omega \times [0, 1]$.

It is easy to check that (4.296) has the exact solution $G(x, \rho, t) = e^{J\rho x t}(t^{3+\alpha} + 1) \sin(x^\gamma) \sin((1-x)^\gamma)$ for $\Omega \times [0, 1]$. The numerical results are given in Tables 4.16 and 4.17.

Tables 4.16 and 4.17 show that the finite difference scheme (4.328) has the global truncation errors $\mathcal{O}(\tau + h^2)$ at time $t = 0.5$. Here, we have used

Table 4.16: The discrete L^2 and maximum errors and their convergence rates to Example 4.10 at $t = 0.5$ for different α, β with $U(x) = x, \lambda = 0.7, \rho = 1, \gamma = 3, \tau = h^2$, and $r_3 = \frac{(\beta-1)(2-\beta)(\beta+3)}{4(\beta+1)(\beta+2)}$.

(α,β)	N	$\|G^M - \widetilde{G}^M\|_\infty$	rate	$\|G^M - \widetilde{G}^M\|$	rate
	16	1.6796e-04	-	1.1172e-4	-
(0.3,1.8)	32	4.2567e-05	1.98	2.9457e-05	1.92
	64	1.0696e-05	1.99	7.5730e-06	1.96
	128	2.6787e-06	2.00	1.9208e-06	1.98
	16	1.1101e-04	-	6.8820e-05	-
(0.8,1.3)	32	3.6655e-05	1.60	2.2172e-05	1.63
	64	1.0432e-05	1.81	6.2746e-06	1.82
	128	2.7754e-06	1.91	1.6675e-06	1.91

Table 4.17: The discrete L^2 and maximum errors and their convergence rates to Example 4.10 at $t = 0.5$ for different α, β with $U(x) = x, \lambda = 0.7, \rho = 1, \gamma = 3, N = 512$, and $r_3 = \frac{(\beta-1)(2-\beta)(\beta+3)}{4(\beta+1)(\beta+2)}$.

(α,β)	M	$\|G^M - \widetilde{G}^M\|_\infty$	rate	$\|G^M - \widetilde{G}^M\|$	rate
	40	2.2683e-05	-	1.5000e-05	-
(0.3,1.8)	80	1.1390e-05	0.99	7.4949e-06	1.00
	120	7.6127e-06	0.98	7.4949e-06	1.00
	160	5.7219e-06	0.98	3.7313e-06	1.01
	40	9.7252e-06	-	6.5348e-06	-
(0.8,1.3)	80	4.9542e-06	0.97	3.2814e-06	0.99
	120	3.3562e-06	0.96	2.1929e-06	0.99
	160	2.5558e-06	0.95	1.6483e-06	0.99

the Gauss quadrature to calculate the left and right fractional derivatives in (4.348). In fact, letting $G(x) \in C^2(a,b)$, it follows that

$$G(x) = G(a) + G'(a)(x-a) + \int_a^x (x-\xi)G^{(2)}(\xi)d\xi, \tag{4.349}$$

$$G(x) = G(b) - G'(b)(b-x) + \int_x^b (\xi-x)G^{(2)}(\xi)d\xi. \tag{4.350}$$

Then we have

$$\begin{aligned} {}_aD_x^\beta G(x) = &\frac{G(a)}{\Gamma(1-\beta)}x^{-\beta} + \frac{G'(a)}{\Gamma(2-\beta)}(x-a)^{1-\beta} \\ &+\frac{1}{\Gamma(2-\beta)}\int_a^x (x-a)^{1-\beta}G^{(2)}(\xi)d\xi, \end{aligned} \tag{4.351}$$

and

$$
{}_xD_b^{\beta}G(x) = \frac{G(b)}{\Gamma(1-\beta)}(b-x)^{-\beta} - \frac{G'(b)}{\Gamma(2-\beta)}(b-x)^{1-\beta} + \frac{1}{\Gamma(2-\beta)}\int_x^b (\xi - x)^{1-\beta}G^{(2)}(\xi)d\xi. \tag{4.352}
$$

Note that

$$
\int_a^x (x-a)^{1-\beta}G^{(2)}(\xi)d\xi = \left(\frac{x-a}{2}\right)^{2-\beta}\int_{-1}^{1}(1-\xi)^{1-\beta}G^{(2)}\left(\frac{x-a}{2}\xi + \frac{x+a}{2}\right)d\xi \tag{4.353}
$$

and

$$
\int_x^b (\xi - x)^{1-\beta}G^{(2)}(\xi)d\xi = \left(\frac{b-x}{2}\right)^{2-\beta}\int_{-1}^{1}(1+\xi)^{1-\beta}G^{(2)}\left(\frac{b-x}{2}\xi + \frac{b+x}{2}\right)d\xi. \tag{4.354}
$$

Therefore, the Jacobi-Gauss quadrature (see Appendix A) with weights $(1-\xi)^{1-\beta}(1+\xi)^0$ and $(1-\xi)^0(1+\xi)^{1-\beta}$ can be used to calculate (4.353) and (4.354), respectively.

Table 4.18: MGM to solve the scheme (4.328) at $t = 1$, where $\lambda = 0.7$, $U(x) = x, \rho = 1, \gamma = 2, \tau = h$, and $r_3 = \frac{(\beta-1)(2-\beta)(\beta+3)}{4(\beta+1)(\beta+2)}$.

N	$\beta = 1.3, \alpha = 0.8$	Iter	CPU(s)	$\beta = 1.8, \alpha = 0.3$	Iter	CPU(s)
16	2.4702e-03	8.0	0.23 s	1.7526e-03	9.0	0.28 s
32	1.2761e-03	7.0	0.72 s	5.1049e-04	9.0	0.82 s
64	6.4040e-04	6.0	2.41 s	1.6173e-04	9.0	2.85 s
128	3.2027e-04	6.0	8.68 s	5.7602e-05	10.0	10.18 s

Finally, the numerical results of V-cycle multigrid method (MGM), with the computational cost $\mathcal{O}(M\log M)$ for every V-cycle [Chen and Deng (2014b); Pang and Sun (2012)], to solve Example 4.10 are listed in Table 4.18, where the parameters of MGM, e.g., 'Iter', 'CPU', etc, are the same as the ones of [Chen and Deng (2014b)]. For the convenience to the readers, the MGM's pseudo codes are added in the Appendix C. The numerical experiments are programmed in Python, and the computations are carried out on a PC with the configuration: Intel(R) Core(TM) i5-3470 3.20 GHZ and 8 GB RAM and a 64 bit Windows 7 operating system.

Chapter 5

Variational numerical method for the space fractional PDEs

Besides the foregoing numerical methods, the methods based on weak formulations are also developed for solving the fractional PDEs, such as the finite element methods [Deng (2008); Ervin and Roop (2006); Ervin *et al.* (2007); Jin *et al.* (2015a); Wang *et al.* (2014)], the spectral Galerkin methods [Li and Xu (2009, 2010)], and the discontinuous finite element methods [Deng and Hesthaven (2013); Qiu *et al.* (2015); Wang *et al.* (2016); Xu and Hesthaven (2014)]. These variational methods are generally classified as the h-type, the p-type, and the hp-type methods [Ern and Guermond (2004)], and a suitable variational formulation is their key part. For the finite element methods and the spectral Galerkin methods, the finite dimensional subspaces are usually required to be conforming, and their convergences are achieved by refining the mesh and increasing the polynomial degree, respectively; on the contrary, the DG methods are non-conformal, for which the degrees of freedom on each element are independent. These three methods have been well developed for solving the integer order PDEs, while for fractional PDEs, the studies on these methods are still on the early stage, and the corresponding theoretical analysis and the potential improvements on the methods seem to be inadequate.

In this chapter, we introduce the appropriate weak formulations and the related theoretical analysis for solving fractional PDEs, including the finite element methods, the spectral Galerkin methods, and the discontinuous finite element methods. We also discuss the efficient implementation techniques of the corresponding algorithms. More specifically, in Sec. 5.1, we introduce the fractional variational spaces, which are the basis for the finite element methods and spectral methods. In Sec. 5.2, we consider numerical schemes for the fractional Fokker-Planck equation, where the finite element method and spectral Galerkin method are discussed. In Sec. 5.3

and Sec. 5.4, we develop the LDG method and the general DG method for solving the fractional PDEs. For the LDG method, the discrete problems can be solved element by element, and for the general DG method, the corresponding algebraic equation needs to be solved as a whole.

Let Ω be a bounded interval (domain) in following sections, and throughout, we denote by $\tilde{u}$ the extension of u by zero to $\mathbb{R}$ (or $\mathbb{R}^2$). By $A \lesssim B$, we mean that A can be bounded by a multiple of B, independent of the parameters they may depend on; and the expression $A \simeq B$ means that $A \lesssim B \lesssim A$. Considering the space-time functions, we also need the following Banach space-valued functions space [Evans (2010)]:

$$L^p(0,T;X) := \left\{ v : [0,T] \to X \middle| \int_0^T \|v(t)\|_X^p \, dt < \infty \right\}, \tag{5.1}$$

where X is a Banach space endowed with the norm $\|\cdot\|_X$. In a similar way we can define $L^\infty(0,T;X)$ and $C^0(0,T;X)$. The Sobolev space $H^1(0,T;X)$ is given as

$$H^1(0,T;X) := \left\{ v \in L^2(0,T;X) \middle| \frac{\partial v}{\partial t} \in L^2(0,T;X) \right\}, \tag{5.2}$$

equipped with the norm

$$\|v\|_{H^1(0,T;X)} = \left(\int_0^T \|v\|_X^2 + \left\| \frac{\partial v}{\partial t} \right\|_X^2 dt \right)^{\frac{1}{2}}. \tag{5.3}$$

5.1 Variational spaces of fractional derivatives

In this section, we give a brief introduction of the fractional variational spaces and their properties; refer to [Ervin and Roop (2006); Li and Xu (2009, 2010)] for details.

For any real number $\mu \geq 0$, let $H^\mu(\mathbb{R})$ be the Sobolev space of order μ on $\mathbb{R}$, i.e.,

$$H^\mu(\mathbb{R}) = \left\{ u(x) \in L^2(R) \,\middle|\, |u|^2_{H^\mu(\mathbb{R})} < \infty \right\} \tag{5.4}$$

endowed with the seminorm

$$|u|^2_{H^\mu(\mathbb{R})} := \int_{\mathbb{R}} |\omega|^{2\mu} \left|\mathscr{F}[u](\omega)\right|^2 d\omega, \tag{5.5}$$

and the norm

$$\begin{aligned} \|u\|^2_{H^\mu(\mathbb{R})} &:= \int_{\mathbb{R}} \left(1 + |\omega|^2\right)^\mu \left|\mathscr{F}[u](\omega)\right|^2 d\omega \\ &\simeq \int_{\mathbb{R}} \left(1 + |\omega|^{2\mu}\right) \left|\mathscr{F}[u](\omega)\right|^2 d\omega; \end{aligned} \tag{5.6}$$

and for $\Omega = (a, b)$, let

$$H^\mu(\Omega) = \{u \in L^2(\Omega) : \exists U \in H^\mu(\mathbb{R}) \text{ such that } U|_\Omega = u\} \tag{5.7}$$

endowed with the norm

$$\|u\|_{H^\mu(\Omega)} := \inf_{U|_\Omega = u} \|U\|_{H^\mu(\mathbb{R})}. \tag{5.8}$$

Note that $C_0^\infty(\mathbb{R})$ is dense in $H^\mu(\mathbb{R})$, but this is not always true for $C_0^\infty(\Omega)$ in $H^\mu(\Omega)$. We usually denote $H_0^\mu(\Omega)$ as the closure of $C_0^\infty(\Omega)$ w.r.t. $\|\cdot\|_{H^\mu(\Omega)}$. There are also some other definitions of the fractional Sobolev space; for the relations between them see Appendix A.

Theorem 5.1. *Let $\mu > 0$ and $u(x) \in C_0^\infty(\mathbb{R})$ be a real valued function. Then*

$$\begin{aligned}({}_{-\infty}D_x^\mu u(x), {}_xD_\infty^\mu u(x)) &= \cos(\pi\mu)\, \|{}_{-\infty}D_x^\mu u(x)\|_{L^2(\mathbb{R})}^2 \\ &= \cos(\pi\mu)\, \|{}_xD_\infty^\mu u(x)\|_{L^2(\mathbb{R})}^2.\end{aligned} \tag{5.9}$$

Proof. Note that

$$\mathscr{F}[{}_{-\infty}D_x^\mu u(x)](\omega) = (i\omega)^\mu \hat{u}(\omega), \tag{5.10}$$

$$\mathscr{F}[{}_xD_\infty^\mu u(x)](\omega) = (-i\omega)^\mu \hat{u}(\omega). \tag{5.11}$$

By the Plancherel theorem (see Lemma A.16), it follows that

$$\|{}_{-\infty}D_x^\mu u(x)\|_{L^2(\mathbb{R})}^2 = \frac{1}{2\pi} \|\mathscr{F}[{}_{-\infty}D_x^\mu u(x)](\omega)\|_{L^2(\mathbb{R})}^2, \tag{5.12}$$

$$\|{}_xD_\infty^\mu u(x)\|_{L^2(\mathbb{R})}^2 = \frac{1}{2\pi} \|\mathscr{F}[{}_xD_\infty^\mu u(x)](\omega)\|_{L^2(\mathbb{R})}^2, \tag{5.13}$$

and

$$\begin{aligned}({}_{-\infty}D_x^\mu u(x), {}_xD_\infty^\mu u(x)) &= \frac{1}{2\pi} \int_{\mathbb{R}} (i\omega)^\mu \hat{u}(\omega) \overline{(-i\omega)^\mu \hat{u}(\omega)} d\omega \\ = \frac{1}{2\pi} \int_{-\infty}^0 |(i\omega)^\mu|^2 |\hat{u}(\omega)|^2 e^{-i\pi\mu} d\omega &+ \frac{1}{2\pi} \int_0^\infty |(i\omega)^\mu|^2 |\hat{u}(\omega)|^2 e^{i\pi\mu} d\omega \\ = \frac{\cos(\pi\mu)}{2\pi} \int_{-\infty}^\infty |\omega|^{2\mu} |\hat{u}(\omega)|^2 d\omega,\end{aligned} \tag{5.14}$$

where in the last step, the property $|\hat{u}(\omega)| = |\hat{u}(-\omega)|$ has been used. Thus, we complete the proof. □

Remark 5.1. In the particular cases $\mu = n + 1/2$, $n \in \mathbb{N}$, the right side of (5.9) will be zero.

Theorem 5.2. *Suppose that* $\mu \in (0, \frac{1}{2}) \cup (\frac{1}{2}, 1)$. *Then the operators* ${}_aD_x^{\mu}u(x)$ *and* ${}_xD_b^{\mu}u(x)$ *defined for* $u \in C_0^{\infty}(\Omega)$ *can be continuously extended to operators from* $H_0^{\mu}(\Omega)$ *to* $L^2(\Omega)$; *and for* $u \in H_0^{\mu}(\Omega)$, *it follows that*

$$\|u\|_{L^2(\Omega)} \lesssim \|{}_aD_x^{\mu}u\|_{L^2(\Omega)} \lesssim \|u\|_{H^{\mu}(\Omega)}, \tag{5.15}$$

$$\|u\|_{L^2(\Omega)} \lesssim \|{}_xD_b^{\mu}u\|_{L^2(\Omega)} \lesssim \|u\|_{H^{\mu}(\Omega)}. \tag{5.16}$$

Proof. Firstly, if $u \in C_0^{\infty}(\Omega)$, then $\tilde{u} \in C_0^{\infty}(\mathbb{R})$. By the proof processes of Theorem 5.1, (A.19), and Proposition A.4, it is easy to see that

$$\|{}_{-\infty}D_x^{\mu}\tilde{u}\|_{L^2(\mathbb{R})}^2 = \|{}_xD_{\infty}^{\mu}\tilde{u}\|_{L^2(\mathbb{R})}^2 \simeq |\tilde{u}|_{H^{\mu}(\mathbb{R})}^2 \lesssim \|u\|_{H^{\mu}(\Omega)}^2. \tag{5.17}$$

Moreover,

$${}_{-\infty}D_x^{\mu}\tilde{u}\,|_{x\in\Omega} = {}_aD_x^{\mu}u, \quad {}_xD_{\infty}^{\mu}\tilde{u}\,|_{x\in\Omega} = {}_xD_b^{\mu}u. \tag{5.18}$$

Therefore,

$$\|{}_aD_x^{\mu}u\|_{L^2(\Omega)} \lesssim \|u\|_{H^{\mu}(\Omega)}, \quad \|{}_xD_b^{\mu}u\|_{L^2(\Omega)} \lesssim \|u\|_{H^{\mu}(\Omega)}. \tag{5.19}$$

Noticing that $C_0^{\infty}(\Omega)$ is dense in $H_0^{\mu}(\Omega)$, then a standard density argument allows us to extend ${}_aD_x^{\mu}u(x)$ and ${}_xD_b^{\mu}u(x)$ to continuous operators from $H^{\beta}(\mathbb{R})$ to $L^2(\mathbb{R})$ with the same bounds as in (5.19).

Secondly, for any $u \in H_0^{\mu}(\Omega)$, there exists a sequence $u_n \in C_0^{\infty}(\Omega)$ such that $\lim_{n\to\infty} \|u - u_n\|_{H^{\mu}(\Omega)} = 0$, and by Proposition B.14, it holds that

$${}_aD_x^{-\mu}\,{}_aD_x^{\mu}u_n = {}_xD_b^{-\mu}\,{}_xD_b^{\mu}u_n = u_n. \tag{5.20}$$

Therefore,

$$\begin{aligned}
&\left\|{}_xD_b^{-\mu}\,{}_xD_b^{\mu}u - u\right\|_{L^2(\Omega)} \\
&\leq \left\|{}_xD_b^{-\mu}\,{}_xD_b^{\mu}(u - u_n)\right\|_{L^2(\Omega)} + \|u_n - u\|_{L^2(\Omega)} \\
&\lesssim \|{}_xD_b^{\mu}(u - u_n)\|_{L^2(\Omega)} + \|u - u_n\|_{L^2(\Omega)} \\
&\lesssim |u - u_n|_{H^{\mu}(\Omega)} + \|u - u_n\|_{L^2(\Omega)} \to 0,
\end{aligned}$$

where the property $\left\|{}_xD_b^{-\mu}u\right\|_{L^2(\Omega)} \lesssim \|u\|_{L^2(\Omega)}$ (see Proposition B.10) has been used. Similarly, one can prove that

$$\left\|{}_aD_x^{-\mu}\,{}_aD_x^{\mu}u - u\right\|_{L^2(\Omega)} = 0. \tag{5.21}$$

Using Proposition B.10 again, one has

$$\|u\|_{L^2(\Omega)} = \left\|{}_aD_x^{-\mu}\,{}_aD_x^{\mu}u\right\|_{L^2\Omega} \lesssim \|{}_aD_x^{\mu}u\|_{L^2(\Omega)}, \tag{5.22}$$

$$\|u\|_{L^2(\Omega)} = \left\|{}_xD_b^{-\mu}\,{}_xD_b^{\mu}u\right\|_{L^2\Omega} \lesssim \|{}_xD_b^{\mu}u\|_{L^2(\Omega)}. \tag{5.23}$$

Thus, we complete the proof. □

Theorem 5.3. *Let $u \in H_0^\mu(\Omega)$ with $\mu \in (0, \frac{1}{2}) \cup (\frac{1}{2}, 1)$. Then*

$$\left|({}_aD_x^\mu u, {}_xD_b^\mu u)_{L^2(\Omega)}\right| \simeq |{}_aD_x^\mu u|_{L^2(\Omega)}^2 \simeq |{}_xD_b^\mu u|_{L^2(\Omega)}^2, \tag{5.24}$$

and all of them are equivalent to $\|u\|_{H^\mu(\Omega)}$.

Proof. It suffices to prove (5.24) for $u \in C_0^\infty(\Omega)$. Since

$$({}_aD_x^\mu u, {}_xD_b^\mu u)_{L^2(\Omega)} = ({}_{-\infty}D_x^\mu \tilde{u}, {}_xD_\infty^\mu \tilde{u})_{L^2(\mathbb{R})},$$

then by (A.19), Proposition A.4, and Theorem 5.1, one has

$$\|u\|_{H^\mu(\Omega)}^2 \lesssim \|{}_{-\infty}D_x^\mu \tilde{u}\|_{L^2(\mathbb{R})}^2 \simeq \left|({}_aD_x^\mu u, {}_xD_b^\mu u)_{L^2(\Omega)}\right|. \tag{5.25}$$

Note that

$$\left|({}_aD_x^\mu u, {}_xD_b^\mu u)_{L^2(\Omega)}\right| \le \|{}_aD_x^\mu u\|_{L^2(\Omega)} \|{}_xD_b^\mu u\|_{L^2(\Omega)}. \tag{5.26}$$

Then combining Theorem 5.2, (5.25), and (5.26), one can complete the proof. □

By Proposition A.6, when $\mu_1, \mu_2 \in (0, 1)$ and $\mu_1 < \mu_2$, for any $u \in H^{\mu_2}(\Omega)$, it holds that

$$\|u\|_{H^{\mu_1}(\Omega)} \lesssim \|u\|_{H^{\mu_2}(\Omega)}. \tag{5.27}$$

Thus, combining Theorem 5.2 with 5.3, we have

$$\|{}_aD_x^{\mu_1} u\|_{L^2(\Omega)} \lesssim \|{}_aD_x^{\mu_2} u\|_{L^2(\Omega)}. \tag{5.28}$$

5.2 Finite element method and spectral method for solving the fractional Fokker-Planck equations

In this section, we focus on developing the variational methods for the space and time fractional Fokker-Planck equation [Magdziarz and Weron (2007); Metzler and Klafter (2000); Metzler and Nonnenmacher (2002)], describing the competition between Lévy flights and traps under the influence of an external potential $U(x)$, given by

$$\frac{\partial}{\partial t} p(x,t) = {}_0D_t^{1-\alpha} \left[\frac{\partial}{\partial x} \frac{U'(x)}{\eta_\alpha} + \kappa_\alpha \nabla^\beta\right] p(x,t), \tag{5.29}$$

where $p(x,t)$ is the probability density, prime stands for the derivative w.r.t. the space coordinate, κ_α denotes the anomalous diffusion coefficient with physical dimension $[m^\beta s^{-\alpha}]$, η_α represents the generalized friction coefficient possessing the dimension $[kgs^{\alpha-2}]$, $\alpha \in (0,1)$, and $\beta \in (1,2)$. Here, the operators

$$_0D_t^{1-\alpha}p(x,t) = \frac{1}{\Gamma(\alpha)}\frac{\partial}{\partial t}\int_0^t (t-\xi)^{\alpha-1}p(x,\xi)d\xi; \tag{5.30}$$

and $\nabla^\beta = \frac{1}{2}\left({}_{-\infty}D_x^\beta + {}_xD_\infty^\beta\right)$ with ${}_{-\infty}D_x^\beta$ and ${}_xD_\infty^\beta$ being the left and right R-L space fractional derivatives of order β, respectively (see formulae (B.13) and (B.14)).

Performing ${}_0D_t^{-(1-\alpha)}$ on both sides of (5.29) and using (B.26), we have the equivalent form of (5.29) as

$${}_0^CD_t^\alpha p(x,t) = \left[\frac{\partial}{\partial x}\frac{U'(x)}{\eta_\alpha} + \kappa_\alpha \nabla^\beta\right] p(x,t), \tag{5.31}$$

where ${}_0^CD_t^\alpha$ is the Caputo fractional derivative being defined by

$${}_0^CD_t^\alpha p(x,t) = \frac{1}{\Gamma(1-\alpha)}\int_0^t (t-\xi)^{-\alpha}\frac{\partial p(x,\xi)}{\partial \xi}d\xi. \tag{5.32}$$

5.2.1 *Variational formulation and time semi-discrete scheme*

Setting $T > 0$, $\Omega = (a,b)$, rewriting (5.31), and making it subject to the given initial and boundary conditions, we have

$${}_0^CD_t^\alpha p(x,t) = \left[\frac{\partial}{\partial x}\frac{U'(x)}{\eta_\alpha} + \kappa_\alpha \nabla^\beta\right] p(x,t), \quad (x,t) \in \Omega \times (0,T], \tag{5.33}$$

with initial-boundary conditions

$$p(x,0) = g(x), \quad x \in \Omega, \tag{5.34}$$
$$p(x,t) = 0, \quad x \in \mathbb{R}\backslash\Omega, \quad 0 \le t \le T. \tag{5.35}$$

Because of the absorbing condition (5.35), for $x \in \Omega$, we can rewrite $\nabla^\beta = \frac{1}{2}\left({}_aD_x^\beta + {}_xD_b^\beta\right)$, and discuss the model in the space $\tilde{H}^{\frac{\beta}{2}}(\mathbb{R}) = \left\{v \in H^{\frac{\beta}{2}}(\mathbb{R}) : v(x) = 0 \text{ for } x \in \mathbb{R}\backslash\Omega\right\}$, which is consistent with $H_0^{\frac{\beta}{2}}(\Omega)$ $(1 < \beta < 2)$ (see Proposition A.3). In the following the model will be considered in $H_0^{\frac{\beta}{2}}(\Omega)$.

Lemma 5.1. *Let $1 < \beta \le 2$. Then for all $u \in H_0^{\frac{\beta}{2}}(\Omega)$ and $w(x) \in C^1(\overline{\Omega})$, the product $uw \in H_0^{\frac{\beta}{2}}(\Omega)$ and $\|uw\|_{H^{\frac{\beta}{2}}(\Omega)} \lesssim \|u\|_{H^{\frac{\beta}{2}}(\Omega)}$.*

Proof. Since $H^{\frac{\beta}{2}}(\Omega) \subset C(\Omega)$, one has $(uw)(a) = (uw)(b) = 0$. We only focus on the case for $1 < \beta < 2$, which is trivial for $\beta = 2$. Note that

$H_0^{\frac{\beta}{2}}(\Omega) = \left[L^2(\Omega), H_0^1(\Omega)\right]_{\frac{\beta}{2}}$. It suffices to show that the linear mappings T_0, T_1 defined by

$$T_0(u) = T_1(u) = wu$$

are bounded. First, we have that $T_0 : L^2(\Omega) \to L^2(\Omega)$ is bounded as

$$\|T_0(u)\|_{L^2(\Omega)}^2 = \|wu\|_{L^2(\Omega)}^2 \leq \|w\|_{L^\infty(\Omega)}^2 \|u\|_{L^2(\Omega)}^2 .$$

Next, we have that $T_1 : H_0^1(\Omega) \to H_0^1(\Omega)$ is bounded as

$$\begin{aligned}
&\|T_1(u)\|_{H^1(\Omega)}^2 \\
&= \|wu\|_{H^1(\Omega)}^2 \\
&= \|wu\|_{L^2(\Omega)}^2 + \|wu'\|_{L^2(\Omega)}^2 + \|w'u\|_{L^2(\Omega)}^2 \\
&\leq \left(\|w\|_{L^\infty(\Omega)}^2 + \|w'\|_{L^\infty(\Omega)}^2\right) \|u\|_{H^1(\Omega)}^2 .
\end{aligned}$$

Therefore, using Proposition A.7, we have

$$\|wu\|_{H^{\frac{\beta}{2}}(\Omega)} \leq \|T_0\|^{1-\frac{\beta}{2}} \|T_1\|^{\frac{\beta}{2}} \|u\|_{H^{\frac{\beta}{2}}(\Omega)} \leq C \|u\|_{H^{\frac{\beta}{2}}(\Omega)} .$$

Thus, we complete the proof. □

Now for $U'(x) \in C^1(\overline{\Omega})$, we consider the weak formulation of problem (5.33) as follows: find $p(\cdot, t) : (0, T] \to H_0^{\frac{\beta}{2}}(\Omega)$ such that for any $v \in H_0^{\frac{\beta}{2}}(\Omega)$, it holds that

$$\begin{cases} \left({}_0^C D_t^\alpha p, v\right) = -B(p, v), \\ (p(\cdot, 0), v) = (g, v), \end{cases} \tag{5.36}$$

where

$$\begin{aligned}
B(p, v) := &-\frac{1}{\eta_\alpha}\left({}_aD_x^{\frac{\beta}{2}}(U'p), {}_xD_b^{1-\frac{\beta}{2}} v\right) \\
&- \frac{\kappa_\alpha}{2}\left({}_aD_x^{\frac{\beta}{2}} p, {}_xD_b^{\frac{\beta}{2}} v\right) - \frac{\kappa_\alpha}{2}\left({}_xD_b^{\frac{\beta}{2}} p, {}_aD_x^{\frac{\beta}{2}} v\right).
\end{aligned} \tag{5.37}$$

In fact, the variational formulation (5.36) can be derived from (5.33) directly for sufficient regularity $p(x,t)$ (w.r.t. x). For example, assuming that $p(\cdot, t) \in C^2(\Omega)$ and $v \in C_0^\infty(\Omega)$, one has

$$\begin{aligned}
&2\left(\nabla^\beta p, v\right) \\
&= -\left({}_aD_x^{-(2-\beta)} p', v'\right) - \left({}_xD_b^{-(2-\beta)} p', v'\right) \\
&= -\left({}_aD_x^{-(1-\frac{\beta}{2})} p', {}_xD_b^{-(1-\frac{\beta}{2})} v'\right) - \left({}_xD_b^{-(1-\frac{\beta}{2})} p', {}_aD_x^{-(1-\frac{\beta}{2})} v'\right) \\
&= \left({}_aD_x^{\frac{\beta}{2}} p, {}_xD_b^{\frac{\beta}{2}} v\right) + \left({}_bD_x^{\frac{\beta}{2}} p, {}_aD_x^{\frac{\beta}{2}} v\right).
\end{aligned}$$

Noting that for $0 < q_1, q_2 < 1$ and $v(a) = 0$, it follows that ${}_aD_x^{q_1+q_2}v(x) = {}_aD_x^{q_1}\,{}_aD_x^{q_2}v(x)$ (see formula (B.13)). Therefore

$$\left(\frac{\partial(U'p)}{\partial x}, v\right) = \left({}_aD_x^{1-\frac{\beta}{2}}\,{}_aD_x^{\frac{\beta}{2}}\left(U'p\right), v\right). \tag{5.38}$$

Denote $w(x,t) = {}_aD_x^{\frac{\beta}{2}}\left(U'p\right)$. Then

$$\begin{aligned}
&\left({}_aD_x^{1-\frac{\beta}{2}}w, v\right) \qquad (5.39)\\
&= \frac{1}{\Gamma(\frac{\beta}{2})}\int_a^b \frac{\partial}{\partial x}\int_a^x \frac{w(\xi,t)}{(x-\xi)^{1-\frac{\beta}{2}}}d\xi v(x)dx\\
&= \frac{-1}{\Gamma(\frac{\beta}{2})}\int_a^b \int_a^x \frac{w(\xi,t)}{(x-\xi)^{1-\frac{\beta}{2}}}d\xi v'(x)dx\\
&= \frac{-1}{\Gamma(\frac{\beta}{2})}\int_a^b \int_\xi^b \frac{v'(x)}{(x-\xi)^{1-\frac{\beta}{2}}}dx w(\xi,t)d\xi\\
&= \left(w(x,t), {}_x^CD_b^{1-\frac{\beta}{2}}v(x)\right)\\
&= \left({}_aD_x^{\frac{\beta}{2}}\left(U'p\right), {}_xD_b^{1-\frac{\beta}{2}}v\right). \qquad (5.40)
\end{aligned}$$

Lemma 5.2. *Let $U''(x) \leq 0$ for any $x \in \Omega$. Then the bilinear $B(\cdot,\cdot)$ is continuous and coercive, i.e., for any $u(x), v(x) \in H_0^{\frac{\beta}{2}}(\Omega)$, there exist constants C_1 and C_2 such that*

$$|B(u,v)| \leq C_1 \|u\|_{\frac{\beta}{2}(\Omega)} \|v\|_{H^{\frac{\beta}{2}}(\Omega)}, \tag{5.41}$$

$$B(u,u) \geq C_2 \|u\|^2_{H^{\frac{\beta}{2}}(\Omega)}. \tag{5.42}$$

Proof. Firstly,

$$\begin{aligned}
&|B(u,v)|\\
&\leq \frac{1}{\eta_\alpha}\left\|{}_aD_x^{\frac{\beta}{2}}(U'u)\right\|_{L^2(\Omega)}\left\|{}_xD_b^{1-\frac{\beta}{2}}v\right\|_{L^2(\Omega)}\\
&\quad + \frac{\kappa_\alpha}{2}\left\|{}_aD_x^{\frac{\beta}{2}}u\right\|_{L^2(\Omega)}\left\|{}_xD_b^{\frac{\beta}{2}}v\right\|_{L^2(\Omega)} + \frac{\kappa_\alpha}{2}\left\|{}_xD_b^{\frac{\beta}{2}}u\right\|_{L^2(\Omega)}\left\|{}_aD_x^{\frac{\beta}{2}}v\right\|_{L^2(\Omega)}\\
&\lesssim \|U'u\|_{H^{\frac{\beta}{2}}(\Omega)}\|v\|_{H^{\frac{\beta}{2}}(\Omega)} + \|u\|_{H^{\frac{\beta}{2}}(\Omega)}\|v\|_{H^{\frac{\beta}{2}}(\Omega)}\\
&\leq C_1\|u\|_{H^{\frac{\beta}{2}}(\Omega)}\|v\|_{H^{\frac{\beta}{2}}(\Omega)}. \qquad (5.43)
\end{aligned}$$

Secondly, since $H_0^\beta(\Omega)$ is the closure of $C_0^\infty(\Omega)$ w.r.t. $\|\cdot\|_{H^\beta(\Omega)}$, there exists a sequence $u_n \in C_0^\infty(\Omega)$ such that $\lim_{n\to\infty} u_n = u$. For $u_n \in C_0^\infty(\Omega)$, it follows that

$$\begin{aligned} &B(u_n, u_n) \\ &= -\frac{1}{2\eta_\alpha}\left(U'', u_n^2\right) - \kappa_\alpha \left({}_aD_x^{\frac{\beta}{2}} u_n, {}_bD_x^{\frac{\beta}{2}} u_n\right) \\ &\ge -\kappa_\alpha \cos\left(\frac{\pi\beta}{2}\right) \left\| {}_aD_x^{\frac{\beta}{2}} u_n \right\|_{L^2(\Omega)} \\ &\ge C_2 \|u_n\|_{H^{\frac{\beta}{2}}(\Omega)}. \end{aligned} \tag{5.44}$$

Using the continuities of $B(\cdot,\cdot)$ and $\|\cdot\|_{H^{\frac{\beta}{2}}(\Omega)}$, we have $B(u,u) \ge C_2 \|u\|_{H^{\frac{\beta}{2}}(\Omega)}$. Thus, we complete the proof. □

Lemma 5.3. *For $U''(x) \le 0$, the solution of (5.36) satisfies*

$$\|p(\cdot,t)\|_{L^2(\Omega)}^2 + \frac{2}{\Gamma(\alpha)}\int_0^t \frac{B(p,p)}{(t-s)^{1-\alpha}} ds \le \|g(x)\|_{L^2(\Omega)}. \tag{5.45}$$

Proof. Choosing $v = p(\cdot,t) \in H_0^{\frac{\beta}{2}}(\Omega)$ in (5.36) yields

$$\left({}_0^C D_t^\alpha p, p\right) + B(p,p) = 0. \tag{5.46}$$

Noting that $\left({}_0^C D_t^\alpha p, p\right) \ge \frac{1}{2}{}_0^C D_t^\alpha \|p\|_{L^2(\Omega)}^2$ (see (3.102) in Chap. 3), it follows that

$${}_0^C D_t^\alpha \|p\|_{L^2(\Omega)}^2 + 2B(p,p) \le 0. \tag{5.47}$$

By performing the operator ${}_0D_t^{-\alpha}$ on both sides of (5.47) and using (B.29), one has

$$\|p(\cdot,t)\|_{L^2(\Omega)}^2 - \|p(\cdot,0)\|_{L^2(\Omega)}^2 + 2\,{}_0D_t^{-\alpha} B(p,p) \le 0. \tag{5.48}$$

Thus, we complete the proof. □

5.2.1.1 *Stability analysis and error estimates for the time semi-discrete scheme*

There are several ways to discrete the time fractional derivative and speed its computation [Alikhanov (2015); Lin *et al.* (2011); Zhang and Deng (2017)]. Here we use the so-called L1 approximation introduced in [Lin *et al.* (2011)].

Letting $0 = t_0 < t_1 < \cdots < t_m < \cdots t_M = T$ be a partition of $[0, T]$, then

$$\begin{aligned} &{}_0^C D_t^\alpha p(x,t) \mid_{t=t_m} \\ &= \frac{1}{\Gamma(1-\alpha)} \int_0^{t_m} \frac{\partial P(x,\xi)}{\partial \xi} (t_m - \xi)^{-\alpha} d\xi \\ &= \frac{1}{\Gamma(1-\alpha)} \sum_{j=0}^{m-1} \int_{t_j}^{t_{j+1}} \frac{\partial P(x,\xi)}{\partial \xi} (t_m - \xi)^{-\alpha} d\xi; \end{aligned} \tag{5.49}$$

replacing $p(x,t)$ at every interval $[t_j, t_{j+1}]$ with first degree polynomial $p(x,t_j)\frac{t_{j+1}-t}{t_{j+1}-t_j} + p(x,t_{j+1})\frac{t_j - t}{t_j - t_{j+1}}$ yields the approximation

$$\begin{aligned} &{}_0^C D_t^\alpha p(x,t) \mid_{t=t_m} \\ &= \frac{1}{\Gamma(1-\alpha)} \sum_{j=0}^{m-1} \frac{p(x,t_{j+1}) - p(x,t_j)}{t_{j+1} - t_j} \int_{t_j}^{t_{j+1}} \frac{d\xi}{(t_m - \xi)^\alpha} + r_\tau^m. \end{aligned} \tag{5.50}$$

If the partition is uniform with the stepsize $\tau = \frac{T}{M}$, we obtain the so-called L1 discretization which can be written in a compact form

$$\begin{aligned} {}_0 D_t^\alpha p(x,t) \mid_{t=t_m} = \frac{\tau^{-\alpha}}{\Gamma(2-\alpha)} \Bigg[&\sum_{j=0}^{m-1} \left(Q_j^\alpha - Q_{j-1}^\alpha \right) p(x, t_{m-j}) \\ &- Q_{m-1}^\alpha p(x, t_0) \Bigg] + r_\tau^m, \end{aligned} \tag{5.51}$$

where $Q_j^\alpha = (j+1)^{1-\alpha} - j^{1-\alpha} = (1-\alpha)\int_j^{j+1} s^{-\alpha} ds$ for $j = 0, 1, \cdots, m-1$ and $Q_{-1}^\alpha := 0$. Furthermore, if $p(x,t) \in C^2[0,T]$ w.r.t. t, then

$$r_\tau^m \le C_p \tau^{2-\alpha}, \tag{5.52}$$

where C_p is a constant depending only on $p(x,t)$. By $Q_j^\alpha = (1 - \alpha)\int_j^{j+1} s^{-\alpha} ds$, it is easy to check that

Proposition 5.1. *For any* $0 < \alpha < 1$, *the coefficients* Q_j^α $(j \in \mathbb{N})$ *satisfy*

$$\begin{aligned} &Q_0^\alpha = 1 > Q_j^\alpha > Q_{j+1}^\alpha > 0, \\ &(1-\alpha)(j+1)^{-\alpha} \le Q_j^\alpha \le j^{-\alpha}(1-\alpha) \text{ for } j = 1, 2, \cdots. \end{aligned}$$

Using p^m as the discrete approximation of $p(x, t_m)$ leads to the following time-discrete scheme of (5.36): find $p^m \in H_0^{\frac{\beta}{2}}(\Omega)$ $(m = 1, 2, \cdots, M)$ such that for any $v \in H_0^{\frac{\beta}{2}}(\Omega)$

$$(p^m, v) + \alpha_0 B(p^m, v) = \sum_{j=1}^{m-1} \left(Q_{j-1}^\alpha - Q_j^\alpha \right) \left(p^{m-j}, v \right) + Q_{m-1}^\alpha \left(p^0, v \right), \tag{5.53}$$

where $p^0 \in H_0^{\frac{\beta}{2}}(\Omega)$ is the approximation of $g(x)$, and $\alpha_0 = \Gamma(2-\alpha)\tau^\alpha$.

Define

$$|||p||| = \left(\|p\|^2_{L^2(\Omega)} + \alpha_0 B(p,p)\right)^{\frac{1}{2}}. \tag{5.54}$$

It is easy to see that $|||\cdot|||$ is equivalent to $\|\cdot\|_{H^{\frac{\beta}{2}}(\Omega)}$ in $H_0^{\frac{\beta}{2}}(\Omega)$.

Theorem 5.4. *Let $U''(x) \leq 0$ for any $x \in \Omega$. The time semi-discrete scheme (5.53) is unconditionally stable in the sense that for any time-step length $\tau > 0$, it holds that*

$$|||p^m||| \leq \left\|p^0\right\|_{L^2(\Omega)}. \tag{5.55}$$

Proof. In order to simply the subsequent analysis, we consider the weak formulation (5.53) as a special case of

$$\begin{aligned}(p^m, v) &+ \alpha_0 B(p^m, v)\\ &= \sum_{j=1}^{m-1}\left(Q^\alpha_{j-1} - Q^\alpha_j\right)\left(p^{m-j}, v\right) + Q^\alpha_{m-1}\left(p^0, v\right) + \alpha_0\left(f^m, v\right),\end{aligned} \tag{5.56}$$

where $f^m \in L^2(\Omega)$.

When $m = 1$, we have

$$(p^1, v) + \alpha_0 B(p^1, v) = (p^0, v) + \alpha_0 (f, v) \quad \forall v \in H_0^{\frac{\beta}{2}}(\Omega). \tag{5.57}$$

Taking $v = p^1 \in H_0^{\frac{\beta}{2}}(\Omega)$ and applying the Cauchy-Schwarz inequality, we obtain

$$|||p^1|||^2 \leq \left\|p^0\right\|_{L^2(\Omega)}\left\|p^1\right\|_{L^2(\Omega)} + \alpha_0 \|f\|_{L^2(\Omega)}\left\|p^1\right\|_{L^2(\Omega)}. \tag{5.58}$$

Applying $\left\|p^1\right\|_{L^2(\Omega)} \leq |||p^1|||$ and $Q_0^\alpha = 1$, we immediately attain

$$|||p^1||| \leq \left\|p^0\right\|_{L^2(\Omega)} + \frac{\alpha_0}{Q_0^\alpha}\left\|f^1\right\|_{L^2(\Omega)}. \tag{5.59}$$

Assume we have proven the following inequality

$$|||p^k||| \leq \left\|p^0\right\|_{L^2(\Omega)} + \frac{\alpha_0}{Q^\alpha_{k-1}} \max_{1\leq l\leq k}\left\|f^l\right\|_{L^2(\Omega)}, \quad k = 1, 2, \cdots, m-1. \tag{5.60}$$

Then we need to prove

$$|||p^m||| \leq \left\|p^0\right\|_{L^2(\Omega)} + \frac{\alpha_0}{Q^\alpha_{m-1}} \max_{1\leq l\leq m}\left\|f^l\right\|_{L^2(\Omega)}.$$

Letting $v = p^m$ in (5.68) gives

$$\begin{aligned}
&(p^m, p^m) + \alpha_0 B(p^m, p^m) \\
&= \sum_{j=1}^{m-1} \left(Q_{j-1}^\alpha - Q_j^\alpha\right)\left(p^{m-j}, p^m\right) + Q_{m-1}^\alpha \left(p^0, p^m\right) + \alpha_0 \left(f^m, p^m\right). \quad (5.61)
\end{aligned}$$

Then

$$\begin{aligned}
|||p^m||| &\leq \sum_{j=1}^{m-1} \left(Q_{j-1}^\alpha - Q_j^\alpha\right) \left\|p^{m-j}\right\|_{L^2(\Omega)} \\
&\quad + Q_{m-1}^\alpha \left(\left\|p^0\right\|_{L^2(\Omega)} + \frac{\alpha_0}{Q_{m-1}^\alpha} \|f^m\|_{L^2(\Omega)} \right) \\
&\leq \left(\sum_{j=1}^{m-1} \left(Q_{j-1}^\alpha - Q_j^\alpha\right) + Q_{m-1}^\alpha \right) \left(\left\|p^0\right\|_{L^2(\Omega)} + \frac{\alpha_0}{Q_{m-1}^\alpha} \max_{1 \leq l \leq m} \left\|f^l\right\|_{L^2(\Omega)} \right) \\
&= \left\|p^0\right\|_{L^2(\Omega)} + \frac{\alpha_0}{Q_{m-1}^\alpha} \max_{1 \leq l \leq m} \left\|f^l\right\|_{L^2(\Omega)}. \quad (5.62)
\end{aligned}$$

Thus, the inequality (5.55) can be obtained by letting $f^l = 0$. □

Theorem 5.5. *Let $U''(x) \leq 0$ for any $x \in \Omega$, $p(x,t)$ be the exact solution of (5.36), and $\{p^m\}$ be the solutions of the time semidiscrete scheme (5.53). Then the error estimate is*

$$|||p(x, t_m) - p^m||| \leq C_p \Gamma(1-\alpha) T^\alpha (b-a)^{\frac{1}{2}} \tau^{2-\alpha}, \quad m = 1, 2, \cdots, M. \quad (5.63)$$

Proof. Let $E^m = p^m - p(x, t_m)$. Then for any $v \in H_0^{\frac{\beta}{2}}(\Omega)$, it holds that

$$\begin{aligned}
&(E^m, v) + \alpha_0 B(E^m, v) \\
&= \sum_{j=1}^{m-1} \left(Q_{j-1}^\alpha - Q_j^\alpha\right)\left(E^{m-j}, v\right) + Q_{m-1}^\alpha \left(E^0, v\right) + \alpha_0 \left(r_\tau^m, v\right) \quad (5.64)
\end{aligned}$$

with $E^0 = 0$. From (5.62), one immediately obtains

$$|||E^m||| \leq \frac{\alpha_0}{Q_{m-1}^\alpha} \max_{1 \leq l \leq m} \left\|r_\tau^l\right\|_{L^2(\Omega)}. \quad (5.65)$$

Then noting that $\alpha_0 = \Gamma(2-\alpha)\tau^\alpha$, $Q_{m-1}^\alpha \geq (1-\alpha)m^{-\alpha}$ and $r_\tau^l \leq C_p \tau^{2-\alpha}$ for $l = 1, 2, \cdots, M$, one can complete the proof. □

Remark 5.2. If $\alpha \to 1$, then $\Gamma(1-\alpha) \to +\infty$. So the estimate (5.63) makes no sense. In this case, we may need an estimate of another form: for $m = 1, 2, \cdots, M$,

$$|||E^m||| \leq C_p \Gamma(2-\alpha)(b-a)^{\frac{1}{2}} T \tau. \quad (5.66)$$

In fact, using the mathematical induction, we can prove that for all $m = 1, 2, \cdots, M$, it follows that

$$|||E^m||| \leq C_p\Gamma(2-\alpha)(b-a)^{\frac{1}{2}} m\tau^2. \tag{5.67}$$

A simple proof of (5.67) can be given as follows: obviously, for $m = 1$, by inequality (5.65), we have

$$|||E^1||| \leq C_p\Gamma(2-\alpha)(b-a)^{\frac{1}{2}} \tau^2.$$

Assume that

$$|||E^k||| \leq C_p\Gamma(2-\alpha)(b-a)^{\frac{1}{2}} k\tau^2, \ k = 1, 2, \cdots, m-1.$$

Then

$$\begin{aligned} |||E^m||| &\leq \sum_{j=1}^{m-1} \left(Q_{j-1}^\alpha - Q_j^\alpha\right) \left\|E^{m-j}\right\|_{L^2(\Omega)} + \alpha_0 \left\|r_\tau^m\right\|_{L^2(\Omega)} \\ &\leq \Big(\sum_{j=1}^{m-1} \left(Q_{j-1}^\alpha - Q_j^\alpha\right)\Big) C_p\Gamma(2-\alpha)(b-a)^{\frac{1}{2}}(m-1)\tau^2 + \alpha_0 \left\|r_\tau^m\right\|_{L^2(\Omega)} \\ &\leq C_p\Gamma(2-\alpha)(b-a)^{\frac{1}{2}} m\tau^2. \end{aligned}$$

To obtain the space discretizations of (5.53) in the Galerkin framework, we should choose an appropriate subspace $S_h \subset H_0^{\frac{\beta}{2}}(\Omega)$, and formulate the full discretization of (5.53) as : find $p_h^m \in S_h$ such that for all $v \in S_h$

$$\begin{aligned} &(p_h^m, v) + \alpha_0 B(p_h^m, v) \\ &\quad = \sum_{j=1}^{m-1} \left(Q_{j-1}^\alpha - Q_j^\alpha\right)\left(p_h^{m-j}, v\right) + Q_{m-1}^\alpha \left(p_h^0, v\right), \end{aligned} \tag{5.68}$$

where $p_h^0 \in S_h$ is an appropriate approximation of $g(x)$. There are two choices for the basis functions: the finite element approximation and the spectral approximation. The former is to use the basis functions which have the compact support, while the ones of the latter are global. In the following two subsections, we will introduce them separately. However, no matter what kind of space discretizations are used, we should note that

(1) similar to the proof of Theorem 5.4, when $U''(x) \leq 0$, the full discretization scheme (5.68) always is unconditionally stable, i.e.,

$$|||p_h^m||| \leq \left\|p_h^0\right\|_{L^2(\Omega)} \quad \text{for } m = 1, 2, \cdots, M; \tag{5.69}$$

(2) the continuity and coercivity of the bilinear $B(\cdot,\cdot)$ (see Lemma 5.2) actually mean that for any $u \in H_0^{\frac{\beta}{2}}(\Omega)$, there exists a unique $R_h u \in S_h$ such that

$$B(R_h u, v) = B(u, v) \quad \forall v \in S_h. \tag{5.70}$$

In addition, since

$$\begin{aligned} &C_2 \|u - R_h u\|^2_{H^{\frac{\beta}{2}}(\Omega)} \\ &\leq B(u - R_h u, u - R_h u) \\ &= B(u - R_h u, u - v) \\ &\leq C_1 \|u - R_h u\|_{H^{\frac{\beta}{2}}(\Omega)} \|u - v\|_{H^{\frac{\beta}{2}}(\Omega)}, \end{aligned}$$

we have

$$\|u - R_h u\|_{H^{\frac{\beta}{2}}(\Omega)} \leq \frac{C_1}{C_2} \inf_{v \in S_h} \|u - v\|_{H^{\frac{\beta}{2}}(\Omega)}. \tag{5.71}$$

5.2.2 *Finite element space discretization and related implementation*

In this subsection, we consider the piecewise polynomial finite element approximation. Let $\{X_h\}$ be a family of partitions of Ω with grid parameter h, and associated with X_h, define the finite dimensional subspace S_h to be the basis of the piecewise polynomial of order $d\,(d \in \mathbb{N})$. Denote $\Pi_h u$ as the Scott-Zhang interpolation of u in S_h [Ciarlet (2013); Scott and Zhang (1990); Ern and Guermond (2004)]. Then for any $u \in H^l(\Omega)$ and $\frac{\beta}{2} \leq l \leq d$ it follows that

$$\|u - \Pi_h u\|_{H^{\frac{\beta}{2}}(\Omega)} \lesssim h^{l-\frac{\beta}{2}} \|u\|_{H^l(\Omega)}. \tag{5.72}$$

Therefore, by (5.71), if $p(\cdot,t) \in H^l(\Omega)$, one has

$$\|p(\cdot,t) - R_h p(\cdot,t)\|_{H^{\frac{\beta}{2}}(\Omega)} \lesssim h^{l-\frac{\beta}{2}} \|p(\cdot,t)\|_{H^l(\Omega)}. \tag{5.73}$$

Theorem 5.6. *Let $U''(x) \leq 0$ for any $x \in \Omega$, $p(x,t)$ be the exact solution of (5.36), and $\{p_h^m\}$ be the solution of (5.53) with the initial condition $p_h^0 = R_h g$. Suppose that $g \in H^l(\Omega)\,(\frac{\beta}{2} \leq l \leq d)$, $p, \frac{\partial p}{\partial t} \in L^\infty(0,T; H_0^{\frac{\beta}{2}}(\Omega) \cap H^l(\Omega))\,(\frac{\beta}{2} \leq l \leq d)$, and $\frac{\partial^2 p}{\partial t^2} \in L^\infty(0,T; L^\infty(\Omega))$. Then*

(1) when $0 < \alpha < 1$, one has

$$\begin{aligned} |||p(x,t_m) - p_h^m||| &\lesssim C_p \Gamma(1-\alpha) T^\alpha (b-a)^{\frac{1}{2}} \tau^{2-\alpha} \\ &+ \frac{T}{1-\alpha} h^{l-\frac{\beta}{2}} \max_{0 \leq t \leq T} \|p_t\|_{H^l(\Omega)} + (1 + \tau^\alpha \Gamma(2-\alpha)) h^{l-\frac{\beta}{2}} \|p\|_{H^l(\Omega)}; \end{aligned}$$

(2) when $\alpha \to 1$, one has

$$|||p(x,t_m) - p_h^m||| \lesssim C_p\Gamma(2-\alpha)T(b-a)^{\frac{1}{2}}\tau + T^2\tau^{\alpha-1}h^{l-\frac{\beta}{2}} \max_{0\le t\le T} \|p_t\|_{H^l(\Omega)} + (1+\tau^\alpha\Gamma(2-\alpha))h^{l-\frac{\beta}{2}} \|p\|_{H^l(\Omega)}.$$

Proof. Let $E_h^m = p_h^m - p(x,t_m) = \theta^m + \eta^m$, where $\theta^m = R_h p(x,t_m) - p(x,t_m)$ and $\eta^m = p_h^m - R_h p(x,t_m)$. Using the property $B(\theta^m, v) = 0, \forall v \in S_h$, it yields the error equation

$$\begin{aligned}(\eta^m, v) &+ \alpha_0 B(\eta^m, v)\\ &= \sum_{j=1}^{m-1} \left(Q_{j-1}^\alpha - Q_j^\alpha\right)\left(\eta^{m-j}, v\right)\\ &\quad + Q_{m-1}^\alpha\left(\eta^0, v\right) + \alpha_0(r_\tau^m + r_h^m),\end{aligned} \tag{5.74}$$

where

$$r_h^m = \frac{\tau^{-\alpha}}{\Gamma(2-\alpha)}\left[\sum_{j=0}^{m-1}\left(Q_{j-1}^\alpha - Q_j^\alpha\right)\theta^{m-j} + Q_{m-1}^\alpha\theta^0\right]. \tag{5.75}$$

Note that $\eta^0 = 0$. Therefore,

(1) when $0 < \alpha < 1$, similar to the inequality (5.65), one has

$$|||\eta^m||| \le \frac{\alpha_0}{Q_{m-1}^\alpha} \max_{1\le l\le m} \left\|r_\tau^l + r_h^m\right\|_{L^2(\Omega)}. \tag{5.76}$$

(2) when $\alpha \to 1$, similar to the proof of the inequality (5.66), one has

$$|||\eta^m||| \le \Gamma(2-\alpha)\tau^\alpha m \max_{1\le l\le m} \left\|r_\tau^l + r_h^m\right\|_{L^2(\Omega)}. \tag{5.77}$$

Moreover, it holds that $\|r_\tau^m\|_{L^2(\Omega)} \le C_p\tau^{2-\alpha}$ for $m = 1, 2, \cdots, M$, and

$$\begin{aligned}&\|r_h^m\|_{L^2(\Omega)}\\ &= \frac{\tau^{-\alpha}}{\Gamma(2-\alpha)}\left\|\sum_{j=0}^{m-1} Q_{m-1-j}^\alpha\left(\theta^{j+1} - \theta^j\right)\right\|_{L^2(\Omega)}\\ &\le \frac{\tau^{-\alpha}}{\Gamma(2-\alpha)}\sum_{j=0}^{m-1} Q_{m-1-j}^\alpha \int_{t_j}^{t_{j+1}} \left\|\frac{\partial\theta}{\partial t}\right\|_{L^2(\Omega)} dt\\ &\le \max_{0\le t\le t_n}\left\|\frac{\partial\theta}{\partial t}\right\|_{L^2(\Omega)} \sum_{j=0}^{m-1} \frac{\tau^{1-\alpha}}{\Gamma(2-\alpha)} Q_{m-1-j}^\alpha\\ &= \frac{t_m^{1-\alpha}}{\Gamma(2-\alpha)} \max_{0\le t\le t_n}\left\|\frac{\partial\theta}{\partial t}\right\|_{L^2(\Omega)}.\end{aligned} \tag{5.78}$$

Combining $\|\theta\|_{L^2(\Omega)} \lesssim \|\theta\|_{H^{\frac{\beta}{2}}(\Omega)} \lesssim h^{l-\frac{\beta}{2}} \|p\|_{H^l(\Omega)}$ and the triangle inequality, one can complete the proof. □

Remark 5.3. If the analytical solution of (5.33) has a low regularity near $t = 0$ [Le *et al.* (2016)], the non-uniform time partition may be used to obtain a better approximation, and the corresponding numerical schemes are still stable [Zhang and Deng (2017)].

One challenge exists in numerical solutions of fractional PDEs because of the non-locality of the underlying fractional integral and derivative operators. Indeed, the finite difference/finite element methods usually lead to dense matrices. It is therefore of importance to construct fast solvers by carefully analysing the structures of the matrices. In the following, we give the effective implementation. Let $a = x_0 < \cdots < x_k < \cdots < x_N = b$ with $x_k = a + kh,\ h = \frac{b-a}{N}$ be the space partition. Then the linear element ($i.e., d = 2$) bases can be given as

$$\phi_{h,k}(x) = \begin{cases} \frac{x-x_k}{\sqrt{h}} & x_k \le x \le x_{k+1}, \\ \frac{x_{k+2}-x}{\sqrt{h}} & x_{k+1} < x \le x_{k+2}, \end{cases} \quad k = 0, 1, \cdots, N-2. \tag{5.79}$$

Denote

$$\phi(x) = \begin{cases} x, & 0 \le x \le 1, \\ 2-x, & 1 < x \le 2. \end{cases} \tag{5.80}$$

It is easy to check that

$$\phi_{h,k}(x) = \phi\left(\frac{x-a}{h} - k\right), \quad k = 0, 1, \cdots, N-2. \tag{5.81}$$

Therefore, $\phi_{h,k}$ is just the dilation and translation of a single function $\phi(x)$. Let $\Phi_h(x) = \{\phi_{h,k}(x)\}_{k=0}^{N-2}$. Then $S_h = \text{span}\{\Phi_h(x)\}$.

Assume $p_h^l = \sum_{k=0}^{N-2} p_{h,k}^l \phi_{h,k}$ for $l = 0, 1, \cdots, M$, and denote

$$A_L = \left({}_aD_x^{\frac{\beta}{2}}\Phi_h, {}_xD_b^{\frac{\beta}{2}}\Phi_h\right), A_R = \left({}_xD_b^{\frac{\beta}{2}}\Phi_h, {}_aD_x^{\frac{\beta}{2}}\Phi_h\right), \tag{5.82}$$

$$B = (\Phi_h, \Phi_h), B_1 = \left(\frac{\partial}{\partial x}(U'\Phi_h), \Phi_h\right), \tag{5.83}$$

$$P^l = \left[p_{h,0}^l, p_{h,1}^l, \cdots, p_{h,N-2}^l\right]^{\mathrm{T}}. \tag{5.84}$$

Then the matrix form of the full discretization scheme (5.68) can be given as

$$\begin{aligned} &\left[B - \frac{\alpha_0}{\eta_\alpha}B_1 - \frac{\alpha_0\kappa_\alpha}{2}(A_L + A_R)\right]P^m \\ &= \sum_{j=1}^{m-1}\left(Q_{j-1}^\alpha - Q_j^\alpha\right)BP^{m-j} + Q_{m-1}^\alpha BP^0. \end{aligned} \tag{5.85}$$

It is easy to see that the main task to solve system (5.85) is to handle A_L and A_R efficiently. Since $A_R = {A_L}^{\mathrm{T}}$, it is enough to consider A_L. To simplify the calculation and reduce the storage, we present the following result.

Lemma 5.4 ([Deng and Zhang (2018)]). *Let* $\mathcal{L}[u](x) = \int_a^x g(x-\xi)u(\xi)d\xi$ *and* $c > 0$. *Assume that there exist* $\phi_1(x)$ *and* $\phi_2(x)$ *with compact supports* $\operatorname{supp}\phi_1(x) = [0, d_1]$ *and* $\operatorname{supp}\phi_2(x) = [0, d_2]$, *such that all the supports of* $\phi_{1,k}(x) = \phi_1(c(x-a)-k)$ *and* $\phi_{2,k}(x) = \phi_2(c(x-a)-k), k = 0, 1, \cdots, M$ *lie in* (a, b), *and define* $\Phi_1(x) = \{\phi_{1,k}\}_{k=0}^M$ *and* $\Phi_2(x) = \{\phi_{2,k}\}_{k=0}^M$. *Then the matrix* $(\mathcal{L}[\Phi_1], \Phi_2)$ *with the element* $(\mathcal{L}[\Phi_1], \Phi_2)_{ij} = (\phi_{1,i}(x), \phi_{2,j}(x))$ *is Toeplitz.*

Proof. Since

$$
\begin{aligned}
&(\mathcal{L}[\phi_{1,k_1}], \phi_{2,k_2}) \\
&= \int_a^b \int_a^x g(x-\xi)\phi_{1,k_1}(\xi)\, d\xi\, \phi_{2,k_2}(x)\, dx \qquad (5.86)\\
&= \int_0^{b-a} \int_0^x g(x-\xi)\phi_1(c\xi - k_1)\, d\xi \phi_2(cx - k_2)\, dx \\
&= \int_{\frac{k_2}{c}}^{\frac{d_2+k_2}{c}} \int_0^x g(x-\xi)\phi_1(c\xi - k_1)\, d\xi \phi_2(cx - k_2)\, dx \\
&= \frac{1}{c}\int_0^{d_2} \int_0^{\frac{x+k_2}{c}} g\left(\frac{k_2+x}{c} - \xi\right) \phi_1(c\xi - k_1)\, d\xi \phi_2(x)\, dx \\
&= \frac{1}{c^2}\int_0^{d_2} \int_0^{\min\{x+k_2-k_1, d_1\}} g\left(\frac{x+k_2-k_1-\xi}{c}\right) \phi_1(\xi)\, d\xi \phi_2(x)\, dx,
\end{aligned}
$$

which just depends on the value of $k_2 - k_1$. Thus, we complete the proof. □

Using the fact that $\Phi_h \in H_0^1(\Omega)$, one has

$$
A_L = \left({}_aD_x^{\frac{\beta}{2}}\Phi_h, {}_xD_b^{\frac{\beta}{2}}\Phi_h\right) = -\left({}_aD_x^{-(2-\beta)}\frac{d\Phi_h}{dx}, \frac{d\Phi_h}{dx}\right). \qquad (5.87)
$$

Obviously, all the elements of the vector function $\frac{d\Phi_h}{dx}$ are still the dilation and translation of a single function $\phi'(x)$. Letting $g(x) = \frac{x^{1-\beta}}{\Gamma(2-\beta)}$ in Lemma 5.4, then matrix A_L is Toeplitz. Combining with the compact support of $\phi_{h,k}$, we only need to calculate and store N elements of A_L. It is also possible to calculate A_L produced with the high-degree elements, with the storage cost only $\mathcal{O}(N)$, instead of $\mathcal{O}(N^2)$. Based on Lemma 5.4, we try to take the bases as the dilations and translations of several known functions. For example, if one defines the compact support functions

$$\phi_1^2(x) := \begin{cases} 2x^2 - x, & 0 \le x < 1, \\ 2x^2 - 7x + 6, & 1 \le x \le 2, \end{cases} \tag{5.88}$$

$$\phi_2^2(x) := -4x^2 + 4x, \quad 0 \le x \le 1, \tag{5.89}$$

and takes

$$\Phi_{h,1}^2 = \left\{ \phi_{h,1,k}^2, k = 0, \cdots, N-2 : \phi_{h,1,k}^2 = \frac{1}{\sqrt{h}} \phi_1^2 \left(\frac{x-a}{h} - k \right) \right\},$$

$$\Phi_{h,2}^2 = \left\{ \phi_{h,2,k}^2, k = 0, \cdots, N-1 : \phi_{h,2,k}^2 = \frac{1}{\sqrt{h}} \phi_2^2 \left(\frac{x-a}{h} - k \right) \right\},$$

then $\Phi^2 = \left\{ \Phi_{h,1}^2, \Phi_{h,2}^2 \right\}$ are the bases of the quadratic element space. By Lemma 5.4, A_L is a block Toeplitz matrix. To compute A_L, we can separate it into four parts, and in total only $4(N+3)$ entries need to be computed and stored.

We can also calculate the fractional derivatives of the bases quickly. Indeed, for $d/c \ge a$ with $c > 0$, $k \in \mathbb{N}^+$, one has

$$_aD_x^{\beta-1} (cx - d)_+^k = c^{\beta-1} \frac{\Gamma(k+1)}{\Gamma(k-\beta+2)} (cx-d)_+^{k-\beta+1}, \tag{5.90}$$

where $x_+^k = (\max\{0, x\})^k$. Since

$$\phi(x) = x_+ - 2(x-1)_+ + (x-2)_+. \tag{5.91}$$

Then

$$\begin{aligned} &_aD_x^{\beta-1} \phi_{h,k}(x) \\ &= \frac{1}{\sqrt{h}} {_aD_x^{\beta-1}} \phi \left(\frac{x-a}{h} - k \right) \\ &= \frac{h^{\frac{1}{2}-\beta}}{\Gamma(3-\beta)} \left(X_+^{2-\beta} - 2(X-1)_+^{2-\beta} + (X-2)_+^{2-\beta} \right), \end{aligned} \tag{5.92}$$

where $X = \frac{x-a}{h} - k$. These techniques can also be applied to the high-degree elements, such as, for the quadratic element, one has

$$\phi_1^2(x) = 2x_+^2 - x_+ - 6(x-1)_+ - 2(x-2)_+^2 - (x-2)_+,$$
$$\phi_2^2(x) = -4x_+^2 + 4x_+ + 4(x-1)_+^2 + 4(x-1)_+.$$

The Toeplitz or block Toeplitz structure allows one to compute the matrix vector product with the cost $\mathcal{O}(N \log N)$. Then the iterative methods and multigrid method work efficiently (see Appendix C).

Remark 5.4. Besides the above classic piecewise polynomial basis functions, one can also use the interval B-spline functions as the basis of the finite element space S_h [Zhang and Deng (2017)]. They are the dilation and translation of a single function, except a few boundary basis functions, and the Teoplitz structure of A_L for high degree element can be kept.

5.2.3 *Spectral Galerkin space discretization and related implementation*

The nonlocality of fractional derivative suggest that global schemes such as spectral method may be appropriate tools for discretizing fractional differential equation. In [Li and Xu (2009)], the authors develop a time-space spectral method for the time-fractional diffusion equation. Spectral collocation methods for fractional problems using classical interpolation basis functions is proposed in [Tian *et al.* (2014)]. However, spectral accuracy requires high regularity of the solution in the whole domain. Due to the possible boundary low regularity [Jin *et al.* (2015a); Wang and Zhang (2015)], the authors in [Chen *et al.* (2016); Zayernouri and Karniadakis (2013, 2014a,b)] approximate the solution of the fractional differential equation with one-sided fractional derivatives by the generalized Jacobi functions, but it seems hard to know how to choose the basis functions appropriately. In the following, we give the usual spectral Galerkin method for solving the Fokker-Planck equation.

For convenience, let $\Omega = (-1,1)$ and $\mathbb{P}_N$ be the space of polynomials on $[-1,1]$ with the degree no greater than N. For any $n \in \mathbb{N}$, let $L_n(x)$ be the nth degree Legendre polynomial on the interval $[-1,1]$, defined by the recurrence relation:

$$L_0(x) = 1, \quad L_1(x) = x,$$
$$L_{n+1}(x) = \frac{2n+1}{n+1} x L_n(x) - \frac{n}{n+1} L_{n-1}(x), \quad n \geq 1.$$

It is well known that Legendre polynomials are orthogonal in $[-1,1]$, i.e.,

$$(L_n(x), L_m(x)) = \frac{2}{2n+1}\delta_{m,n}, \tag{5.93}$$

and it holds that

$$L'_{n+1}(x) - L'_{n-1}(x) = (2n+1)L_n(x), \tag{5.94}$$

$$L'_n(x) = \frac{1}{2}(n+1)J^{1,1}_{n-1}(x), \, n \geq 1, \quad L_n(\pm 1) = (\pm 1)^n. \tag{5.95}$$

Let

$$\phi_n(x) = L_n(x) - L_{n+2}(x). \tag{5.96}$$

Then the space $S_N = \{v \in \mathbb{P}_N[-1,1] : v(-1) = v(1) = 0\} \subset H_0^{\frac{\beta}{2}}(\Omega)$ can be generated by $\phi_n(x)\,(n = 0, \cdots, N-2)$ [Shen and Tang (2006), p. 110], and the spectral Galerkin scheme can be formulated as: find $p_N^m \in S_N$, such that for all $v \in S_N$

$$\begin{aligned}&(p_N^m, v) + \alpha_0 B(p_N^m, v)\\ &= \sum_{j=1}^{m-1} \left(Q_{j-1}^{\alpha} - Q_j^{\alpha}\right)\left(p_N^{m-j}, v\right) + Q_{m-1}^{\alpha}\left(p_N^0, v\right),\end{aligned} \tag{5.97}$$

where $p_N^0 \in S_N$ is an appropriate approximation of $g(x)$.

In order to perform the error estimate of the spectral Galerkin scheme, we define the $H_0^1(\Omega)$-orthogonal project operator $\Pi_N : H_0^1(\Omega) \to S_N$ by

$$((\Pi_N u - u)', v') = 0, \quad \forall v \in S_N, \tag{5.98}$$

then if $0 \le \mu \le 1 \le l$ and $u \in H_0^1(\Omega) \cap H^l(\Omega)$, it holds that [Guo (1998), Theorem 2.11]

$$\|u - \Pi_N u\|_{H^{\mu}(\Omega)} \lesssim N^{\mu - l} \|u\|_{H^l(\Omega)}. \tag{5.99}$$

Lemma 5.5. *Let $R_N : H_0^{\frac{\beta}{2}}(\Omega) \to S_N$ be the R_h defined in (5.70). Then if $p(\cdot,t) \in H_0^{\frac{\beta}{2}}(\Omega) \cap H^l(\Omega)$ with $l \ge \frac{\beta}{2}$, one has*

$$\|p(\cdot,t) - R_N p(\cdot,t)\|_{H^{\frac{\beta}{2}}(\Omega)} \lesssim N^{\frac{\beta}{2} - l} \|p(\cdot,t)\|_{H^l(\Omega)}. \tag{5.100}$$

Proof. If $l = \frac{\beta}{2}$, by (5.71), we obviously have

$$\|p(\cdot,t) - R_N p(\cdot,t)\|_{H^{\frac{\beta}{2}}(\Omega)} \lesssim \|p(\cdot,t)\|_{H^{\frac{\beta}{2}}(\Omega)}. \tag{5.101}$$

If $l \ge 1$, making use of (5.71) and (5.99), we also have

$$\|p(\cdot,t) - R_N p(\cdot,t)\|_{H^{\frac{\beta}{2}}(\Omega)} \lesssim N^{\frac{\beta}{2} - l} \|p(\cdot,t)\|_{H^l(\Omega)}. \tag{5.102}$$

In the following, we give the proof for $\frac{\beta}{2} < l < 1$ by space interpolation. Let us define an operator, $\mathcal{T}$, that maps $p(\cdot,t)$ to the error $p(\cdot,t) - R_N p(\cdot,t)$, that is,

$$\mathcal{T} p(\cdot,t) := p(\cdot,t) - R_N p(\cdot,t). \tag{5.103}$$

Estimate (5.101) implies that $\mathcal{T}$ maps $H^{\frac{\beta}{2}}(\Omega) \cap H_0^{\frac{\beta}{2}}(\Omega)$ to $H_0^{\frac{\beta}{2}}(\Omega)$, with

$$\|\mathcal{T}\|_{H^{\frac{\beta}{2}}(\Omega) \cap H_0^{\frac{\beta}{2}}(\Omega) \to H_0^{\frac{\beta}{2}}(\Omega)} \lesssim 1; \tag{5.104}$$

and (5.102) implies that $\mathcal{T}$ maps $H^1(\Omega) \cap H_0^{\frac{\beta}{2}}(\Omega)$ to $H_0^{\frac{\beta}{2}}(\Omega)$, with

$$\|\mathcal{T}\|_{H_0^{\frac{\beta}{2}}(\Omega)\cap H^1(\Omega)\to H_0^{\frac{\beta}{2}}(\Omega)} \lesssim N^{\frac{\beta}{2}-1}. \tag{5.105}$$

In addition, for $\frac{\beta}{2} < l < 1$, we have [Guo (1998), p. 14]

$$H_0^{\frac{\beta}{2}}(\Omega) \cap H^l(\Omega) = \left[H^{\frac{\beta}{2}}(\Omega) \cap H_0^{\frac{\beta}{2}}(\Omega), H^1(\Omega) \cap H_0^{\frac{\beta}{2}}(\Omega)\right]_{\frac{l-\frac{\beta}{2}}{1-\frac{\beta}{2}}}. \tag{5.106}$$

Therefore, by Proposition A.7, we have

$$\|\mathcal{T}\|_{H_0^{\frac{\beta}{2}}(\Omega)\cap H^l(\Omega)\to H_0^{\frac{\beta}{2}}(\Omega)} \lesssim N^{\frac{\beta}{2}-l}. \tag{5.107}$$

Thus, we complete the proof. □

Now, similar to the proof of Theorem 5.6, one has

Theorem 5.7. *Let $U''(x) \leq 0$ for any $x \in \Omega$, $p(x,t)$ be the exact solution of (5.36), and $\{p_N^m\}$ be the solution of (5.97) with the initial condition $p_N^0 = R_N g$. Suppose that $g \in H^l(\Omega)$ $(\frac{\beta}{2} \leq l \leq d)$, $p, \frac{\partial p}{\partial t} \in L^\infty(0,T; H_0^{\frac{\beta}{2}}(\Omega)\cap H^l(\Omega))$ $(\frac{\beta}{2} \leq l \leq d)$, and $\frac{\partial^2 p}{\partial t^2} \in L^\infty(0,T; L^\infty(\Omega))$. Then*

(1) when $0 < \alpha < 1$, one has

$$\begin{aligned}&|||p(x,t_m) - p_N^m||| \lesssim C_p\Gamma(1-\alpha)T^\alpha(b-a)^{\frac{1}{2}}\tau^{2-\alpha}\\&+\frac{T}{1-\alpha}N^{\frac{\beta}{2}-l}\max_{0\leq t\leq T}\|p_t\|_{H^l(\Omega)} + (1+\tau^\alpha\Gamma(2-\alpha))N^{\frac{\beta}{2}-l}\|p\|_{H^l(\Omega)};\end{aligned}$$

(2) when $\alpha \to 1$, one has

$$\begin{aligned}&|||p(x,t_m) - p_N^m||| \lesssim C_p\Gamma(2-\alpha)T(b-a)^{\frac{1}{2}}\tau\\&+T^2\tau^{\alpha-1}N^{\frac{\beta}{2}-l}\max_{0\leq t\leq T}\|p_t\|_{H^l(\Omega)} + (1+\tau^\alpha\Gamma(2-\alpha))N^{\frac{\beta}{2}-l}\|p\|_{H^l(\Omega)}.\end{aligned}$$

Remark 5.5. Theorem 5.7 indicates that the convergence of numerical solution can be exponential if the exact solution is smooth. However, the exponential convergence is difficult to achieve because the smoothness of the coefficients and right-hand side does not ensure the regularity of the true solutions to fractional diffusion equations [Jin *et al.* (2015a)]. This is the primary difference when using spectral method to solve the fractional PDEs and the integer order PDEs.

Because of (5.93) and (5.94), matrices B and B_1 can be quickly generated. In fact, B is a penta-diagonal matrix, and if $U'(x)$ is constant, B_1 is a three diagonal matrix. In the following, we discuss how to produce the fractional differential matrix $A_L = \left({}_{-1}D_x^{\frac{\beta}{2}} \Phi_N, {}_xD_1^{\frac{\beta}{2}} \Phi_N \right)$ efficiently, where $\Phi_N = \{\phi_k(x)\}_{k=0}^{N-2}$. Unlike the finite element method, generally A_L has no special structure in the spectral method, so we need to compute each element of it directly.

Let $J_n^{\beta_1,\beta_2}(x)$ $(\beta_1, \beta_2 > -1)$ be the Jacobi polynomials given in Appendix A. We have

Lemma 5.6. *For $\gamma > 0, \beta_1 > -1, \beta_2 > -1$, and $\forall x \in [-1,1]$, we have*

$$\begin{aligned} &{}_{-1}D_x^{-\gamma}\left((1+x)^{\beta_2} J_n^{\beta_1,\beta_2}(x)\right) \\ &\quad = \frac{\Gamma(n+\beta_2+1)}{\Gamma(n+\beta_2+\gamma+1)}(1+x)^{\beta_2+\gamma} J_n^{\beta_1-\gamma,\beta_2+\gamma}(x). \end{aligned} \tag{5.108}$$

$$\begin{aligned} &{}_xD_1^{-\gamma}\left((1-x)^{\beta_1} J_n^{\beta_1,\beta_2}(x)\right) \\ &\quad = \frac{\Gamma(n+\beta_1+1)}{\Gamma(n+\beta_1+\gamma+1)}(1-x)^{\beta_1+\gamma} J_n^{\beta_1+\gamma,\beta_2-\gamma}(x). \end{aligned} \tag{5.109}$$

Now, let $\gamma = \frac{\beta}{2}$. Performing the fractional operator ${}_{-1}D_x^{\frac{\beta}{2}}$ on both sides of (5.108) and choosing $\beta_1 = \frac{\beta}{2}, \beta_2 = -\frac{\beta}{2}$, we obtain

$${}_{-1}D_x^{\frac{\beta}{2}} L_n(x) = \frac{\Gamma(n+1)}{\Gamma(n-\frac{\beta}{2}+1)}(1+x)^{-\frac{\beta}{2}} J_n^{\frac{\beta}{2},-\frac{\beta}{2}}(x), \quad x \in [-1,1].$$

Similarly, we have

$${}_xD_1^{\frac{\beta}{2}} L_n(x) = \frac{\Gamma(n+1)}{\Gamma(n-\frac{\beta}{2}+1)}(1-x)^{-\frac{\beta}{2}} J_n^{-\frac{\beta}{2},\frac{\beta}{2}}(x), \quad x \in [-1,1].$$

Therefore,

$$\begin{aligned} &{}_{-1}D_x^{\frac{\beta}{2}} \phi_i(x) = \\ &\quad -\left[\frac{\Gamma(i+3)}{\Gamma(i-\frac{\beta}{2}+3)} J_{i+2}^{\frac{\beta}{2},-\frac{\beta}{2}}(x) - \frac{\Gamma(i+1)}{\Gamma(i-\frac{\beta}{2}+1)} J_i^{\frac{\beta}{2},-\frac{\beta}{2}}(x)\right](1+x)^{-\frac{\beta}{2}}, \\ &{}_xD_1^{\frac{\beta}{2}} \phi_k(x) = \\ &\quad -\left[\frac{\Gamma(k+3)}{\Gamma(k-\frac{\beta}{2}+3)} J_{k+2}^{\frac{-\beta}{2},\frac{\beta}{2}}(x) - \frac{\Gamma(k+1)}{\Gamma(k-\frac{\beta}{2}+1)} J_k^{-\frac{\beta}{2},\frac{\beta}{2}}(x)\right](1-x)^{-\frac{\beta}{2}}. \end{aligned}$$

Then

$$(A_L)_{i,k} = \left({}_{-1}D_x^{\frac{\beta}{2}} \phi_k(x), {}_xD_1^{\frac{\beta}{2}} \phi_i(x) \right)$$

can be calculated exactly by the Gauss Jacobi quadrature with the weight $(1-x)^{-\frac{\beta}{2}}(1+x)^{-\frac{\beta}{2}}$ (see Appendix A).

5.2.4 Numerical results

Example Consider

$$ {}_0^C D_t^\alpha p(x,t) = \left[\frac{\partial}{\partial x} U'(x) + \nabla^\beta\right] p(x,t) + f(x,t) \tag{5.110}$$

$\times (0,T]$ with $U(x) = 3x$,

$$p(x,t) = 0, \quad (x,t) \in \mathbb{R}\backslash(0,1) \times [0,T], \tag{5.111}$$

$$p(x,0) = 0, \quad x \in (0,1), \tag{5.112}$$

and

$$\begin{aligned} f(x,t) =& \frac{2\Gamma(2)}{\Gamma(3-\alpha)} t^{2-\alpha} x^2 (1-x)^2 - 6t^2 \left(x(x-1)^2 + x^2(1-x)\right) \\ & - \frac{t^2}{2}\Big(\frac{\Gamma(5)}{\Gamma(5-\beta)}\left(x^{4-\beta} + (1-x)^{4-\beta}\right) - \frac{2\Gamma(4)}{\Gamma(4-\beta)}\left(x^{3-\beta} + (1-x)^{3-\beta}\right) \\ & + \frac{\Gamma(3)}{\Gamma(3-\beta)}\left(x^{2-\beta} + (1-x)^{2-\beta}\right)\Big), \quad (x,t) \in (0,1) \times (0,T]. \end{aligned}$$

For $(x,t) \in [0,1] \times [0,T]$, the exact solution is $p(x,t) = t^2x^2(1-x)^2$. The numerical results with the finite element approximation are listed in Table 5.1.

Table 5.1: The L^2 errors (i.e., $\|p(x,T) - p_h^M\|_{L^2(\Omega)}$) and convergence rates to Example 5.1 approximated by the finite element scheme at $t = T = 0.5$ for different α, β with $\tau = h^2$.

	$(\alpha,\beta) = (0.5, 1.3)$		$(\alpha,\beta) = (0.7, 1.8)$		$(\alpha,\beta) = (0.3, 1.7)$	
h	L^2-Err	rate	L^2-Err	rate	L^2-Err	rate
$1/2^5$	2.1800e-05	-	1.6706e-05	-	1.7272e-05	-
$1/2^6$	4.9992e-06	2.12	4.0323e-06	2.05	4.0919e-06	2.08
$1/2^7$	1.1350e-06	2.14	9.7382e-07	2.05	9.6803e-07	2.08

Example 5.2. Consider (5.110) in $(0,1) \times (0,T]$ with $U(x) = 3x$,

$$p(x,t) = 0, \quad (x,t) \in \mathbb{R}\backslash(0,1) \times [0,T], \tag{5.113}$$

$$p(x,0) = 0, \quad x \in (0,1); \tag{5.114}$$

and the source term $f(x,t)$ is derived from the exact solution

$$p(x,t) = \frac{t^2 x^{n+\beta/2}(1-x)^{n+\beta/2}}{\Gamma(\beta+1+2n)}, \quad n = 0, 1.$$

More specifically, if $n=0$,

$$f(x,t)=\frac{1}{\Gamma(\beta+1)}\Big(\frac{\Gamma(3)}{\Gamma(3-\alpha)}t^{2-\alpha}x^{\beta/2}(1-x)^{\beta/2}$$
$$-\frac{3\beta}{2}t^2x^{\beta/2-1}(1-x)^{\beta/2-1}(1-2x)\Big)-\cos\left(\frac{\pi\beta}{2}\right)t^2,$$

and if $n=1$,

$$f(x,t)=\frac{1}{\Gamma(\beta+3)}\Big(\frac{\Gamma(3)}{\Gamma(3-\alpha)}t^{2-\alpha}x^{1+\beta/2}(1-x)^{1+\beta/2}$$
$$-3\Big(1+\frac{\beta}{2}\Big)t^2x^{\beta/2}(1-x)^{\beta/2}(1-2x)\Big)$$
$$-\cos\left(\frac{\pi\beta}{2}\right)t^2\left(\frac{\beta+3}{(2\beta+2)(12+4\beta)}-\frac{3+\beta}{24+8\beta}(2x-1)^2\right),$$

where the result [Podlubny (1999), Theorem 6.5]

$$\frac{d^2}{dx^2}\int_{-1}^{1}\frac{(1-\xi^2)^{\beta/2}J_i^{\beta/2,\beta/2}(\xi)}{|x-\xi|^{2-\beta}}$$
$$=\frac{2\cos(\beta\pi/2)\Gamma(\beta+i+1)}{\Gamma(2-\beta)\Gamma(i+1)}J_i^{\beta/2,\beta/2}(x),\quad i=0,1,2,\cdots$$

has been used to calculate $\nabla^\beta p(x,t)$ explicitly.

The numerical results with the finite element approximation are listed in Table 5.2, which show that the convergence rate is also limited by the regularity of the exact solution.

Table 5.2: The L^2 errors (i.e., $\|p(x,T)-p_h^M\|_{L^2(\Omega)}$) and convergence rates to Example 5.2 approximated by the finite element scheme at $t=T=0.5$ for different β with $\alpha=0.4,\tau=h^2$.

	$\beta=1.4$		$\beta=1.7$		$\beta=1.9$	
$(n,\ h)$	L^2-Err	rate	L^2-Err	rate	L^2-Err	rate
$(0,\ 1/2^5)$	1.0724e-03	-	1.6466e-04	-	3.9083e-05	-
$(0,\ 1/2^6)$	5.3267e-04	1.01	6.9339e-05	1.24	1.2815e-05	1.61
$(0,\ 1/2^7)$	2.6293e-04	1.02	3.0636e-05	1.18	4.5923e-06	1.48
$(1,\ 1/2^5)$	4.4580e-06	-	1.5597e-06	-	9.1396e-07	-
$(1,\ 1/2^6)$	1.1777e-06	1.92	3.9010e-07	2.00	2.2690e-07	2.01
$(1,\ 1/2^7)$	3.0393e-07	1.95	9.7151e-08	2.01	5.6365e-08	2.01

Example 5.3. Consider (5.110) in $(-1,1)\times(0,T]$ with $U(x)=3x$,

$$p(x,t)=0,\quad (x,t)\in\mathbb{R}\backslash(-1,1)\times[0,T],$$
$$p(x,0)=0,\qquad x\in(-1,1),$$

and

$$f(x,t) = \frac{\Gamma(3)}{\Gamma(3-\alpha)} t^{2-\alpha} \sin(\pi x) - 3\pi t^2 \cos(\pi x)$$
$$- \frac{1}{2}\left({}_{-1}D_x^{\beta} \sin(\pi x) + {}_xD_1^{\beta} \sin(\pi x)\right), \quad (x,t) \in (-1,1) \times (0,T].$$

For $(x,t) \in [-1,1] \times [0,T]$, the exact solution is $p(x,t) = t^2 \sin(\pi x)$. The numerical results with the spectral Galerkin approximation are given in Tables 5.3 and 5.4, where the convergence rates in Table 5.3 are calculated by

$$\text{rate} = \frac{\log\left(\|p(x,T) - p_N^{M_1}\|_{L^2(\Omega)} / \|p(x,T) - p_N^{M_2}\|_{L^2(\Omega)}\right)}{\log(M_2/M_1)}, \quad (5.115)$$

and the convergence rates in Table 5.4 are calculated by

$$\text{rate} = \frac{\log\left(\|p(x,T) - p_{N_1}^{M}\|_{L^2(\Omega)} / \|p(x,T) - p_{N_2}^{M}\|_{L^2(\Omega)}\right)}{|N_2 - N_1|}. \quad (5.116)$$

Moreover, to calculate the integrals involving $f(x,t)$, noting that

$$\begin{aligned}
&\int_{-1}^{1} {}_{-1}D_x^{\beta} \sin(\pi x)\, \phi_i(x) dx \\
&\quad = -\int_{-1}^{1} {}_{-1}D_x^{\beta-1} \sin(\pi x)\, \phi_i'(x) dx \\
&\quad = \frac{-2^{\beta-2}\pi}{\Gamma(2-\beta)} \int_{-1}^{1} (x+1)^{2-\beta} \phi_i'(x) \eta_1(x)\, dx, \qquad (5.117)
\end{aligned}$$

and

$$\begin{aligned}
&\int_{-1}^{1} {}_{-1}D_x^{\beta} \sin(\pi x)\, \phi_i(x) dx \\
&\quad = -\int_{-1}^{1} {}_xD_1^{\beta-1} \sin(\pi x)\, \phi_i'(x) dx \\
&\quad = \frac{-2^{\beta-2}\pi}{\Gamma(2-\beta)} \int_{-1}^{1} (1-x)^{2-\beta} \phi_i'(x) \eta_2(x)\, dx, \qquad (5.118)
\end{aligned}$$

where

$$\eta_1(x) = \int_{-1}^{1} (1-\xi)^{1-\beta} \cos\left(\frac{(x+1)\xi + x - 1}{2}\pi\right) d\xi, \quad (5.119)$$

$$\eta_2(x) = \int_{-1}^{1} (1+\xi)^{1-\beta} \cos\left(\frac{(1-x)\xi + x + 1}{2}\right) d\xi, \quad (5.120)$$

then all the integrals in (5.117)-(5.120) can be calculated by the Jacobi-Gauss quadratures with the corresponding weights functions (see Appendix A). The results show that the convergence rate in time is $\mathcal{O}(\tau^{2-\alpha})$, while the exponential convergence in spatial direction are not well demonstrated (see Table 5.4), due to the limited machine accuracy and the error brought by the time discretization.

Table 5.3: The L^2 errors (i.e., $\|p(x,T)-p_h^M\|_{L^2(\Omega)}$) and convergence rates to Example 5.3 approximated by the spectral Galerkin method at $t = T = 0.5$ for different α, β with $N = 25$.

	$(\alpha,\beta) = (0.6, 1.5)$		$(\alpha,\beta) = (0.8, 1.5)$		$(\alpha,\beta) = (0.3, 1.8)$	
M	L^2-Err	rate	L^2-Err	rate	L^2-Err	rate
2^4	5.5569e-04	-	1.6095e-03	-	7.5748e-05	-
2^5	2.1307e-04	1.38	7.0471e-04	1.19	2.4276e-05	1.64
2^6	8.1350e-05	1.39	3.0769e-04	1.20	7.7118e-06	1.65
2^7	3.0976e-05	1.39	1.3415e-04	1.20	2.4335e-06	1.66

Table 5.4: The L^2 errors (i.e., $\|p(x,T)-p_h^M\|_{L^2(\Omega)}$) and convergence rates to Example 5.3 approximated by the spectral Galerkin method at $t = T = 0.5$ for different β with $\alpha = 0.3$ and $\tau = 1/2^{18}$.

	$\beta = 1.3$		$\beta = 1.5$		$\beta = 1.8$	
N	L^2-Err	rate	L^2-Err	rate	L^2-Err	rate
8	1.8333e-04	-	1.0075e-04	-	7.8358e-05	-
11	3.0566e-08	2.90	2.8789e-08	2.72	2.9270e-08	2.63
14	7.4912e-10	1.24	3.6612e-10	1.45	3.4953e-10	1.48
17	2.8296e-11	1.09	2.5139e-11	0.89	2.0290e-11	0.95

5.3 Local discontinuous Galerkin methods for one dimensional fractional diffusion equations

In this section, we introduce the LDG methods for the fractional diffusion problems. This development is based on the extensive work on LDG for problems founded in classic calculus [Cockburn and Shu (1998); Castillo *et al.* (2001); Hesthaven and Warburton (2004, 2008); Yan and Shu. (2002)]. In particular, we consider the extension of the LDG method [Cockburn and Shu (1998)], based on previous work in [Bassi and Rebay (1997)], to problems containing fractional spatial derivatives. We find that almost all the advantages or characteristics [Hesthaven and Warburton (2008)] of LDG methods when used to solve classical PDEs carries over to fractional PDEs, i.e., the methods are naturally formulated for any order of accuracy in each element, there is flexibility in choosing element sizes in different places, and the mass matrix is local and easily invertible, leading an explicit formulation for time dependent problems. We shall also discuss, however, the choice of the numerical flux is essential to ensure the accuracy and stability of the scheme.

We shall concern ourselves with the model

$$\begin{cases} \frac{\partial u(x,t)}{\partial t} = d \ {}_{-\infty}D_x^{\beta} u(x,t) + f(x,t), \quad (x,t) \in \Omega \times (0,T], \\ u(x,0) = u_0(x), \quad x \in \Omega, \\ u(x,t) = 0, \quad x \in \mathbb{R} \backslash \Omega \times [0,T], \end{cases} \tag{5.121}$$

where $\Omega = (a,b)$, $d \in \mathbb{R}^+$, $f(x,t)$ is a source term, and ${}_{-\infty}D_x^{\beta} u(x,t)$ $(1 < \beta < 2)$ denotes the left R-L fractional derivatives. In fact, when $\beta = 1$ and 2, ${}_{-\infty}D_x^{\beta} u(x,t)$ still makes sense and recovers exactly the first order and second order classical derivatives, respectively, at this moment, d becomes a velocity ($\beta = 1$) or a pure diffusion coefficient ($\beta = 2$). Hence, within this framework we can unify the classical and fractional calculus and recover known formulations in special cases. We point out that the suggested scheme also apply to the models with the mixed (left and right) R-L fractional derivatives.

5.3.1 *Negative fractional norms*

Let $\Omega = (a,b)$ and

$${}_aD_x^{-\alpha} v(x) := \frac{1}{\Gamma(\alpha)} \int_a^x (x-\xi)^{\alpha-1} v(\xi) d\xi, \ x > a, \tag{5.122}$$

$${}_xD_b^{-\alpha} v(x) := \frac{1}{\Gamma(\alpha)} \int_x^b (\xi-x)^{\alpha-1} v(\xi) d\xi, \ x < b. \tag{5.123}$$

If $v(x) \in L^2(\Omega)$, by Proposition B.10 in Appendix B, it holds that

$$\left\| {}_aD_x^{-\alpha} v \right\|_{L_2(\Omega)} \leq \frac{(b-a)^{\alpha}}{\Gamma(\alpha+1)} \|v\|_{L_2(\Omega)}, \tag{5.124}$$

$$\left\| {}_xD_b^{-\alpha} v \right\|_{L_2(\Omega)} \leq \frac{(b-a)^{\alpha}}{\Gamma(\alpha+1)} \|v\|_{L_2(\Omega)}. \tag{5.125}$$

Moreover, combining (B.10), (B.11) and (B.12), we have

$$\left({}_aD_x^{-\alpha} u, v \right)_{L^2(\Omega)} = \left({}_aD_x^{-\frac{\alpha}{2}} u, {}_xD_b^{-\frac{\alpha}{2}} v \right)_{L^2(\Omega)} = \left(u, {}_xD_b^{-\alpha} v \right)_{L^2(\Omega)}. \tag{5.126}$$

For $\alpha > 0$ and $v(x) \in L^2(\Omega)$, by (5.124) and (5.125), we define

$$\|v(x)\|_{J_L^{-\alpha}(\Omega)} := \|{}_aD_x^{-\alpha}v\|_{L^2(\Omega)}, \tag{5.127}$$

and

$$\|v(x)\|_{J_R^{-\alpha}(\Omega)} := \|{}_xD_b^{-\alpha}v\|_{L^2(\Omega)}. \tag{5.128}$$

Theorem 5.8. *Let $-\alpha_2 < -\alpha_1 < 0$ and $v(x) \in L^2(\Omega)$. Then*

$$\|v(x)\|_{J_L^{-\alpha_2}(\Omega)} \lesssim \|v(x)\|_{J_L^{-\alpha_1}(\Omega)}, \tag{5.129}$$

$$\|v(x)\|_{J_R^{-\alpha_2}(\Omega)} \lesssim \|v(x)\|_{J_R^{-\alpha_1}(\Omega)}. \tag{5.130}$$

Proof. By (B.10) and (5.124), we have

$${}_aD_x^{-\alpha_2}v = {}_aD_x^{-(\alpha_2-\alpha_1)}{}_aD_x^{-\alpha_1}v \quad \text{and} \quad {}_aD_x^{-\alpha_1}v \in L^2(\Omega).$$

Therefore, using (5.124) again, it follows that

$$\|v(x)\|_{J_L^{-\alpha_2}(\Omega)} \le \frac{(b-a)^{\alpha_2-\alpha_1}}{\Gamma(\alpha_2-\alpha_1+1)} \|v(x)\|_{J_L^{-\alpha_1}(\Omega)}. \tag{5.131}$$

The proof for (5.130) is similar. □

Theorem 5.9. *Suppose that $\alpha \in (0,1)$, and $v(x) \in L^2(\Omega)$ is a real function. Then*

$$\left({}_aD_x^{-\frac{\alpha}{2}}v, {}_xD_b^{-\frac{\alpha}{2}}v\right)_{L^2(\Omega)} \simeq \|v(x)\|_{J_L^{-\frac{\alpha}{2}}(\Omega)} \simeq \|v(x)\|_{J_R^{-\frac{\alpha}{2}}(\Omega)}. \tag{5.132}$$

Proof. By Cauchy-Schwarz's inequality, we have

$$\left({}_aD_x^{-\frac{\alpha}{2}}v, {}_xD_b^{-\frac{\alpha}{2}}v\right)_{L^2(\Omega)} \le \|v\|_{J_L^{-\frac{\alpha}{2}}(\Omega)} \|v\|_{J_R^{-\frac{\alpha}{2}}(\Omega)}. \tag{5.133}$$

Moreover, note that

$$\left({}_aD_x^{-\frac{\alpha}{2}}v, {}_xD_b^{-\frac{\alpha}{2}}v\right)_{L^2(\Omega)} = \left({}_{-\infty}D_x^{-\frac{\alpha}{2}}\tilde{v}, {}_xD_\infty^{-\frac{\alpha}{2}}\tilde{v}\right)_{L^2(\mathbb{R})}, \tag{5.134}$$

where $\tilde{v}$ is the zero extension of v. Using the Fourier transform of the left and right R-L fractional integrals (see Proposition B.8) and the Plancherel theorem (see Lemma A.16), we have

$$\begin{aligned}
&\left({}_{-\infty}D_x^{-\frac{\alpha}{2}}\tilde{v}, {}_xD_\infty^{-\frac{\alpha}{2}}\tilde{v}\right)_{L^2(\mathbb{R})} \\
&= \frac{1}{2\pi}\int_{\mathbb{R}} (i\omega)^{-\frac{\alpha}{2}}\hat{\tilde{v}}(\omega)\overline{(-i\omega)^{-\frac{\alpha}{2}}\hat{\tilde{v}}(\omega)}d\omega \\
&= \frac{\cos(\alpha\pi/2)}{2\pi}\int_{\mathbb{R}} |\pm i\omega|^{-\alpha}\left|\hat{\tilde{v}}(\omega)\right|^2 d\omega \\
&= \cos(\alpha\pi/2)\int_{\mathbb{R}} \left|{}_{-\infty}D_x^{-\frac{\alpha}{2}}\tilde{v}(x)\right|^2 dx \\
&= \cos(\alpha\pi/2)\int_{\mathbb{R}} \left|{}_xD_\infty^{-\frac{\alpha}{2}}\tilde{v}(x)\right|^2 dx.
\end{aligned} \tag{5.135}$$

Therefore,

$$\left({}_aD_x^{-\frac{\alpha}{2}}v, {}_xD_b^{-\frac{\alpha}{2}}v\right)_{L^2(\Omega)} \geq \cos(\alpha\pi/2)\|v\|^2_{J_L^{-\frac{\alpha}{2}}(\Omega)}, \tag{5.136}$$

$$\left({}_aD_x^{-\frac{\alpha}{2}}v, {}_xD_b^{-\frac{\alpha}{2}}v\right)_{L^2(\Omega)} \geq \cos(\alpha\pi/2)\|v\|^2_{J_R^{-\frac{\alpha}{2}}(\Omega)}. \tag{5.137}$$

Combining (5.133), (5.134), (5.136), and (5.137), we can complete the proof. □

Since $\|\cdot\|_{J_L^{-\frac{\alpha}{2}}(\Omega)} = 0$ means that ${}_aD_x^{-\frac{\alpha}{2}}v(x) = 0$ *a.e.*, then $v(x) = {}_aD_x^{\frac{\alpha}{2}}\left({}_aD_x^{-\frac{\alpha}{2}}v(x)\right) = 0$ *a.e.*. Therefore, $\|\cdot\|_{J_L^{-\frac{\alpha}{2}}(\Omega)}$ is also a norm on $L^2(\Omega)$. For $\|\cdot\|_{J_R^{-\frac{\alpha}{2}}(\Omega)}$ is similar.

5.3.2 *Derivation of the numerical schemes*

5.3.2.1 *Weak formulation*

Noting that ${}_{-\infty}D_x^{\beta}u(x,t) = {}_aD_x^{\beta}u(x,t)$ for $x \in \Omega$, and by Proposition B.15, we have ${}_aD_x^{\beta-1}u(x,t) = {}_a^CD_x^{\beta-1}u(x,t)$. These allow us to more clearly describe the equation in (5.121) as

$$\frac{\partial u(x,t)}{\partial t} = \frac{\partial}{\partial x} d\, {}_aD_x^{-(2-\beta)} \frac{\partial}{\partial x} u(x,t) + f(x,t), \quad \text{in } \Omega_T, \tag{5.138}$$

where $\Omega_T := \Omega \times (0,T]$. Then following the standard approach for the development of LDG methods for problems with higher derivatives [Cockburn and Shu (1998); Yan and Shu. (2002); Hesthaven and Warburton (2008)], we introduce the auxiliary variables p and q, and rewrite (5.121) as

$$\begin{cases} \dfrac{\partial u(x,t)}{\partial t} - \sqrt{d}\dfrac{\partial q(x,t)}{\partial x} = f(x,t), \text{ in } \Omega_T, \\ q - {}_aD_x^{-(2-\beta)}p(x,t) = 0, & \text{in } \Omega_T, \\ p - \sqrt{d}\dfrac{\partial u(x,t)}{\partial x} = 0, & \text{in } \Omega_T, \\ u(x,0) = u_0(x), & \text{on } \Omega, \\ u(x,t) = 0, & \text{on } \mathbb{R}\backslash\Omega \times [0,T]. \end{cases} \tag{5.139}$$

Using standard notation, given the nodes $a = x_0 < x_1 < \cdots < x_{M-1} < x_M = b$, we define the mesh $\mathcal{T} = \{I_j = (x_{j-1}, x_j),\, j = 1,\ldots,M\}$ and set $h_j := |I_j| = x_j - x_{j-1}$; and $h := \max_{j=1}^M h_j$. Associated with the mesh $\mathcal{T}$, we define the broken Sobolev spaces

$$L^2(\Omega,\mathcal{T}) := \{v : \Omega \to \mathbb{R} \mid v|_{I_j} \in L^2(I_j),\, j = 1,\ldots,M\}$$

and

$$H^1(\Omega,\mathcal{T}) := \{v : \Omega \to \mathbb{R} \mid v|_{I_j} \in H^1(I_j),\ j = 1,\dots,M\}.$$

Note that for these broken spaces, the corresponding norms should be understood as the broken norms [Hesthaven and Warburton (2008)].

Denote the one-sided limits of $v(x)$ at the nodes $\{x_j\}$ by

$$v^{\pm}(x_j) = v(x_j^{\pm}) := \lim_{x\to x_j^{\pm}} v(x). \tag{5.140}$$

We assume that the exact solution $\mathbf{w} = (u,p,q)$ of (5.139) belongs to

$$H^1(0,T;H^1(\Omega,\mathcal{T})) \times L^2(0,T;L^2(\Omega,\mathcal{T})) \times L^2(0,T;H^1(\Omega,\mathcal{T})), \tag{5.141}$$

and require that $\mathbf{w}$ satisfies

$$\left(\frac{\partial u(x,t)}{\partial t}, v\right)_{I_j} + \sqrt{d}\left(q(x,t), \frac{\partial v}{\partial x}\right)_{I_j} - \sqrt{d}q(x,t)v\Big|_{x_{j-1}^+}^{x_j^-} = (f,v)_{I_j}, \tag{5.142}$$

$$(q,w)_{I_j} - \left({}_aD_x^{-(2-\beta)}p(x,t), w\right)_{I_j} = 0, \tag{5.143}$$

$$(p,z)_{I_j} + \sqrt{d}\left(u(x,t), \frac{\partial z}{\partial x}\right)_{I_j} - \sqrt{d}u(x,t)z\Big|_{x_{j-1}^+}^{x_j^-} = 0, \tag{5.144}$$

$$(u(\cdot,0),v)_{I_j} = (u_0(\cdot),v)_{I_j}, \tag{5.145}$$

for all test functions $w \in L^2(\Omega,\mathcal{T})$, and $v,z \in H^1(\Omega,\mathcal{T})$, and for $j = 1,\dots,M$. Here the time derivative is understood in the weak sense and $(u,v)_I = \int_I u(x)v(x)dx$ is the standard inner product over the element.

5.3.2.2 *Numerical schemes*

In the following we shall propose numerical schemes for (5.121) based on the equations (5.142)–(5.145). The global nature of the fractional derivative is reflected in (5.143) while (5.142) and (5.144) remain local as in a more traditional DG formulations.

We now restrict the trial and test functions v, w, and z to the finite dimensional subspaces $V \subset H^1(\Omega,\mathcal{T})$, and choose V to be the space of discontinuous, piecewise polynomial functions

$$V = \{v : \Omega \to \mathbb{R} \mid v|_{I_j} \in \mathcal{P}^k(I_j),\ j = 1,\dots,M\},$$

where $\mathcal{P}^k(I_j)$ denotes the set of all polynomials of degree less than or equal k ($\geqslant 1$) on I_j. Furthermore, we define U, P, and Q as the approximations of u, p, and q, respectively, in the space V. We then need to find $(U,P,Q) \in H^1(0,T;V) \times L^2(0,T;V) \times L^2(0,T;V)$ such that for all v, w, and $z \in V$, and for $j = 1, \ldots, M$ the following holds:

$$\left(\frac{\partial U(x,t)}{\partial t}, v\right)_{I_j} + \sqrt{d}\left(Q(x,t), \frac{\partial v}{\partial x}\right)_{I_j} - \sqrt{d}\hat{Q}(x,t)v\Big|_{x_{j-1}^+}^{x_j^-} = (f,v)_{I_j}, \quad (5.146)$$

$$(Q,w)_{I_j} - \left({}_aD_x^{-(2-\beta)}P(x,t), w\right)_{I_j} = 0, \quad (5.147)$$

$$(P,z)_{I_j} + \sqrt{d}\left(U(x,t), \frac{\partial z}{\partial x}\right)_{I_j} - \sqrt{d}\hat{U}(x,t)z\Big|_{x_{j-1}^+}^{x_j^-} = 0, \quad (5.148)$$

$$(U(\cdot,0),v)_{I_j} = (u_0(\cdot),v)_{I_j}. \quad (5.149)$$

To complete the formulation of the numerical schemes, we must define the numerical fluxes $\hat{Q}(x,t)$ and $\hat{U}(x,t)$. As with traditional LDG methods, this choice is the most delicate one as it determines not only locality but also consistency, stability, and order of convergence of the scheme. Seeking inspiration in the mixed formulation for the heat equation we use the 'alternating principle' [Yan and Shu. (2002)] in choosing the numerical fluxes for (5.146) and (5.148), that is, we choose

$$\hat{Q}(x_j,t) = Q^+(x_j,t), \quad \hat{U}(x_j,t) = U^-(x_j,t); \quad (5.150)$$

or

$$\hat{Q}(x_j,t) = Q^-(x_j,t), \quad \hat{U}(x_j,t) = U^+(x_j,t); \quad (5.151)$$

at all interior boundaries. At the external boundaries we use

$$\hat{Q}(a,t) = Q^+(a,t) = Q^-(a,t), \quad \hat{Q}(b,t) = Q^-(b,t) = Q^+(b,t), \quad (5.152)$$

and

$$\hat{U}(a,t) = 0, \quad \hat{U}(b,t) = 0, \quad (5.153)$$

reflecting the boundary conditions.

As we shall discuss in more detail shortly, this scheme, (5.146)–(5.149) with fluxes (5.150)–(5.153), is stable for any $\beta \in [1,2]$; and it has the optimal convergence order $k+1$ for $\beta \in (1,2]$ and suboptimal convergence order k for $\beta \to 1$. In fact, when $\beta = 1$ the numerical dissipation term disappears. We therefore introduce local dissipation to enhance the stability of the scheme at $\beta = 1$ by adding a penalty term in (5.147). This suggests a new scheme with

$$(Q,w)_{I_j} - \left({}_aD_x^{-(2-\beta)}p(x,t), w\right)_{I_j} + L(h,\beta)\cdot\left(\hat{P}(x,t)w\right)\Big|_{x_{j-1}^+}^{x_j^-} = 0, \quad (5.154)$$

where

$$\hat{P}(x_j^-, t) = [P(x_j, t)] := P(x_j^+, t) - P(x_j^-, t),$$
$$\hat{P}(x_{j-1}^+, t) = [P(x_{j-1}, t)] := P(x_{j-1}^+, t) - P(x_{j-1}^-, t), \tag{5.155}$$
$$\hat{P}(x_0^+, t) = \hat{P}(x_M^-, t) = 0.$$

Here $L(h, \beta)$ is a constant depending on β and h, the local cell size. As we shall discuss, this term recovers the optimal $k+1$ order of convergence for any $\beta \in [1, 2]$. If the value of $L(h, \beta)$ is at least of order h, the order of convergence is $k+1$ for $\beta \in (1, 2]$ (see Theorem 4.3). In the computational examples we take $L(h, \beta) = h^\beta$ in agreement with the scaling of the global operator.

5.3.3 *Stability and error estimate*

The LDG schemes can be expressed as: find $(U, P, Q) \in H^1(0, T; V) \times L^2(0, T; V) \times L^2(0, T; V)$ such that for all $(v, w, z) \in H^1(0, T; V) \times L^2(0, T; V) \times L^2(0, T; V)$, the following holds

$$B(U, P, Q;\ v, w, z) = \mathcal{L}(v, w, z). \tag{5.156}$$

Here $(U(\cdot, 0), v(\cdot, 0)) = (u_0(\cdot), v(\cdot, 0))$ and the discrete bilinear form B is defined as

$$\begin{aligned}
&B(U, P, Q;\ v, w, z) \\
&:= \int_0^T \left(\frac{\partial U(\cdot, t)}{\partial t}, v(\cdot, t) \right) dt - \int_0^T \left({}_aD_x^{-(2-\beta)} p(x, t), w(x, t) \right) dt \\
&\quad + \int_0^T (Q(\cdot, t), w(\cdot, t))\, dt + \int_0^T (P(\cdot, t), z(\cdot, t)) dt \\
&\quad + \sqrt{d} \int_0^T \left(U(x, t), \frac{\partial z(x, t)}{\partial x} \right) dt + \sqrt{d} \int_0^T \left(Q(x, t), \frac{\partial v(x, t)}{\partial x} \right) dt \\
&\quad + \sqrt{d} \int_0^T \sum_{j=1}^{M-1} \hat{Q}(x_j, t)[v](x_j, t) dt + \sqrt{d} \int_0^T \sum_{j=1}^{M-1} \hat{U}(x_j, t)[z](x_j, t) dt \\
&\quad + \sqrt{d} \int_0^T (Q^+(a, t) v^+(a, t) - Q^-(b, t) v^-(b, t)) dt \\
&\quad - \chi \cdot L(h, \beta) \int_0^T \sum_{j=1}^{M-1} [P](x_j, t)[w](x_j, t) dt,
\end{aligned} \tag{5.157}$$

where $\chi = 0$ or 1; when $\chi = 0$ denotes the penalty term in (5.154) is not added. The discrete linear form $\mathcal{L}$ is given by

$$\mathcal{L}(v, w, z) = \int_0^T (f(\cdot, t), v(\cdot, t))dt. \tag{5.158}$$

Since the scheme is consistent with (5.139), if the exact solution (u, p, q) of (5.139) is regular enough, we have

$$B(u, p, q;\ v, w, z) = \mathcal{L}(v, w, z) \tag{5.159}$$

for all $(v, w, z) \in H^1(0, T; V) \times L^2(0, T; V) \times L^2(0, T; V)$.

5.3.3.1 *Numerical stability*

Let $(\widetilde{U}, \widetilde{P}, \widetilde{Q}) \in H^1(0, T; V) \times L^2(0, T; V) \times L^2(0, T; V)$ be the perturbed solution of (U, P, Q); i.e.,$(\widetilde{U}, \widetilde{P}, \widetilde{Q})$ and (U, P, Q) satisfy (5.156) with different initial condition. We denote $e_U := \widetilde{U} - U$, $e_P := \widetilde{P} - P$, and $e_Q := \widetilde{Q} - Q$ as the errors.

Theorem 5.10. *Schemes (5.156) are L^2 stable, and for all $t \in [0, T]$ their solutions satisfy*

$$\begin{aligned}\|e_U(\cdot, t)\|^2_{L^2(\Omega)} = \|e_U(\cdot, 0)\|^2_{L^2(\Omega)} - 2\int_0^T \left({}_aD_x^{-(2-\beta)} e_P(\cdot, t), e_P(\cdot, t)\right) dt \\ -2\chi \cdot L(h, \beta) \int_0^t \sum_{j=1}^{M-1} [e_P]^2(x_j, t)dt.\end{aligned} \tag{5.160}$$

Remark 5.6. If the unstabilized scheme (i.e., $\chi = 0$) is considered in the limit of $\beta = 1$, since $\left({}_aD_x^{-(2-\beta)} e_P(\cdot, t),\ e_P(\cdot, t)\right) = 0$ (see (5.126) and (5.135)), then $\|e_U(\cdot, t)\|^2_{L^2(\Omega)} = \|e_U(\cdot, 0)\|^2_{L^2(\Omega)}$, indicating the numerical dissipation disappears.

Proof. It suffices to prove the result for the case $t = T$. From (5.156) we recover the perturbation equation

$$B_2(e_U, e_P, e_Q;\ v, w, z) = 0, \tag{5.161}$$

for all $(v, w, z) \in H^1(0, T; V) \times L^2(0, T; V) \times L^2(0, T; V)$. Taking $v = e_U$, $w = -e_P$, $z = e_Q$, we obtain

$$
\begin{aligned}
0 &= B(e_U, e_P, e_Q; e_U, -e_P, e_Q) \\
&= \frac{1}{2}\int_0^T \frac{\partial}{\partial t}\|e_U(\cdot,t)\|^2_{L^2(\Omega)}dt + \int_0^T \left({}_aD_x^{-(2-\beta)}e_P(\cdot,t), e_P(\cdot,t)\right)dt \\
&\quad +\chi\cdot L(h,\beta)\int_0^T\sum_{j=1}^{M-1}[e_P]^2(x_j,t)dt + \sqrt{d}\int_0^T\int_a^b \frac{\partial(e_U\cdot e_Q)}{\partial x}dxdt \\
&\quad +\sqrt{d}\int_0^T\sum_{j=1}^{M-1}\hat{e}_Q(x_j,t)[e_U](x_j,t)dt + \sqrt{d}\int_0^T\sum_{j=1}^{M-1}\hat{e}_U(x_j,t)[e_Q](x_j,t)dt \\
&\quad +\sqrt{d}\int_0^T \left(e_Q^+(a,t)e_U^+(a,t) - e_Q^-(b,t)e_U^-(b,t)\right)dt.
\end{aligned}
\tag{5.162}
$$

In (5.162),

$$
\begin{aligned}
\int_0^T\int_a^b \frac{\partial(e_U\cdot e_Q)}{\partial x}dxdt &= \int_0^T \left(-e_Q^+(a,t)e_U^+(a,t) + e_Q^-(b,t)e_U^-(b,t)\right)dt \\
&\quad -\int_0^T\sum_{j=1}^{M-1}[e_U\cdot e_Q](x_j,t),
\end{aligned}
\tag{5.163}
$$

and when $\hat{e}_Q = e_Q^+$ and $\hat{e}_U = e_U^-$ (or $\hat{e}_Q = e_Q^-$ and $\hat{e}_U = e_U^+$),

$$
\begin{aligned}
&\int_0^T\sum_{j=1}^{M-1}\hat{e}_Q(x_j,t)[e_U](x_j,t) + \sum_{j=1}^{M-1}\hat{e}_U(x_j,t)[e_Q](x_j,t)dt \\
&= \int_0^T\sum_{j=1}^{M-1}[e_U\cdot e_Q](x_j,t)dt.
\end{aligned}
\tag{5.164}
$$

Combining (5.162) and (5.163)–(5.164), the desired result is obtained. □

5.3.3.2 *Error estimate*

For the error analysis, we define the projection operators $\mathrm{P}^{\pm}$, S, and S′ from $H^1(\Omega,\mathcal{T})$ to V. For intervals $I_j = (x_{j-1}, x_j)$, $j = 1, 2, \cdots, M$, and any sufficiently regular function u, $\mathrm{P}^{\pm}$ are defined to satisfy the $k+1$ conditions:

$$
\begin{aligned}
&(\mathrm{P}^{\pm}u - u, v)_{I_j} = 0 \qquad \forall v \in \mathcal{P}^{k-1}(I_j), \text{ if } k > 0, \\
&\mathrm{P}^{-}u(x_j) = u^-(x_j) \quad \mathrm{P}^{+}u(x_{j-1}) = u^+(x_{j-1}).
\end{aligned}
\tag{5.165}
$$

S and S′ are the standard L^2-projections, which are defined, respectively, as

$$(\mathrm{S}u - u, v)_{I_j} = 0 \quad \forall v \in \mathcal{P}^k(I_j), \tag{5.166}$$

$$(\mathrm{S}'u - u, v)_{I_j} = 0 \quad \forall v \in \mathcal{P}^{k-1}(I_j), \text{ if } k > 0. \tag{5.167}$$

We are now ready to state our results and then prove them. These results is under the assumption that the corresponding analytical solutions are sufficiently regular as functions of x.

Theorem 5.11 (Error estimate). *The error for the scheme (5.146), (5.154) (or (5.147)), (5.148), and (5.149) with fluxes (5.150)–(5.153) and (5.155) applied to the model (5.121) satisfies*

$$\sqrt{\int_a^b (u(x,t) - U(x,t))^2 dx} \leqslant \begin{cases} \left(\sqrt{\frac{C(h,\beta)}{h}}c + c(\beta)\right) h^{k+1} & \text{for } 1 < \beta \leqslant 2, \\ ch^k & \text{for } \beta = 1, \end{cases} \tag{5.168}$$

where $c(\beta)$ and c depend on $\frac{\partial^{k+1}U(x,t)}{\partial x^{k+1}}$, $\frac{\partial^{k+\beta-1}U(x,t)}{\partial x^{k+\beta-1}}$, $\frac{\partial^{k+\beta}U(x,t)}{\partial x^{k+\beta}}$, and t.

Proof. We denote

$$e_u = u(x,t) - U(x,t), \quad e_p = p(x,t) - P(x,t), \quad e_q = q(x,t) - Q(x,t).$$

From (5.156) and (5.159), we recover the error equation

$$B(e_u, e_p, e_q; v, w, z) = 0 \tag{5.169}$$

for all $(v, w, z) \in H^1(0,T;V) \times L^2(0,T;V) \times L^2(0,T;V)$. Take

$$v = \mathrm{P}^{\pm}u - U, \quad w = P - \mathrm{S}p, \quad z = \mathrm{P}^{\mp}q - Q$$

in (5.169). After rearranging terms, we obtain

$$B(v, -w, z; v, w, z) = B(v^e, -w^e, z^e; v, w, z), \tag{5.170}$$

where v^e, w^e, and z^e are given as

$$v^e = \mathrm{P}^{\pm}u - u, \quad w^e = p - \mathrm{S}p, \quad z^e = \mathrm{P}^{\mp}q - q.$$

Following the discussion in the proof of Theorem 5.10 the left hand side of (5.170) becomes

$$\begin{aligned} & B(v, -w, z; v, w, z) \\ & = \frac{1}{2}\int_0^T \frac{\partial}{\partial t}\|v(\cdot,t)\|^2_{L^2(\Omega)} dt + \int_0^T \left({}_aD_x^{-(2-\beta)} e_P(\cdot,t), e_P(\cdot,t)\right) dt \\ & \quad + \mathcal{VI}', \end{aligned} \tag{5.171}$$

where

$$\mathcal{VI}' = \chi \cdot L(h,\beta) \int_0^T \sum_{j=1}^{M-1} [\omega]^2(x_j,t)dt. \tag{5.172}$$

Using the notation in [Yan and Shu. (2002)], the right hand side of (5.170) can be expressed as

$$B(v^e, -w^e, z^e;\, v, w, z) = \mathcal{I} + \mathcal{II} + \mathcal{III} + \mathcal{IV} + \mathcal{V} + \mathcal{VI}, \tag{5.173}$$

where

$$\mathcal{I} = \int_0^T \left(\frac{\partial v^e(\cdot,t)}{\partial t}, v(\cdot,t) \right) dt, \tag{5.174}$$

$$\begin{aligned}\mathcal{II} = &\sqrt{d} \int_0^T \left(z^e(x,t), \frac{\partial v(x,t)}{\partial x} \right) dt + \sqrt{d} \int_0^T \left(v^e(x,t), \frac{\partial z(x,t)}{\partial x} \right) dt \\ &- \int_0^T (w^e(\cdot,t), z(\cdot,t))dt, \end{aligned}\tag{5.175}$$

$$\mathcal{III} = \sqrt{d} \int_0^T \sum_{j=1}^{M-1} \hat{z}^e(x_j,t)[v](x_j,t)dt + \sqrt{d} \int_0^T \sum_{j=1}^{M-1} \hat{v}^e(x_j,t)[z](x_j,t)dt, \tag{5.176}$$

$$\mathcal{IV} = \sqrt{d} \int_0^T ((z^e)^+(a,t)v^+(a,t) - (z^e)^-(b,t)v^-(b,t))dt, \tag{5.177}$$

$$\mathcal{V} = \int_0^T (z^e(\cdot,t), w(\cdot,t))dt + \int_0^T \left({}_aD_x^{-(2-\beta)} p(x,t), w(x,t) \right) dt, \tag{5.178}$$

and

$$\mathcal{VI} = \chi \cdot L(h,\beta) \int_0^T \sum_{j=1}^{M-1} [w^e](x_j,t)[w](x_j,t)dt. \tag{5.179}$$

Using standard approximation theory [Ciarlet (1975)], we obtain

$$\begin{aligned}\mathcal{I} &\leqslant \frac{1}{2} \int_0^T \int_a^b \left(\frac{\partial v^e(x,t)}{\partial t} \right)^2 dxdt + \int_0^T \int_a^b \left(\frac{v^2(x,t)}{2} \right) dxdt \\ &\leqslant ch^{2k+2} + \frac{1}{2} \int_0^T \|v(\cdot,t)\|^2_{L^2(\Omega)} dt, \end{aligned}\tag{5.180}$$

where c is a constant.

All the terms in $\mathcal{II}$ are vanish due to Galerkin orthogonality, i.e., $p - \mathrm{S}p$ is orthogonal to all polynomials of degree up to k, and $\mathrm{P}^{\pm}u - u$ and $\mathrm{P}^{\pm}q - q$ to $k-1$. For the terms in $\mathcal{III}$, when taking $\hat{z}^e = (z^e)^-$ and $\hat{v}^e = (v^e)^+$, we use $z^e = \mathrm{P}^- q - q$ and $v^e = \mathrm{P}^+ u - u$; and when giving $\hat{z}^e = (z^e)^+$ and

$\hat{v}^e = (v^e)^-$, we choose $z^e = \mathrm{P}^+q - q$ and $v^e = \mathrm{P}^-u - u$. Hence, both terms in $\mathcal{III}$ are zero and so is $\mathcal{III}$. An application of the inequality $xy \leqslant \frac{1}{2}(x^2 + y^2)$, the standard approximation on point values of z^e, and the equivalence of norms in finite dimensional spaces implies

$$
\begin{aligned}
\mathcal{IV} &\leqslant \frac{\sqrt{d}}{2}\int_0^T \left(((z^e)^+(a,t))^2 + ((z^e)^-(b,t))^2 + (v^+(a,t))^2 + (v^-(b,t))^2\right) dt \\
&\leqslant ch^{2k+2} + c\int_0^T \|v(\cdot,t)\|^2_{L^2(\Omega)} dt.
\end{aligned} \tag{5.181}
$$

One of the terms in $\mathcal{IV}$ is exactly zero. When $z^e = \mathrm{P}^+q - q$, the first term is zero; and when $z^e = \mathrm{P}^-q - q$, the second term is zero. For $\mathcal{V}$, we recover two different kinds of estimates corresponding to $\beta = 1$ and $\beta \in (1,2]$ respectively. Both of them use (5.131).

When $\beta = 1$, further applying the operators $\mathcal{S}$ and $\mathcal{S}'$ defined in (5.166) and (5.167) yields

$$
\begin{aligned}
\mathcal{V} &= \int_0^T (z^e(\cdot,t), w(\cdot,t) - \mathcal{S}'w(\cdot,t)) + (z^e(\cdot,t), \mathcal{S}'w(\cdot,t))dt \\
&\quad + \int_0^T \left(w^e(\cdot,t), \int_x^b w(\xi,t)d\xi - \mathcal{S}\int_x^b w(\xi,t)d\xi\right) \\
&\quad + \left(w^e(\cdot,t), \mathcal{S}\int_x^b w(\xi,t)d\xi\right) dt \\
&= \int_0^T (z^e(\cdot,t), w(\cdot,t) - \mathcal{S}'w(\cdot,t)) \\
&\quad + \left(w^e(\cdot,t), \int_x^b w(\xi,t)d\xi - \mathcal{S}\int_x^b w(\xi,t)d\xi\right) dt \\
&\leqslant \int_0^T\int_a^b \left(\frac{(z^e(x,t))^2}{2}\right) dxdt + \int_0^T\int_a^b \left(\frac{(w(x,t) - \mathcal{S}'w(x,t))^2}{2}\right) dxdt \\
&\quad + \frac{1}{2}\int_0^T\int_a^b \left(\frac{(w^e(x,t))^2}{2} + \left(\int_x^b w(\xi,t)dx - \mathcal{S}\int_x^b w(\xi,t)d\xi\right)^2\right) dxdt \\
&\leqslant ch^{2k+2} + ch^{2k} + ch^{2k+2} + ch^{2k+2} \\
&\leqslant ch^{2k}.
\end{aligned} \tag{5.182}
$$

When $\beta \in (1,2]$, further using the inequality $xy \leqslant \frac{x^2}{2\epsilon} + \frac{\epsilon y^2}{2}$ and the norm-equivalence gives

$$\begin{aligned}
\mathcal{V} &\leqslant \int_0^T \int_a^b \frac{(z^e(x,t))^2}{2\epsilon} dxdt + \frac{\epsilon}{2} \int_0^T \|w(\cdot,t)\|^2_{L^2(\Omega)} dt \\
&\quad + \int_0^T \int_a^b \frac{(w^e(x,t))^2}{2\epsilon} dxdt + \frac{\epsilon}{2} \int_0^T \|w(\cdot,t)\|^2_{J_R^{-(2-\beta)}(\Omega)} dt \\
&\leqslant (c/\epsilon)h^{2k+2} + c\epsilon \int_0^T \|w(\cdot,t)\|^2_{J_R^{-(1-\frac{\beta}{2})}(\Omega)} dt,
\end{aligned} \tag{5.183}$$

where ϵ is a small number, chosen such that $c\epsilon \int_0^T \|w(\cdot,t)\|^2_{J_R^{-(1-\frac{\beta}{2})}(\Omega)} dt \leq \int_0^T \left({}_aD_x^{-(2-\beta)} w(\cdot,t), w(\cdot,t)\right) dt$ (see (5.171) and Theorem 5.9).

The estimate of $\mathcal{VI}$ is

$$\begin{aligned}
\mathcal{VI} &\leqslant \chi \cdot L(h,\beta) \int_0^T \frac{1}{2} \sum_{j=1}^{M-1} \left([w^e]^2(x_j,t) + [w]^2(x_j,t)\right) dt \\
&\leqslant \chi \cdot L(h,\beta) c h^{2k+1} + \frac{1}{2}\mathcal{VI}'.
\end{aligned} \tag{5.184}$$

In summary, we have the final estimate:

(1) for $\beta \in (1,2]$,

$$\begin{aligned}
&\frac{1}{2}\|v(\cdot,T)\|^2_{L^2(\Omega)} + \chi \cdot \frac{L(h,\beta)}{2} \int_0^T \sum_{j=1}^{M-1} [w]^2(x_j,t) dt \\
&\leqslant \chi \cdot L(h,\beta) c h^{2k+1} + (c/\epsilon) h^{2k+2} + c \int_0^T \|v(\cdot,t)\|^2_{L^2(\Omega)} dt \\
&\leqslant \left(\frac{L(h,\beta)}{h} c + (c/\epsilon)\right) h^{2k+2} + c \int_0^T \|v(\cdot,t)\|^2_{L^2(\Omega)} dt;
\end{aligned} \tag{5.185}$$

(2) for the special case of $\beta = 1$,

$$\frac{1}{2}\|v(\cdot,T)\|^2_{L^2(\Omega)} + \chi \cdot \frac{L(h,\beta)}{2} \int_0^T \sum_{j=1}^{M-1} [w]^2(x_j,t)dt$$

$$\leqslant \chi \cdot L(h,\beta)ch^{2k+1} + ch^{2k} + c\int_0^T \|v(\cdot,t)\|^2_{L^2(\Omega)}dt \tag{5.186}$$

$$\leqslant ch^{2k} + c\int_0^T \|v(\cdot,t)\|^2_{L^2(\Omega)}dt.$$

According to the continuous Grönwall lemma (see Appendix A) and the standard approximation on $v^e = \mathrm{P}u - u$, the desired estimate (5.168) is obtained. □

Remark 5.7. The introduction of the terms $\mathcal{VI}'$ and $\mathcal{VI}$ does not appear to add any additional freedom to improve the error estimates ($\mathcal{VI}'$ can't control $\|w\|^2$, so we can't improve the order of convergence when $\beta = 1$). On the contrary, we have to estimate the trace errors of w. It seems hard to choose a special projection to eliminate the traces errors, since at one point we have two traces of w. It is not clear whether the error estimate (5.168) is sharp. The numerical computations below show that the order of convergence is $k+1$ for any $\beta \in [1, 1+\varepsilon]$, where ε may be 0.01 or 0.001.

Remark 5.8. The developed LDG schemes can be directly extended to the model including mixed fractional derivatives, that is, the model

$$\frac{\partial u(x,t)}{\partial t} = \theta\, {}_{-\infty}D_x^{\beta} u(x,t) + (1-\theta)\, {}_xD_{\infty}^{\beta} u(x,t) + f,\ \theta \in [0,1] \tag{5.187}$$

in $\Omega \times (0, T]$, with

$$u(x,0) = u_0(x), \quad x \in \Omega, \tag{5.188}$$

$$u(x,t) = 0, \quad (x,t) \in \mathbb{R}\backslash\Omega \times [0,T]. \tag{5.189}$$

In fact, we can rewrite (5.187) as

$$\begin{cases} \frac{\partial u(x,t)}{\partial t} - \frac{\partial q(x,t)}{\partial x} = f(x,t), & \text{in } \Omega_T, \\ q - \left(\theta\, {}_aD_x^{-(2-\beta)} + (1-\theta)\, {}_xD_b^{-(2-\beta)}\right) p(x,t) = 0, & \text{in } \Omega_T, \\ p - \frac{\partial u(x,t)}{\partial x} = 0, & \text{in } \Omega_T, \\ u(x,0) = u_0(x), & \text{on } \Omega, \\ u(x,t) = 0, & \text{on } \mathbb{R}\backslash\Omega \times [0,T]. \end{cases}$$

By Theorem 5.9, the numerical fluxes, penalty terms, and theoretical analysis developed for model (5.121) can be directly used here. A slightly different LDG method can also be found in [Xu and Hesthaven (2014)].

5.3.4 *Numerical results*

Let us now offer some numerical results to validate analysis. We mainly focus on the accuracy and stability of the spatial approximation although we shall also numerically study the CFL condition. We use a fourth order explicit Runge-Kutta method [Canuto *et al.* (2006); Hesthaven and Warburton (2008)] to solve the method-of-line fractional PDE, i.e., the classical ODE system. To ensure the overall error is dominated by space error, small time steps are used. Noting that with the alternating fluxes, we can compute P, Q, U in (5.146)–(5.148) element by element.

Example 5.4. Consider

$$\frac{\partial u(x,t)}{\partial t} = \frac{\Gamma(6-\beta)}{\Gamma(6)} \frac{\partial^\beta u(x,t)}{\partial x^\beta} + f(x,t), \quad \beta \in [1,2] \tag{5.190}$$

in $\Omega = (0,1)$, with the source term

$$f(x,t) = -e^{-t}\left(x^5 - x^3 + x^{5-\beta} - \frac{(5-\beta)(4-\beta)}{20} x^{3-\beta}\right), \tag{5.191}$$

and the initial-boundary conditions

$$u(x,0) = x^5 - x^3, \ x \in \Omega, \quad \text{and } u(x,t) = 0, \ x \in \mathbb{R}\backslash\Omega \times [0,T]. \tag{5.192}$$

Its exact solution is $u(x,t) = e^{-t}(x^5 - x^3)$. In the following, we measure the errors corresponding to the exact solution $u(x,t)$ in the broken L^2 norm, i.e.,

$$\|u(\cdot,t) - U(\cdot,t)\|_{L^2(\Omega,\mathcal{T})} := \left(\sum_{j=1}^{M} \|u(\cdot,t) - U(\cdot,t)\|^2_{L^2(I_j)}\right)^{\frac{1}{2}}. \tag{5.193}$$

When $t = T = 1$, Tables 5.5–5.7 demonstrate the errors and order of convergence of the scheme and confirm optimality for $\beta \in [1+\varepsilon, 2.0]$ with suboptimal convergence for $\beta = 1.0$ as predicted by Theorem 5.11.

However, the results also clearly demonstrate that the optimal asymptotic order of convergence is recovered as $\beta \to 1$; as to the stabilized scheme (i.e., $\chi = 1$), the optimal asymptotic order of convergence is recovered uniformly for $\beta \in [1,2]$.

Table 5.5: The error and order of convergence for first order polynomial approximation ($k = 1$). M denotes the number of elements. For $\beta > 1.1$ the results are independent of whether stabilization with $L(h, \beta) = h^\beta$ is included. Recovery of the optimal order of convergence is possible only with the stabilization as illustrated in the last two rows which show results with the stabilized scheme.

β	$M = 2^6$	$M = 2^7$		$M = 2^8$		$M = 2^9$	
	error	error	rate	error	rate	error	rate
2.00	7.82e-05	2.04e-05	1.94	5.20e-06	1.97	1.31e-06	1.99
1.80	9.37e-05	2.44e-05	1.94	6.14e-06	1.99	1.52e-06	2.01
1.50	1.27e-04	3.23e-05	1.98	7.93e-06	2.03	1.95e-06	2.02
1.20	3.17e-04	8.32e-05	1.93	2.02e-05	2.05	5.09e-06	1.99
1.10	1.90e-04	5.22e-05	1.86	1.32e-05	1.98	3.31e-06	2.00
1.09	2.09e-04	5.96e-05	1.81	1.53e-05	1.96	3.86e-06	1.99
1.08	2.31e-04	6.85e-05	1.76	1.78e-05	1.94	4.53e-06	1.98
1.07	2.56e-04	7.94e-05	1.69	2.12e-05	1.90	5.40e-06	1.98
1.06	2.82e-04	9.26e-05	1.61	2.57e-05	1.85	6.52e-06	1.98
1.05	3.11e-04	1.09e-04	1.52	3.17e-05	1.78	8.11e-06	1.97
1.04	3.43e-04	1.29e-04	1.41	4.01e-05	1.68	1.04e-05	1.95
1.03	3.75e-04	1.52e-04	1.30	5.18e-05	1.55	1.45e-05	1.84
1.02	4.10e-04	1.80e-04	1.19	6.89e-05	1.39	2.16e-05	1.67
1.01	4.46e-04	2.12e-04	1.07	9.31e-05	1.19	3.44e-05	1.44
1.00	4.83e-04	2.49e-04	0.95	1.27e-04	0.97	6.38e-05	0.99
1.01	5.38e-04	1.43e-04	1.91	3.71e-05	1.95	9.45e-06	1.97
1.00	5.38e-04	1.43e-04	1.91	3.71e-05	1.95	9.45e-06	1.97

Table 5.6: The error and order of convergence for second order polynomial approximation ($k = 2$). M denotes the number of elements. For $\beta > 1.1$ the results are independent of whether stabilization with $L(h, \beta) = h^\beta$ is included. Recovery of the optimal order of convergence is possible only with stabilization as illustrated in the last two rows which show results with the stabilized scheme.

β	$M = 2^3$	$M = 2^4$		$M = 2^5$		$M = 2^6$	
	error	error	rate	error	rate	error	rate
2.00	4.92e-04	6.62e-05	2.89	8.61e-06	2.94	1.09e-06	2.98
1.80	5.19e-04	6.76e-05	2.94	8.50e-06	2.99	1.07e-06	2.99
1.50	5.53e-04	7.27e-05	2.93	9.08e-06	3.00	1.12e-06	3.02
1.20	6.04e-04	7.02e-05	3.11	8.73e-06	3.01	1.10e-06	2.99
1.10	3.04e-04	8.28e-05	1.88	9.80e-06	3.08	9.74e-07	3.33
1.09	2.98e-04	8.20e-05	1.86	1.01e-05	3.01	8.97e-07	3.50
1.08	2.92e-04	8.10e-05	1.85	1.05e-05	2.95	7.99e-07	3.71
1.07	2.86e-04	7.99e-05	1.84	1.08e-05	2.88	6.80e-07	4.00
1.06	2.80e-04	7.87e-05	1.83	1.12e-05	2.82	5.51e-07	4.34
1.05	2.74e-04	7.75e-05	1.82	1.15e-05	2.75	4.36e-07	4.72
1.04	2.69e-04	7.62e-05	1.82	1.18e-05	2.69	3.63e-07	5.02
1.03	2.63e-04	7.48e-05	1.81	1.21e-05	2.63	3.69e-07	5.03
1.02	2.57e-04	7.34e-05	1.81	1.23e-05	2.58	4.28e-07	4.85
1.01	2.51e-04	7.19e-05	1.81	1.25e-05	2.53	5.08e-07	4.62
1.00	2.46e-04	7.04e-05	1.81	1.26e-05	2.48	5.79e-07	4.45
1.01	1.50e-05	1.43e-06	3.41	1.10e-07	3.70	1.07e-08	3.36
1.00	1.52e-05	1.43e-06	3.39	1.10e-07	3.71	1.06e-08	3.36

Table 5.7: The error and order of convergence for third order polynomial approximation ($k = 3$). M denotes the number of elements. For $\beta > 1.1$ the results are independent of whether stabilization with $L(h, \beta) = h^\beta$ is included. Recovery of the optimal order of convergence is possible only with the stabilization as illustrated in the last two rows which show results with the stabilized scheme.

β	$M = 4$	$M = 8$		$M = 12$	
	error	error	rate	error	rate
2.0	1.49e-04	1.12e-05	3.74	2.51e-06	3.69
1.8	1.56e-04	1.13e-05	3.79	2.48e-06	3.73
1.5	2.00e-04	1.37e-05	3.87	2.74e-06	3.97
1.2	3.89e-04	2.98e-05	3.71	5.16e-06	4.33
1.10	5.45e-04	5.37e-05	3.34	1.04e-05	4.06
1.09	5.65e-04	5.78e-05	3.29	1.14e-05	4.00
1.08	5.85e-04	6.22e-05	3.23	1.26e-05	3.94
1.07	6.05e-04	6.72e-05	3.17	1.40e-05	3.89
1.06	6.26e-04	7.28e-05	3.11	1.56e-05	3.80
1.05	6.48e-04	7.89e-05	3.04	1.74e-05	3.72
1.04	6.71e-04	8.57e-05	2.97	1.96e-05	3.64
1.03	6.94e-04	9.32e-05	2.90	2.22e-05	3.54
1.02	7.18e-04	1.01e-04	2.82	2.51e-05	3.45
1.01	7.43e-04	1.11e-04	2.75	2.86e-05	3.34
1.00	7.70e-04	1.21e-04	2.67	3.27e-05	3.22
1.01	3.25e-06	1.38e-06	3.84	6.79e-07	3.89
1.00	3.25e-06	1.38e-06	3.84	6.79e-07	3.89

We finally consider the question of how the spectral radius scales with β and the spatial resolution. Illustrated in Table 5.8, we observe a fully discrete scaling for solving equation (5.121) as

$$\Delta t \sim \left(\frac{h}{k^2}\right)^\beta, \tag{5.194}$$

where Δt is the time step size and k still is the order of the approximation polynomial. Provided the stabilization is taken as $L(h, \beta) = (k^2/h)^{\beta-1}$, this does not impact the scaling. This is in agreement with expectations based on the experience for integer values of β [Hesthaven and Warburton (2008)].

We also note that in agreement with the theoretical analysis, both fluxes (5.150) and (5.151) work well for any $\beta \in [1, 2]$. This may seem counter-intuitive when $\beta = 1$ since in this case one of the flux choices reflect down winding. This illustrates that when $\beta = 1$ the single equation is equivalent to the system but in doing numerical computations a new mechanism (some

Table 5.8: The time step sizes required with a fixed space step length $h = 0.01$ for different values of β first and second order polynomial approximations. Here 's' denotes stable and 'u' unstable.

Δt (1st order)	Δt (2nd order)	$\beta = 1.0$	$\beta = 1.2$	$\beta = 1.5$	$\beta = 1.8$	$\beta = 2.0$
2.0e - 02	1.0e - 02	s	u	u	u	u
1.0e - 02	5.0e - 03	s	s	u	u	u
1.0e - 03	1.0e - 03	s	s	s	u	u
5.0e - 04	1.0e - 04	s	s	s	s	u
1.0e - 04	1.0e - 05	s	s	s	s	s

kind of symmetry) is introduced when writing the pure diffusion equation as a system due to the introduction of the fluxes.

5.4 Discontinuous Galerkin method for two dimensional fractional convection-diffusion equations

In this section, we consider the time dependent space fractional convection-diffusion equation in $\Omega \times (0, T]$

$$\frac{\partial u(x,y,t)}{\partial t} + \boldsymbol{b} \cdot \nabla u - {}_{-\infty}D_x^{\beta} u(x,y,t) - {}_{-\infty}D_y^{\beta_1} u(x,y,t) = f(x,y,t) \quad (5.195)$$

with the initial-boundary conditions

$$u(x,y,0) = u_0(x,y), \quad (x,y) \in \Omega, \quad (5.196)$$

$$u(x,y,t) = 0, \quad (x,y,t) \in \mathbb{R}^2 \backslash \Omega \times [0,T], \quad (5.197)$$

where $\Omega = (a,b) \times (c,d)$, ${}_{-\infty}D_x^{\beta} u(x,y,t)$ and ${}_{-\infty}D_y^{\beta_1} u(x,y,t)$ $(1 < \beta, \beta_1 < 2)$ denote the left R-L fractional derivatives w.r.t. x and y, respectively, the convection coefficient $\boldsymbol{b} = (b_1(x,y,t), b_2(x,y,t)) \in L^{\infty}([0,T]; W^{1,\infty}(\Omega)^2)$, the source term $f(x,y,t) \in L^2([0,T]; L^2(\Omega))$, and the initial function $u_0(x,y) \in L^2(\Omega)$.

We shall design a stable and accurate DG method for (5.195)–(5.197). This development is built on the extension of the previous section, where a qualitative study of the high-order LDG methods is discussed and some theoretical results are offered in one space dimension. In order to perform the error analysis, some special projection operators are defined to prove the convergence results. Unfortunately, the defined projection operators can not be easily extended to two dimensional case. Hence, to avoid this difficulty, a generalized DG method is developed in this section by carefully choosing the numerical fluxes and adding penalty terms. We present the full stability and convergence analyses. Of course, both the numerical schemes and theoretical analysis can be extended directly to the higher dimension problems or the models with the mixed (i.e., left and right) R-L fractional derivatives.

5.4.1 *Derivation of the numerical schemes*

We first give some notations, and then focus on deriving the numerical scheme of (5.195)–(5.197).

5.4.1.1 *Notations*

Let the domain Ω be subdivided into elements E. Here E is a triangle in 2D. We assume that the intersection of two elements is either empty, or an edge (2D). The mesh is called regular if

$$\forall E \in \mathscr{E}_h, \quad \frac{h_E}{\rho_E} \leq C,$$

where $\mathscr{E}_h$ is the subdivision of Ω, C a constant, h_E the diameter of the element E, and ρ_E the diameter of the inscribed circle in element E. The set of edges of the subdivision $\mathscr{E}_h$ is denoted by $\mathscr{E}_h^B$. Denote $\mathscr{E}_h^i$ as the set of interior edges, and $\mathscr{E}_h^b = \mathscr{E}_h^B \backslash \mathscr{E}_h^i$ the set of edges on $\partial\Omega$. Throughout this section $h = \max_{E\in\mathscr{E}_h} h_E$.

For $s \geq 0$, we introduce the fragment smooth function space

$$H^s(\mathscr{E}_h) = \left\{v \in L^2(\Omega) : v|_E \in H^s(E), \ \forall E \in \mathscr{E}_h\right\}$$

equipped with the broken Sobolev norm

$$\|v\|_{H^s(\Omega,\mathscr{E}_h)} = \left(\sum_{E\in\mathscr{E}_h} \| v \|^2_{H^s(E)}\right)^{\frac{1}{2}}.$$

Obviously, $L^2(\mathscr{E}_h) = H^0(\mathscr{E}_h)$. It is well known if $v \in H^s(\Omega, \mathscr{E}_h)\,(s > \frac{1}{2})$, by trace theorem, $v \in L^2(\mathscr{E}_h^B)$[Guo (1998); Lions and Magenes (1972); Quarteroni and Valli (2008)]. In the following, we retain the same style with the traditional Sobolev space and rewrite $\|v\|_{H^s(\Omega,\mathscr{E}_h)}$ as $\|v\|_{H^s(\Omega)}$ when no misunderstanding is possible.

Let E_1 and E_2 be two adjacent elements of $\mathscr{E}_h$; denote (x, y) as an arbitrary point of the set $e = \partial E_1 \cap \partial E_2$, and $\boldsymbol{n}_{E_1}$ and $\boldsymbol{n}_{E_2}$ the corresponding outward unit normals at that point. Assuming that $(\boldsymbol{\sigma}, u)$ are smooth inside each element E_1 and E_2, and denote by $(\boldsymbol{\sigma}_1, u_1)$ and $(\boldsymbol{\sigma}_2, u_2)$ the trace of $(\boldsymbol{\sigma}, u)$ on e from the interior of E_1 and E_2, respectively. Then for $e \in \mathscr{E}_h^i$, we define the mean values $\{\cdot\}$ and jumps $[\![\cdot]\!]$ at $x \in e$ as

$$\{u\} = \frac{1}{2}(u_1 + u_2), \quad [\![u]\!] = u_1 \boldsymbol{n}_{E_1} + u_2 \boldsymbol{n}_{E_2},$$
$$\{\boldsymbol{\sigma}\} = \frac{1}{2}(\boldsymbol{\sigma}_1 + \boldsymbol{\sigma}_2), \quad [\![\boldsymbol{\sigma}]\!] = \boldsymbol{\sigma}_1 \cdot \boldsymbol{n}_{E_1} + \boldsymbol{\sigma}_2 \cdot \boldsymbol{n}_{E_2};$$

for $e \in \mathscr{E}_h^b$, we define

$$\{u\} = u, \ [\![u]\!] = u\,\boldsymbol{n}; \quad \{\boldsymbol{\sigma}\} = \boldsymbol{\sigma}, \ [\![\boldsymbol{\sigma}]\!] = \boldsymbol{\sigma} \cdot \boldsymbol{n},$$

where $\boldsymbol{n}$ is the outward unit normal to $\partial\Omega$. Note that the jump in u is a vector and the jump in $\boldsymbol{\sigma}$ is a scalar which only involves the normal component of $\boldsymbol{\sigma}$.

5.4.1.2 *The DG scheme for space discretization*

Let $\psi(\mathbf{x}, t) = (1 + |\boldsymbol{b}(\mathbf{x}, t)|^2)^{\frac{1}{2}}$, where $\mathbf{x} = (x, y)$, $|\boldsymbol{b}(\mathbf{x}, t)|^2 = b_1^2 + b_2^2$. Hence, the characteristic direction associated with $\partial_t u + \boldsymbol{b} \cdot \nabla u$ is denoted by $\partial_\tau = \frac{\partial_t}{\psi} + \frac{\boldsymbol{b} \cdot \nabla}{\psi}$. Now, we need to rewrite (5.195) as low order system. In fact, similarly to the previous section, using (see Proposition B.15)

$$_{-\infty}D_x^\beta u(\cdot, y, t) = \frac{\partial}{\partial x}\, {}_aD_x^{-(2-\beta)}\left(\frac{\partial}{\partial x} u(\cdot, y, t)\right), \ (x, y) \in \Omega, \quad (5.198)$$
$$_{-\infty}D_y^\beta u(x, \cdot, t) = \frac{\partial}{\partial y}\, {}_cD_y^{-(2-\beta_1)}\left(\frac{\partial}{\partial y} u(x, \cdot, t)\right), \quad (x, y) \in \Omega, \quad (5.199)$$

and introducing two auxiliary vector functions $\boldsymbol{p} = (p_x, p_y)$, $\boldsymbol{\sigma} = (\sigma_x, \sigma_y)$, we have

$$\begin{cases} \psi \partial_\tau u - \nabla \cdot \boldsymbol{\sigma} = f, & (x, y, t) \in \Omega \times (0, T], \\ \boldsymbol{\sigma} - I_{\mathbf{x}}^{\bar{\boldsymbol{\alpha}}} \boldsymbol{p} = 0, & (x, y, t) \in \Omega \times (0, T], \\ \boldsymbol{p} - \nabla u = 0, & (x, y, t) \in \Omega \times (0, T], \\ u(x, y, 0) = u_0(x, y), & (x, y) \in \Omega, \\ u(x, y, t) = 0, & (x, y, t) \in \mathbb{R}^2 \backslash \Omega \times [0, T], \end{cases} \quad (5.200)$$

where $I_{\mathbf{x}}^{\bar{\boldsymbol{\alpha}}} := \left({}_aD_x^{-(2-\beta)},\ {}_cD_y^{-(2-\beta_1)}\right)$ and correspondingly,

$$I_{\mathbf{x}}^{\bar{\boldsymbol{\alpha}}} \boldsymbol{p} := \left({}_aD_x^{-(2-\beta)} p_x,\ {}_cD_y^{-(2-\beta_1)} p_y\right). \quad (5.201)$$

For an arbitrary subset $E \in \mathscr{E}_h$, we multiply the first, second, and the third equation of (5.200) by the smooth test functions $v, \boldsymbol{\omega} = (\omega_x, \omega_y)$, and

$\boldsymbol{q} = (q_x, q_y)$, respectively, and integrate by parts over the element E to obtain

$$\begin{cases} \int_E \psi \partial_\tau u v d\mathbf{x} + \int_E \boldsymbol{\sigma} \cdot \nabla v d\mathbf{x} - \int_{\partial E} \boldsymbol{\sigma} \cdot \boldsymbol{n}_E v ds = \int_E f v d\mathbf{x}, \\ \int_E \boldsymbol{\sigma} \cdot \boldsymbol{\omega} d\mathbf{x} - \int_E I_{\mathbf{x}}^{\bar{\alpha}} \boldsymbol{p} \cdot \boldsymbol{\omega} d\mathbf{x} = 0, \\ \int_E \boldsymbol{p} \cdot \boldsymbol{q} d\mathbf{x} + \int_E u \nabla \cdot \boldsymbol{q} d\mathbf{x} - \int_{\partial E} u \boldsymbol{n}_E \cdot \boldsymbol{q} ds = 0, \end{cases} \tag{5.202}$$

where $\boldsymbol{n}_E$ is the outward unit normal to ∂E, $d\mathbf{x} := dxdy$. Note that the above equations are well defined for the functions $(u, \boldsymbol{\sigma}, \boldsymbol{p})$ belong to

$$H^1\left(0, T; H^1(\mathscr{E}_h)\right) \times \left(L^2(0, T; H^1(\mathscr{E}_h))\right)^2 \times \left(L^2(0, T; L^2(\mathscr{E}_h))\right)^2, \tag{5.203}$$

and $(v, \boldsymbol{q}, \boldsymbol{\omega})$ belong to $H^1(\mathscr{E}_h) \times \left(H^1(\mathscr{E}_h)\right)^2 \times \left(L^2(\mathscr{E}_h)\right)^2$. Here $\left(L^2(\mathscr{E}_h)\right)^2$ and $\left(H^1(\mathscr{E}_h)\right)^2$, etc. denote the vector-valued function spaces.

To complete the DG scheme for space discretization, we introduce the finite dimensional subspace

$$\mathbb{V}_h = \left\{ u_h \in L^2(\Omega) : u_h|_E \in P^k(E), \ \ \forall E \in \mathscr{E}_h \right\} \subset H^1(\mathscr{E}_h), \tag{5.204}$$

where $P^k(E)$ denotes the set of polynomials of degree less than or equal to k. We seek to approximate $(u, \boldsymbol{\sigma}, \boldsymbol{p})$ with functions $(u_h, \boldsymbol{\sigma}_h, \boldsymbol{p}_h)$ in the finite element spaces $H^1(0, T; V_h) \times \left(L^2(0, T; V_h)\right)^2 \times \left(L^2(0, T; V_h)\right)^2$, such that for any $(v, \boldsymbol{q}, \boldsymbol{\omega}) \in V_h \times (V_h)^2 \times (V_h)^2$,

$$\begin{cases} \int_E \psi \partial_\tau u_h v d\mathbf{x} + \int_E \boldsymbol{\sigma}_h \cdot \nabla v d\mathbf{x} - \int_{\partial E} \widehat{\boldsymbol{\sigma}_h} \cdot \boldsymbol{n}_E v ds = \int_E f v d\mathbf{x}, \\ \int_E \boldsymbol{\sigma}_h \cdot \boldsymbol{\omega} d\mathbf{x} - \int_E I_{\mathbf{x}}^{\bar{\alpha}} \boldsymbol{p}_h \cdot \boldsymbol{\omega} d\mathbf{x} = 0, \\ \int_E \boldsymbol{p}_h \cdot \boldsymbol{q} d\mathbf{x} + \int_E u_h \nabla \cdot \boldsymbol{q} d\mathbf{x} - \int_{\partial E} \widehat{u}_h \boldsymbol{n}_E \cdot \boldsymbol{q} ds = 0, \end{cases} \tag{5.205}$$

where the numerical fluxes $\widehat{\boldsymbol{\sigma}_h}$ and $\widehat{u}_h$ need to be carefully chosen for ensuring the stability and the accuracy of the method. Here we use

$$\widehat{u}_h = \{u_h\} - \epsilon_2 [\![\boldsymbol{\sigma}_h]\!], \ \ \widehat{u}_h|_{\partial\Omega} = 0; \quad \widehat{\boldsymbol{\sigma}_h} = \{\boldsymbol{\sigma}_h\} - \epsilon_1 [\![u_h]\!], \tag{5.206}$$

where the nonnegative parameters ϵ_1, ϵ_2 are usually chosen as

$$\epsilon_1 = \mathcal{O}\left(h_e^{\gamma_1}\right), \ \epsilon_2 = \mathcal{O}\left(h_e^{\gamma_2}\right), \quad -1 \le \gamma_1, \gamma_2 \le 1.$$

Obviously, the fluxes $\widehat{\boldsymbol{\sigma}_h}$ and $\widehat{u}_h$ are consistent. Substituting the flux $\widehat{u}_h$ and $\widehat{\boldsymbol{\sigma}_h}$ into (5.205) and summing over all the elements, we get the DG scheme as follows:

$$\begin{cases} (\psi\partial_\tau u_h, v) + (\boldsymbol{\sigma}_h, \nabla v) - (\{\boldsymbol{\sigma}_h\}, [\![v]\!])_{\mathscr{E}_h^B} + \epsilon_1([\![u_h]\!], [\![v]\!])_{\mathscr{E}_h^B} = (f, v), \\ (\boldsymbol{\sigma}_h, \boldsymbol{\omega}) - (I_{\mathbf{x}}^{\bar{\boldsymbol{\alpha}}}\boldsymbol{p}_h, \boldsymbol{\omega}) = 0, \\ (\boldsymbol{p}_h, \boldsymbol{q}) - (\nabla u_h, \boldsymbol{q}) + ([\![u_h]\!], \{\boldsymbol{q}\})_{\mathscr{E}_h^B} + \epsilon_2([\![\boldsymbol{\sigma}_h]\!], [\![\boldsymbol{q}]\!])_{\mathscr{E}_h^i} = 0, \end{cases} \tag{5.207}$$

where

$$(w, v) := \sum_{E\in\mathscr{E}_h} \int_E w \cdot v d\mathbf{x},$$

$$(w, v)_{\mathscr{E}_h^B} := \sum_{e\in\mathscr{E}_h^B} \int_e w \cdot v ds, \quad (w, v)_{\mathscr{E}_h^i} := \sum_{e\in\mathscr{E}_h^i,} \int_e w \cdot v ds,$$

and the relations

$$(u_h, \nabla \cdot \boldsymbol{q}) = \sum_{E\in\mathscr{E}_h} \int_{\partial E} u_h \boldsymbol{q} \cdot \boldsymbol{n}_E ds - (\nabla u_h, \boldsymbol{q}),$$

$$\sum_{E\in\mathscr{E}_h} \int_{\partial E} u_h \boldsymbol{q} \cdot \boldsymbol{n}_E ds = (\{\boldsymbol{q}\}, [\![u_h]\!])_{\mathscr{E}_h^B} + (\{u\}, [\![\boldsymbol{q}]\!])_{\mathscr{E}_h^i}$$

have been used to derive the third equations of (5.207).

Remark 5.9. When $\epsilon_2 = 0$, the LDG scheme is obtained. This means that the auxiliary variables $\boldsymbol{\sigma}_h$ and $\boldsymbol{p}_h$ can be locally solved in terms of u_h by using the second and third equations of (5.205) and then easily eliminated from the first equation of (5.205); u_h is the only variable for the resulting system. In addition, if $\epsilon_1 = 0$ further, $\widehat{u}_h$ and $\widehat{\boldsymbol{\sigma}}_h$ will reduce to the center fluxes, and LDG scheme is still stable (see Theorem 5.12). When $\epsilon_2 \neq 0$, the third equation of (5.205) makes the DG method loose its locality, since $\boldsymbol{p}_h$ is a function of u_h and $\boldsymbol{\sigma}_h$, $\boldsymbol{p}_h$ can not be eliminated from the third equation. So we have to simultaneously solve the three unknowns u_h, p_{xh}, p_{yh}. We will see that, although the extra unknowns can not be eliminated in the this case, the error analysis is available.

5.4.1.3 *Dealing with time*

After performing the DG space approximation, we discretize the time derivative with the characteristic method. For the given positive integer N, let $0 = t^0 < t^1 < \cdots < t^N = T$ be a partition of $[0, T]$ into subintervals $J^n = (t^{n-1}, t^n]$ with uniform mesh and the interval length $\Delta t = t^n - t^{n-1}, 1 \leq n \leq N$. The characteristic tracing back along the field

$\boldsymbol{b}$ of a point $\mathbf{x} = (x, y) \in \Omega$ at time t^n to t^{n-1} is approximated by [Chen (2002); Chen *et al.* (2003); Douglas and Russell (1982)]

$$\check{\mathbf{x}}(\mathbf{x}, t^{n-1}) = \mathbf{x} - \boldsymbol{b}(\mathbf{x}, t^n)\Delta t.$$

Therefore, the approximation for the hyperbolic part of (5.195) at time t^n can be given as

$$\psi^n \partial_\tau u^n \approx \frac{u^n - \check{u}^{n-1}}{\Delta t},$$

where $u^n = u(\mathbf{x}, t^n)$, $\check{u}^{n-1} = u(\check{\mathbf{x}}(\mathbf{x}, t^{n-1}), t^{n-1})$, and $\check{u}^0 = u^0(\mathbf{x})$.

Remark 5.10 ([Douglas and Russell (1982)]). Assume that the solution u of (5.195) is sufficiently regular. Under the assumption of the function $\boldsymbol{b}$, we have

$$\begin{aligned} &\left\| \psi^n \partial_\tau u^n - \frac{u^n - \check{u}^{n-1}}{\Delta t} \right\|^2_{L^2(\Omega)} \\ &\qquad \leq C \parallel \psi^{(4)} \parallel_{L^\infty(J^n; L^\infty(\Omega))} \parallel \partial_{\tau\tau} u \parallel^2_{L^2(J^n; L^2(\Omega))} \Delta t. \end{aligned}$$

Thus, the fully discrete scheme corresponding to the variational formulation (5.207) is to find $(u_h^n, \boldsymbol{\sigma}_h^n, \boldsymbol{p}_h^n) \in \mathbb{V}_h \times (\mathbb{V}_h)^2 \times (\mathbb{V}_h)^2$, such that for any $(v, \boldsymbol{\tau}, \boldsymbol{q}) \in \mathbb{V}_h \times (\mathbb{V}_h)^2 \times (\mathbb{V}_h)^2$

$$\begin{cases} \left(\frac{u_h^n - \check{u}_h^{n-1}}{\Delta t}, v\right) + (\boldsymbol{\sigma}_h^n, \nabla v) - (\{\boldsymbol{\sigma}_h^n\}, [\![v]\!])_{\mathscr{E}_h^B} + \epsilon_1([\![u_h^n]\!], [\![v]\!])_{\mathscr{E}_h^B} = (f^n, v), \\ (\boldsymbol{\sigma}_h^n, \boldsymbol{\omega}) - (I_{\mathbf{x}}^{\bar{\alpha}} \boldsymbol{p}_h^n, \boldsymbol{\omega}) = 0, \\ (\boldsymbol{p}_h^n, \boldsymbol{q}) - (\nabla u_h^n, \boldsymbol{q}) + ([\![u_h^n]\!], \{\boldsymbol{q}\})_{\mathscr{E}_h^B} + \epsilon_2([\![\boldsymbol{\sigma}_h^n]\!], [\![\boldsymbol{q}]\!])_{\mathscr{E}_h^i} = 0, \end{cases} \tag{5.208}$$

where $\check{u}_h^{n-1} = u_h(\check{\mathbf{x}}(\mathbf{x}, t^{n-1}), t^{n-1}), \check{u}_h^0 = u^0$.

Define the bilinear forms by

$$\mathbb{a}(\boldsymbol{\sigma}_h^n, v) := (\boldsymbol{\sigma}_h^n, \nabla v) - (\{\boldsymbol{\sigma}_h^n\}, [\![v]\!])_{\mathscr{E}_h^B}, \quad \mathbb{c}(\boldsymbol{p}_h^n, \boldsymbol{q}) := (\boldsymbol{p}_h^n, \boldsymbol{q}),$$

$$\mathbb{d}(u_h^n, v) := \epsilon_1([\![u_h^n]\!], [\![v]\!])_{\mathscr{E}_h^B}, \quad \mathbb{e}(\boldsymbol{\sigma}_h^n, \boldsymbol{q}) := \epsilon_2([\![\boldsymbol{\sigma}_h^n]\!], [\![\boldsymbol{q}]\!])_{\mathscr{E}_h^i},$$

and the linear form

$$\mathcal{F}(v) := (f^n, v) \quad \forall v \in \mathbb{V}_h.$$

We can rewrite (5.208) as a compact formulation: find $(u_h^n, \boldsymbol{\sigma}_h^n, \boldsymbol{p}_h^n) \in \mathbb{V}_h \times (\mathbb{V}_h)^2 \times (\mathbb{V}_h)^2$ at time $t = t^n$, such that

$$\begin{cases} \left(\frac{u_h^n - \check{u}_h^{n-1}}{\Delta t}, v\right) + \mathbb{a}(\boldsymbol{\sigma}_h^n, v) + \mathbb{d}(u_h^n, v) = \mathcal{F}(v) \quad \forall v \in \mathbb{V}_h, \\ \mathbb{c}(\boldsymbol{\sigma}_h^n, \boldsymbol{\omega}) - \mathbb{c}(I_{\mathbf{x}}^{\bar{\alpha}} \boldsymbol{p}_h^n, \boldsymbol{\omega}) = 0 \quad \forall \boldsymbol{\omega} \in (\mathbb{V}_h)^2, \\ \mathbb{c}(\boldsymbol{p}_h^n, \boldsymbol{q}) - \mathbb{a}(\boldsymbol{q}, u_h^n) + \mathbb{e}(\boldsymbol{\sigma}_h^n, \boldsymbol{q}) = 0 \quad \forall \boldsymbol{q} \in (\mathbb{V}_h)^2. \end{cases} \tag{5.209}$$

5.4.2 *Stability analysis and error estimate*

In this subsection, we focus on providing the proof of the unconditional stability and the convergence of the schemes. To ensure the stability, we only require $\epsilon_1, \epsilon_2 \geq 0$. However, to perform the convergence analysis, we require $\epsilon_1, \epsilon_2 > 0$. In the following, C indicates a generic constant independent of h and Δt, which may take different values in different occurrences.

Lemma 5.7 ([Chen (2002)]). *If $\boldsymbol{b} \in L^\infty([0,T]; W^{1,\infty}(\Omega)^2)$, for any function $v \in L^2(\Omega)$ and each n, there is*

$$\| \check{v} \|_{L^2(\Omega)}^2 - \| v \|_{L^2(\Omega)}^2 \leq C\Delta t \| v \|_{L^2(\Omega)}^2, \tag{5.210}$$

where $\check{v}(\mathbf{x}) = v(\mathbf{x} - \boldsymbol{b}(\mathbf{x}, t^n)\Delta t)$.

Theorem 5.12 (Numerical stability). *If $\boldsymbol{b} \in L^\infty([0,T]; W^{1,\infty}(\Omega)^2)$, the full discretization scheme (5.208) is stable, i.e., for any integer $N = 1, 2, \cdots$, there is*

$$\| u_h^N \|_{L^2(\Omega)}^2 + 2\Delta t \sum_{n=1}^{N} \left|(u_h^n, \boldsymbol{\sigma}_h^n, \boldsymbol{p}_h^n)\right|_{\mathcal{A}}^2 \tag{5.211}$$

$$\leq C\Delta t \sum_{n=1}^{N} \| f^n \|_{L^2(\Omega)}^2 + C \| u^0 \|_{L^2(\Omega)}^2, \tag{5.212}$$

where $u_h^0 = u^0$, and the semi-norm $|\cdot|_{\mathcal{A}}$ is defined as

$$\begin{aligned} &\left|(u_h^n, \boldsymbol{\sigma}_h^n, \boldsymbol{p}_h^n)\right|_{\mathcal{A}}^2 = \mathbb{d}(u_h^n, u_h^n) + \mathbb{c}(I_{\mathbf{x}}^{\bar{\alpha}} \boldsymbol{p}_h^n, \boldsymbol{p}_h^n) + \mathbb{e}(\boldsymbol{\sigma}_h^n, \boldsymbol{\sigma}_h^n) \\ &= \int_c^d \left({}_aD_x^{-(2-\beta)} p_{xh}(\cdot, y), p_{xh}(\cdot, y)\right)_{L^2(a,b)} dy + \epsilon_1 \sum\nolimits_{e \in \mathscr{E}_h^B} \| [\![u_h^n]\!] \|_{L^2(e)}^2 \\ &+ \int_a^b \left({}_cD_y^{-(2-\beta_1)} p_{yh}(x, \cdot), p_{yh}(x, \cdot)\right)_{L^2(c,d)} dx + \epsilon_2 \sum\nolimits_{e \in \mathscr{E}_h^i} \| [\![\boldsymbol{\sigma}_h^n]\!] \|_{L^2(e)}^2 . \end{aligned} \tag{5.213}$$

Proof. Let $v = 2\Delta t u_h^n$, $\boldsymbol{\omega} = -2\Delta t \boldsymbol{p}_h^n$, $\boldsymbol{q} = 2\Delta t \boldsymbol{\sigma}_h^n$ in the equations of (5.209), respectively. By the symmetry of the bilinear formulae, adding the above equations, we obtain

$$2\Delta t \mathcal{F}(u_h^n) = 2\Delta t \mathbb{c}(I_{\mathbf{x}}^{\bar{\alpha}} \boldsymbol{p}_h^n, \boldsymbol{p}_h^n) + 2\Delta t \mathbb{e}(\boldsymbol{\sigma}_h^n, \boldsymbol{\sigma}_h^n)$$
$$+ 2\big(u_h^n - \breve{u}_h^{n-1}, u_h^n\big) + 2\Delta t \mathbb{d}(u_h^n, u_h^n).$$

Following from

$$2\big(u_h^n - \breve{u}_h^{n-1}, u_h^n\big) \geq \| u_h^n \|_{L^2(\Omega)}^2 - \| \breve{u}_h^{n-1} \|_{L^2(\Omega)}^2,$$

the Young inequality, the definition of $\mathcal{F}$ and $|\cdot|_{\mathcal{A}}$, and Lemma 5.7, we have

$$\| u_h^n \|_{L^2(\Omega)}^2 - \| u_h^{n-1} \|_{L^2(\Omega)}^2 + 2\Delta t \big|(u_h^n, \boldsymbol{\sigma}_h^n, \boldsymbol{p}_h^n)\big|_{\mathcal{A}}^2$$
$$\leq C\Delta t \| u_h^{n-1} \|_{L^2(\Omega)}^2 + \Delta t\big(\| u_h^n \|_{L^2(\Omega)}^2 + \| f^n \|_{L^2(\Omega)}^2 \big).$$

Summing from $n = 1$ to N, we get

$$\| u_h^N \|_{L^2(\Omega)}^2 + 2\Delta t \sum_{n=1}^{N} \big|(u_h^n, \boldsymbol{\sigma}_h^n, \boldsymbol{p}_h^n)\big|_{\mathcal{A}}^2$$
$$\leq C\Delta t \sum_{n=1}^{N} \| u_h^n \|_{L^2(\Omega)}^2 + (1 + C\Delta t) \| u_h^0 \|_{L^2(\Omega)}^2 + \Delta t \sum_{n=1}^{N} \| f^n \|_{L^2(\Omega)}^2 .$$

Using the discrete Grönwall inequality (see Appendix A), with $C\Delta t < 1$, $\forall N \geq 1$, there is

$$\| u_h^N \|_{L^2(\Omega)}^2 + 2\Delta t \sum_{n=1}^{N} \big|(u_h^n, \boldsymbol{\sigma}_h^n, \boldsymbol{p}_h^n)\big|_{\mathcal{A}}^2$$
$$\leq C \| u_h^0 \|_{L^2(\Omega)}^2 + C\Delta t \sum_{n=1}^{N} \| f^n \|_{L^2(\Omega)}^2 .$$

Thus, we complete the proof. □

5.4.2.1 *Error estimates*

For the error estimate, we define the orthogonal projection operators, $\Pi : H^1(\mathscr{E}_h) \to V_h$, $\boldsymbol{\Pi} = (\pi_x, \pi_y) : \big(H^1(\mathscr{E}_h)\big)^2 \to (V_h)^2$, and $\boldsymbol{Q} = (Q_x, Q_y) : \big(L^2(\mathscr{E}_h)\big)^2 \to (V_h)^2$. For all the elements $E \subset \mathscr{E}_h$, the operators Π, $\boldsymbol{\Pi}$, $\boldsymbol{Q}$ are defined to satisfy

$$(\Pi u - u, v)_E = 0 \quad \forall v \in V_h, \tag{5.214}$$
$$(\boldsymbol{\Pi}\boldsymbol{\sigma} - \boldsymbol{\sigma}, \boldsymbol{q})_E = 0 \quad \forall \boldsymbol{q} \in (V_h)^2, \tag{5.215}$$
$$(\boldsymbol{Q}\boldsymbol{p} - \boldsymbol{p}, \boldsymbol{\omega})_E = 0 \quad \forall \boldsymbol{\omega} \in (V_h)^2 . \tag{5.216}$$

As usual, we express the errors $(e_u^n, \boldsymbol{e}_{\boldsymbol{\sigma}}^n, e_{\boldsymbol{p}}^n) = (u^n - u_h^n, \boldsymbol{\sigma}^n - \boldsymbol{\sigma}_h^n, \boldsymbol{p}^n - \boldsymbol{p}_h^n)$ as the sum

$$(e_u^n, \boldsymbol{e}_{\boldsymbol{\sigma}}^n, e_{\boldsymbol{p}}^n) = (u^n - \Pi u^n, \boldsymbol{\sigma}^n - \boldsymbol{\Pi}\boldsymbol{\sigma}^n, \boldsymbol{p}^n - \boldsymbol{Q}\boldsymbol{p}^n) + (\Pi e_u^n, \boldsymbol{\Pi e}_{\boldsymbol{\sigma}}^n, \boldsymbol{Q} e_{\boldsymbol{p}}^n).$$

From (5.209), we obtain the compact form

$$\left(\frac{u_h^n - \breve{u}_h^{n-1}}{\Delta t}, v\right) + \mathcal{A}(u_h^n, \boldsymbol{\sigma}_h^n, \boldsymbol{p}_h^n; v, \boldsymbol{\omega}, \boldsymbol{q}) = \mathcal{F}(v), \tag{5.217}$$

where

$$\begin{aligned}&\mathcal{A}(u_h^n, \boldsymbol{\sigma}_h^n, \boldsymbol{p}_h^n; v, \boldsymbol{\omega}, \boldsymbol{q}) = \mathbb{a}(\boldsymbol{\sigma}_h^n, v) + \mathbb{d}(u_h^n, v)\\ &\quad + \mathbb{c}(\boldsymbol{\sigma}_h^n, \boldsymbol{\omega}) - \mathbb{c}(I_{\mathbf{x}}^{\bar{\alpha}} \boldsymbol{p}_h^n, \boldsymbol{\omega}) + \mathbb{c}(\boldsymbol{p}_h^n, \boldsymbol{q}) - \mathbb{a}(\boldsymbol{q}, u_h^n) + \mathbb{e}(\boldsymbol{\sigma}_h^n, \boldsymbol{q}).\end{aligned} \tag{5.218}$$

Lemma 5.8. *Assume that the solution u of problem (5.195) is sufficiently regular. Then*

$$\begin{aligned}&\left(\psi^n \partial_\tau u^n - \frac{u_h^n - \breve{u}_h^{n-1}}{\Delta t}, \Pi e_u^n\right) + \left|(\Pi e_u^n, \boldsymbol{\Pi e}_{\boldsymbol{\sigma}}^n, \boldsymbol{Q} e_{\boldsymbol{p}}^n)\right|_{\mathcal{A}}^2\\ &\quad = \mathcal{A}(\Pi u^n - u^n, \boldsymbol{\Pi}\boldsymbol{\sigma}^n - \boldsymbol{\sigma}^n, \boldsymbol{Q}\boldsymbol{p}^n - \boldsymbol{p}^n; \Pi e_u^n, -\boldsymbol{Q} e_{\boldsymbol{p}}^n, \boldsymbol{\Pi e}_{\boldsymbol{\sigma}}^n).\end{aligned} \tag{5.219}$$

Proof. By the consistency of the numerical fluxes, the exact solution $(u, \boldsymbol{\sigma}, \boldsymbol{p})$ satisfies (5.207). Taking $v = \Pi e_u^n, \boldsymbol{\omega} = -\boldsymbol{Q} e_{\boldsymbol{p}}^n, \boldsymbol{q} = \boldsymbol{\Pi e}_{\boldsymbol{\sigma}}^n$ and subtracting (5.208) from (5.207) yield

$$\left(\psi^n \partial_\tau u^n - \frac{u_h^n - \breve{u}_h^{n-1}}{\Delta t}, \Pi e_u^n\right) + \mathcal{A}(e_u^n, \boldsymbol{e}_{\boldsymbol{\sigma}}^n, e_{\boldsymbol{p}}^n; \Pi e_u^n, -\boldsymbol{Q} e_{\boldsymbol{p}}^n, \boldsymbol{\Pi e}_{\boldsymbol{\sigma}}^n) = 0 \tag{5.220}$$

and

$$\left|(\Pi e_u^n, \boldsymbol{\Pi e}_{\boldsymbol{\sigma}}^n, \boldsymbol{Q} e_{\boldsymbol{p}}^n)\right|_{\mathcal{A}}^2 = \mathcal{A}(\Pi e_u^n, \boldsymbol{\Pi e}_{\boldsymbol{\sigma}}^n, \boldsymbol{Q} e_{\boldsymbol{p}}^n; \Pi e_u^n, -\boldsymbol{Q} e_{\boldsymbol{p}}^n, \boldsymbol{\Pi e}_{\boldsymbol{\sigma}}^n). \tag{5.221}$$

By the Galerkin orthogonality, there is

$$\begin{aligned}&\mathcal{A}(e_u^n, \boldsymbol{e}_{\boldsymbol{\sigma}}^n, e_{\boldsymbol{p}}^n; \Pi e_u^n, -\boldsymbol{Q} e_{\boldsymbol{p}}^n, \boldsymbol{\Pi e}_{\boldsymbol{\sigma}}^n)\\ &\quad = \mathcal{A}(\Pi e_u^n, \boldsymbol{\Pi e}_{\boldsymbol{\sigma}}^n, \boldsymbol{Q} e_{\boldsymbol{p}}^n; \Pi e_u^n, -\boldsymbol{Q} e_{\boldsymbol{p}}^n, \boldsymbol{\Pi e}_{\boldsymbol{\sigma}}^n)\\ &\quad\quad - \mathcal{A}(\Pi u^n - u^n, \boldsymbol{\Pi}\boldsymbol{\sigma}^n - \boldsymbol{\sigma}^n, \boldsymbol{Q}\boldsymbol{p}^n - \boldsymbol{p}^n; \Pi e_u^n, -\boldsymbol{Q} e_{\boldsymbol{p}}^n, \boldsymbol{\Pi e}_{\boldsymbol{\sigma}}^n).\end{aligned} \tag{5.222}$$

Substituting the equalities (5.221) and (5.222) into (5.220) leads to the desired result. □

The following two lemmas [Castilo *et al.* (2000)] contain all the information we actually use about our finite element. The first one is the standard approximation result for any linear continuous operator Π from $H^{s+1}(E)$

onto $V_h(E) = \{v; v\big|_E \in P^k(E)\}$ satisfying $\Pi v = v$ for any $v \in P^k(E)$. The second one is the standard trace inequality.

Lemma 5.9. *Let $v \in H^{s+1}(E)$, $s \geq 0$. Π is a linear continuous operator from $H^{s+1}(E)$ onto $V_h(E)$ such that $\Pi v = v$ for any $v \in P^k(E)$. Then, for $m = 0, 1$,*

$$\begin{cases} \big|v - \Pi v\big|_{H^m(E)} \leq C h_E^{min\{s,k\}+1-m} \parallel v \parallel_{H^{s+1}(E)}, \\ \parallel v - \Pi v \parallel_{L^2(\partial E)} \leq C h_E^{min\{s,k\}+\frac{1}{2}} \parallel v \parallel_{H^{s+1}(E)} . \end{cases} \tag{5.223}$$

Lemma 5.10. *There exists a generic constant C being independent of h_E, for any $v \in V_h(E)$, such that*

$$\parallel v \parallel_{L^2(\partial E)} \leq C h_E^{-\frac{1}{2}} \parallel v \parallel_{L^2(E)} . \tag{5.224}$$

We will estimate the first left hand side term of (5.219).

Lemma 5.11 ([Chen (2002)]). *If $\boldsymbol{b} \in L^\infty([0,T]; W^{1,\infty}(\Omega)^2)$, for any function $v \in H^1(\Omega)$ and each n,*

$$\parallel v - \breve{v} \parallel_{L^2(\Omega)} \leq C\Delta t \parallel \nabla v \parallel_{L^2(\Omega)}, \tag{5.225}$$

where $\breve{v} = v(\breve{\mathbf{x}}) = v(\mathbf{x} - \boldsymbol{b}^n \Delta t)$.

The following result is a straightforward consequence of the estimate of the first left-side term of (5.219).

Theorem 5.13. *Assume that the solution u of problem (5.195) is sufficiently smooth and u_h^n satisfies (5.208). If $\mathbf{b} \in L^\infty([0,T]; W^{1,\infty}(\Omega)^2)$, we have*

$$\begin{aligned} &\left(\psi^n \partial_\tau u^n - \frac{u_h^n - \breve{u}_h^{n-1}}{\Delta t}, \Pi e_u^n\right) \\ &\geq \frac{1}{2\Delta t}\left(\parallel \Pi e_u^n \parallel_{L^2(\Omega)}^2 - \parallel \Pi e_u^{n-1} \parallel_{L^2(\Omega)}^2 \right) - C \parallel \Pi e_u^{n-1} \parallel_{L^2(\Omega)}^2 \\ &\quad - C\Delta t \parallel \partial_{\tau\tau} u \parallel_{L^2(J^n;L^2(\Omega))}^2 - \frac{C}{\Delta t} \parallel \partial_t(\Pi u - u) \parallel_{L^2(J^n;L^2(\Omega))}^2 \\ &\quad - C \parallel \nabla(\Pi u^{n-1} - u^{n-1}) \parallel_{L^2(\Omega)}^2 - C \parallel \Pi e_u^n \parallel_{L^2(\Omega)}^2 . \end{aligned} \tag{5.226}$$

Proof. From (5.219), it can be noted that

$$\begin{aligned}
&\left(\psi^n \partial_\tau u^n - \frac{u_h^n - \check{u}_h^{n-1}}{\Delta t}, \Pi e_u^n\right) \\
&= \left(\frac{\Pi e_u^n - \Pi \check{e}_u^{n-1}}{\Delta t}, \Pi e_u^n\right) + \left(\psi^n \partial_\tau u^n - \frac{u^n - \check{u}^{n-1}}{\Delta t}, \Pi e_u^n\right) \\
&- \left(\frac{(\Pi u^n - u^n) - (\Pi \check{u}^{n-1} - \check{u}^{n-1})}{\Delta t}, \Pi e_u^n\right) = \sum_{i=1}^{3} \mathcal{B}_i.
\end{aligned} \tag{5.227}$$

Using Lemma 5.7, we obtain

$$\begin{aligned}
\mathcal{B}_1 &= \left(\frac{\Pi e_u^n - \Pi \check{e}_u^{n-1}}{\Delta t}, \Pi e_u^n\right) \\
&= \frac{1}{2\Delta t}\left(\| \Pi e_u^n \|_{L^2(\Omega)}^2 - \| \Pi \check{e}_u^{n-1} \|_{L^2(\Omega)}^2 + \| \Pi e_u^n - \Pi \check{e}_u^{n-1} \|_{L^2(\Omega)}^2 \right) \\
&\geq \frac{1}{2\Delta t}\left(\| \Pi e_u^n \|_{L^2(\Omega)}^2 - \| \Pi \check{e}_u^{n-1} \|_{L^2(\Omega)}^2 \right) \\
&\geq \frac{1}{2\Delta t}\left(\| \Pi e_u^n \|_{L^2(\Omega)}^2 - \| \Pi e_u^{n-1} \|_{L^2(\Omega)}^2 \right) - C \| \Pi e_u^{n-1} \|_{L^2(\Omega)}^2,
\end{aligned}$$

where $\Pi \check{e}_u^{n-1} = \Pi \check{u}^{n-1} - \check{u}_h^{n-1}$. Also by the Taylor expansion and the Hölder inequality, there are

$$\begin{aligned}
| \mathcal{B}_2 | &= \left| \left(\psi^n \partial_\tau u^n - \frac{u^n - \check{u}^{n-1}}{\Delta t}, \Pi e_u^n\right) \right| \\
&\leq C\Delta t \| \partial_{\tau\tau} u \|_{L^2(J^n; L^2(\Omega))}^2 + C \| \Pi e_u^n \|_{L^2(\Omega)}^2
\end{aligned}$$

and

$$\begin{aligned}
-\mathcal{B}_3 &= \left(\frac{(\Pi u^n - u^n) - (\Pi \check{u}^{n-1} - \check{u}^{n-1})}{\Delta t}, \Pi e_u^n\right) \\
&= \left(\frac{(\Pi u^n - u^n) - (\Pi u^{n-1} - u^{n-1})}{\Delta t}, \Pi e_u^n\right) \\
&+ \left(\frac{(\Pi u^{n-1} - u^{n-1}) - (\Pi \check{u}^{n-1} - \check{u}^{n-1})}{\Delta t}, \Pi e_u^n\right) \\
&= \mathcal{S}_1 + \mathcal{S}_2,
\end{aligned}$$

where

$$\begin{aligned}\mathcal{S}_1 &= \Big(\frac{(\Pi u^n - u^n) - (\Pi u^{n-1} - u^{n-1})}{\Delta t}, \Pi e_u^n\Big)\\ &\le \frac{1}{\Delta t} \parallel \Pi e_u^n \parallel_{L^2(\Omega)} \int_{t^{n-1}}^{t^n} \parallel \partial_t(\Pi u - u) \parallel_{L^2(\Omega)} dt\\ &\le C \parallel \Pi e_u^n \parallel_{L^2(\Omega)}^2 + \frac{C}{\Delta t} \parallel \partial_t(\Pi u - u) \parallel_{L^2(J^n;L^2(\Omega))}^2,\end{aligned}$$

and

$$\begin{aligned}\mathcal{S}_2 &= \Big(\frac{(\Pi u^{n-1} - u^{n-1}) - (\Pi \breve{u}^{n-1} - \breve{u}^{n-1})}{\Delta t}, \Pi e_u^n\Big)\\ &\le C \parallel \Pi e_u^n \parallel_{L^2(\Omega)}^2 + C \parallel \nabla(\Pi u^{n-1} - u^{n-1}) \parallel_{L^2(\Omega)}^2,\end{aligned}$$

following from Cauchy-Schwarz's inequality, Young's inequality and Lemma 5.11. Substituting $\mathcal{B}_1, \mathcal{B}_2, \mathcal{B}_3$ into (5.227), the desired result is reached. □

In the following, we use the general analytic methods to get the bound of the right side term of (5.219).

Theorem 5.14. *Let u be sufficiently smooth solution of (5.200). $(\Pi u^n, \boldsymbol{\Pi\sigma}^n, \boldsymbol{Qp}^n)$ are standard L^2-projection operators of $(u^n, \boldsymbol{\sigma}^n, \boldsymbol{p}^n)$, and $(u_h^n, \boldsymbol{\sigma}_h^n, \boldsymbol{p}_h^n)$ solve (5.208). If $\boldsymbol{b} \in L^\infty([0,T]; W^{1,\infty}(\Omega)^2)$, we have*

$$\begin{aligned}&\big|\mathcal{A}(\Pi u^n - u^n, \boldsymbol{\Pi\sigma}^n - \boldsymbol{\sigma}^n, \boldsymbol{Qp}^n - \boldsymbol{p}^n; \Pi e_u^n, -\boldsymbol{Qe}_{\boldsymbol{p}}^n, \boldsymbol{\Pi e}_{\boldsymbol{\sigma}}^n)\big|\\ &\le C\epsilon_{\alpha_1} \textstyle\int_c^d \parallel Q_x e_{p_x}^n(\cdot, y) \parallel_{J_{R,0}^{-\alpha_1/2}(a,b)}^2 dy + \Big(\frac{C}{\epsilon_1} + C\epsilon_1\Big) h^{2k+1} + \frac{C}{\epsilon_{\alpha_1}} h^{2k+2}\\ &\quad + C\epsilon_{\alpha_2} \textstyle\int_a^b \parallel Q_y e_{p_y}^n(x, \cdot) \parallel_{J_{R,0}^{-\alpha_2/2}(c,d)}^2 dx + \Big(\frac{C}{\epsilon_2} + C\epsilon_2\Big) h^{2k+1} + \frac{C}{\epsilon_{\alpha_2}} h^{2k+2}\\ &\quad + \frac{\epsilon_1}{2} \textstyle\sum_{e\in\mathscr{E}_h^B} \parallel [\![\Pi e_u^n]\!] \parallel_{L^2(e)}^2 + \frac{\epsilon_2}{2} \sum_{e\in\mathscr{E}_h^i} \parallel [\![\boldsymbol{\Pi e}_{\boldsymbol{\sigma}}^n]\!] \parallel_{L^2(e)}^2,\end{aligned} \tag{5.228}$$

where $\alpha_1 = 2 - \beta$, $\alpha_2 = 2 - \beta_1$.

Proof. From the definition of $\mathcal{A}$, we have

$$\begin{aligned}&\mathcal{A}(\Pi u^n - u^n, \boldsymbol{\Pi\sigma}^n - \boldsymbol{\sigma}^n, \boldsymbol{Qp}^n - \boldsymbol{p}^n; \Pi e_u^n, -\boldsymbol{Qe}_{\boldsymbol{p}}^n, \boldsymbol{\Pi e}_{\boldsymbol{\sigma}}^n)\\ &\le \big|\mathbb{a}(\boldsymbol{\Pi\sigma}^n - \boldsymbol{\sigma}^n, \Pi e_u^n)\big| + \big|\mathbb{c}(I_{\mathbf{x}}^{\bar{\alpha}}(\boldsymbol{Qp}^n - \boldsymbol{p}^n), \boldsymbol{Qe}_{\boldsymbol{p}}^n)\big|\\ &\quad + \big|\mathbb{a}(\boldsymbol{\Pi e}_{\boldsymbol{\sigma}}^n, \Pi u^n - u^n)\big| + \big|\mathbb{d}(\Pi u^n - u^n, \Pi e_u^n)\big|\\ &\quad + \big|\mathbb{e}(\boldsymbol{\Pi\sigma}^n - \boldsymbol{\sigma}^n, \boldsymbol{\Pi e}_{\boldsymbol{\sigma}}^n)\big| + \big|\mathbb{c}(\boldsymbol{\Pi\sigma}^n - \boldsymbol{\sigma}^n, -\boldsymbol{Qe}_{\boldsymbol{p}}^n)\big|\\ &\quad + \big|\mathbb{c}(\boldsymbol{Qp}^n - \boldsymbol{p}^n, \boldsymbol{\Pi e}_{\boldsymbol{\sigma}}^n)\big| = \sum_{i=1}^{7} T_i.\end{aligned} \tag{5.229}$$

Using Hölder's, Young's inequalities and Lemma 5.9, we obtain

$$
\begin{aligned}
T_1 &= \left|\mathrm{a}(\mathbf{\Pi}\boldsymbol{\sigma}^n - \boldsymbol{\sigma}^n, \Pi e_u^n)\right| = \left|(\{\mathbf{\Pi}\boldsymbol{\sigma}^n - \boldsymbol{\sigma}^n\}, [\![\Pi e_u^n]\!])_{\mathscr{E}_h^B}\right| \\
&\leq \sum_{e\in\mathscr{E}_h^B} \| \{\mathbf{\Pi}\boldsymbol{\sigma}^n - \boldsymbol{\sigma}^n\} \|_{L^2(e)} \| [\![\Pi e_u^n]\!] \|_{L^2(e)} \\
&\leq \sum_{e\in\mathscr{E}_h^B} \left(\frac{1}{\epsilon_1} \| \{\mathbf{\Pi}\boldsymbol{\sigma}^n - \boldsymbol{\sigma}^n\} \|_{L^2(e)}^2 + \frac{\epsilon_1}{4} \| [\![\Pi e_u^n]\!] \|_{L^2(e)}^2 \right) \\
&\leq \frac{C}{\epsilon_1} h^{2k+1} + \frac{\epsilon_1}{4} \sum_{e\in\mathscr{E}_h^B} \| [\![\Pi e_u^n]\!] \|_{L^2(e)}^2 .
\end{aligned}
$$

From (5.128), (5.126), and Lemma 5.9, we have

$$
\begin{aligned}
T_2 &= \left|\mathrm{c}(I_{\mathbf{x}}^{\tilde{\alpha}}(\boldsymbol{Q}\boldsymbol{p}^n - \boldsymbol{p}^n), \boldsymbol{Q}\boldsymbol{e}_{\boldsymbol{p}}^n)\right| \\
&= \left|(Q_x p_x^n - p_x^n, {}_xD_b^{-(2-\beta)} \Pi e_{p_x}^n) + (Q_y p_y^n - p_y^n, {}_xD_b^{-(2-\beta_1)} \Pi e_{p_y}^n)\right| \\
&\leq \| Q_x p_x^n - p_x^n \|_{L^2(\Omega)} \left(\int_c^d \| Q_x e_{p_x}^n(\cdot, y) \|_{J_{R,0}^{-\alpha_1}(a,b)}^2 \, dy\right)^{\frac{1}{2}} \\
&\quad + \| Q_y p_y^n - p_y^n \|_{L^2(\Omega)} \left(\int_a^b \| Q_y e_{p_y}^n(x, \cdot) \|_{J_{R,0}^{-\alpha_2}(c,d)}^2 \, dx\right)^{\frac{1}{2}} \\
&\leq C \| Q_x p_x^n - p_x^n \|_{L^2(\Omega)} \left(\int_c^d \| Q_x e_{p_x}^n(\cdot, y) \|_{J_{R,0}^{-\alpha_1/2}(a,b)}^2 \, dy\right)^{\frac{1}{2}} \\
&\quad + C \| Q_y p_y^n - p_y^n \|_{L^2(\Omega)} \left(\int_a^b \| Q_y e_{p_y}^n(x, \cdot) \|_{J_{R,0}^{-\alpha_2/2}(c,d)}^2 \, dx\right)^{\frac{1}{2}} \\
&\leq \frac{C}{\epsilon_{\alpha_1}} h^{2k+2} + C\epsilon_{\alpha_1} \int_c^d \| Q_x e_{p_x}^n(\cdot, y) \|_{J_{R,0}^{-\alpha_1/2}(a,b)}^2 \, dy \\
&\quad + \frac{C}{\epsilon_{\alpha_2}} h^{2k+2} + C\epsilon_{\alpha_2} \int_a^b \| Q_y e_{p_y}^n(x, \cdot) \|_{J_{R,0}^{-\alpha_2/2}(c,d)}^2 \, dx.
\end{aligned}
$$

According to (5.126) and Theorem 5.9, we can choose ϵ_{α_1} and ϵ_{α_2} small enough such that

$$
\begin{aligned}
&\int_c^d \left({}_aD_x^{-(2-\beta)} Q_x e_{p_x}^n(\cdot, y), {}_x e_{p_x}^n(\cdot, y)\right)_{L^2(a,b)} dy \\
&\qquad \geq C\epsilon_{\alpha_1} \int_c^d \| Q_x e_{p_x}^n(\cdot, y) \|_{J_{R,0}^{-\alpha_1/2}(a,b)}^2 \, dy, \qquad (5.230)
\end{aligned}
$$

$$\int_a^b \left({}_cD_y^{-(2-\beta_1)} Q_y e_{p_y}^n(x,\cdot),\, Q_y e_{p_y}^n(x,\cdot)\right)_{L^2(c,d)} dx$$
$$\geq C\epsilon_{\alpha_2} \int_a^b \| Q_y e_{p_y}^n(x,\cdot) \|^2_{J_{R,0}^{-\alpha_2/2}(c,d)} dx. \tag{5.231}$$

Integrating the first term of $\mathbb{a}(\mathbf{\Pi e}_{\boldsymbol{\sigma}}^n, \Pi u^n - u^n)$ by parts, and using the orthogonal property of projection operator $\mathbf{\Pi}$, we get

$$\begin{aligned}
T_3 &= \left|\mathbb{a}(\mathbf{\Pi e}_{\boldsymbol{\sigma}}^n, \Pi u^n - u^n)\right| \\
&= \left|(\mathbf{\Pi e}_{\boldsymbol{\sigma}}^n, \nabla(\Pi u^n - u^n)) - (\{\mathbf{\Pi e}_{\boldsymbol{\sigma}}^n\}, [\![\Pi u^n - u^n]\!])_{\mathscr{E}_h^B}\right| \\
&= \left|([\![\mathbf{\Pi e}_{\boldsymbol{\sigma}}^n]\!], \{\Pi u^n - u^n\})_{\mathscr{E}_h^i}\right| \\
&\leq \sum_{e\in\mathscr{E}_h^i} \| [\![\mathbf{\Pi e}_{\boldsymbol{\sigma}}^n]\!] \|_{L^2(e)} \| \{\Pi u^n - u^n\} \|_{L^2(e)} \\
&\leq \sum_{e\in\mathscr{E}_h^i} \left(\frac{1}{\epsilon_2} \| \{\Pi u^n - u^n\} \|^2_{L^2(e)} + \frac{\epsilon_2}{4} \| [\![\mathbf{\Pi e}_{\boldsymbol{\sigma}}^n]\!] \|^2_{L^2(e)}\right) \\
&\leq \frac{C}{\epsilon_2} h^{2k+1} + \frac{\epsilon_2}{4} \sum_{e\in\mathscr{E}_h^i} \| [\![\mathbf{\Pi e}_{\boldsymbol{\sigma}}^n]\!] \|^2_{L^2(e)} .
\end{aligned}$$

With the same deduction of T_1, there is

$$\begin{aligned}
T_4 &= \left|\mathbb{d}(\Pi u^n - u^n, \Pi e_u^n)\right| \\
&\leq \epsilon_1 \sum_{e\in\mathscr{E}_h^B} \| [\![\Pi u^n - u^n]\!] \|_{L^2(e)} \| [\![\Pi e_u^n]\!] \|_{L^2(e)} \\
&\leq \epsilon_1 \sum_{e\in\mathscr{E}_h^B} \left(\| [\![\Pi u^n - u^n]\!] \|^2_{L^2(e)} + \frac{1}{4} \| [\![\Pi e_u^n]\!] \|^2_{L^2(e)}\right) \\
&\leq \epsilon_1 C h^{2k+1} + \frac{\epsilon_1}{4} \sum_{e\in\mathscr{E}_h^B} \| [\![\Pi e_u^n]\!] \|^2_{L^2(e)} .
\end{aligned}$$

By Lemma 5.9, we get

$$\begin{aligned}
T_5 &= \left|\mathbb{e}(\mathbf{\Pi}\boldsymbol{\sigma}^n - \boldsymbol{\sigma}^n, \mathbf{\Pi e}_{\boldsymbol{\sigma}}^n)\right| \\
&\leq \epsilon_2 \sum_{e\in\mathscr{E}_h^i} \| [\![\mathbf{\Pi}\boldsymbol{\sigma}^n - \boldsymbol{\sigma}^n]\!] \|_{L^2(e)} \| [\![\mathbf{\Pi e}_{\boldsymbol{\sigma}}^n]\!] \|_{L^2(e)} \\
&\leq \epsilon_2 \sum_{e\in\mathscr{E}_h^i} \left(\| [\![\mathbf{\Pi}\boldsymbol{\sigma}^n - \boldsymbol{\sigma}^n]\!] \|^2_{L^2(e)} + \frac{1}{4} \| [\![\mathbf{\Pi e}_{\boldsymbol{\sigma}}^n]\!] \|^2_{L^2(e)}\right) \\
&\leq C\epsilon_2 h^{2k+1} + \frac{\epsilon_2}{4} \sum_{e\in\mathscr{E}_h^i} \| [\![\mathbf{\Pi e}_{\boldsymbol{\sigma}}^n]\!] \|^2_{L^2(e)} .
\end{aligned}$$

Note that T_6 and T_7 vanish because of the orthogonal property of the projection $\mathbf{\Pi}$. Substituting $T_i, i = 1, \cdots, 7$ into (5.229), the desired result is obtained. □

Assuming that the solution of (5.195) is sufficiently regular, we have the following error estimates.

Theorem 5.15. *Let $(u^n, \boldsymbol{\sigma}^n, \boldsymbol{p}^n)$ be the exact solution of (5.200), $(u_h^n, \boldsymbol{\sigma}_h^n, \boldsymbol{p}_h^n)$ the numerical solution of the fully discrete discontinuous Galerkin scheme (5.208). If $\boldsymbol{b} \in L^\infty([0,T]; W^{1,\infty}(\Omega)^2)$, for any integer $N = 1, 2, \cdots$, there is*

$$\begin{aligned}
&\|u^N - u_h^N\|^2_{L^2(\Omega)} + \sum_{n=1}^{N} \Delta t \epsilon_1 \sum_{e \in \mathscr{E}_h^B} \|[\![u^n - u_h^n]\!]\|^2_{L^2(e)} \\
&+ \sum_{n=1}^{N} \Delta t \epsilon_2 \sum_{e \in \mathscr{E}_h^i} \|[\![\boldsymbol{\sigma}^n - \boldsymbol{\sigma}_h^n]\!]\|^2_{L^2(e)} \\
&\le C(\Delta t)^2 \sum_{n=1}^{N} \|\partial_{\tau\tau} u\|^2_{L^2(J^n; L^2(\Omega))} + C \sum_{n=1}^{N} \|\partial_t(\Pi u - u)\|^2_{L^2(J^n; L^2(\Omega))} \\
&+ C_\epsilon h^{2k+1} + C\Delta t \sum_{n=1}^{N} |\, \Pi u^{n-1} - u^{n-1} \,|^2_{H^1(\Omega)} \cdot
\end{aligned} \tag{5.232}$$

Proof. Substituting the results of Theorem 5.13 and Theorem 5.14 into (5.219), and using (5.230) and (5.231), there holds

$$\begin{aligned}
&\frac{1}{2\Delta t}\left(\| \Pi e_u^n \|^2_{L^2(\Omega)} - \| \Pi e_u^{n-1} \|^2_{L^2(\Omega)} \right) + \frac{\epsilon_1}{2} \sum_{e \in \mathscr{E}_h^B} \| [\![\Pi e_u^n]\!] \|^2_{L^2(e)} \\
&+ \frac{\epsilon_2}{2} \sum_{e \in \mathscr{E}_h^i} \| [\![\mathbf{\Pi} \boldsymbol{e}_\sigma^n]\!] \|^2_{L^2(e)} \\
&\le C \| \Pi e_u^{n-1} \|^2_{L^2(\Omega)} + C \| \Pi e_u^n \|^2_{L^2(\Omega)} + C\Delta t \| \partial_{\tau\tau} u \|^2_{L^2(J^n; L^2(\Omega))} \\
&+ \frac{C}{\Delta t} \|\partial_t(\Pi u - u)\|^2_{L^2(J^n; L^2(\Omega))} + C \,|\, \Pi u^{n-1} - u^{n-1} \,|^2_{H^1(\Omega)} + C_\epsilon h^{2k+1}.
\end{aligned}$$

With $\Pi e_u^0 = 0$, multiplying the above inequality by $2\Delta t$ on both sides, summing over n from 1 to N, and using the discrete Grönwall inequality, there is

$$\| \Pi e_u^N \|_{L^2(\Omega)}^2 + \Delta t \sum_{n=1}^{N} \left(\epsilon_1 \sum_{e \in \mathscr{E}_h^B} \| [\![\Pi e_u^n]\!] \|_{L^2(e)}^2 + \epsilon_2 \sum_{e \in \mathscr{E}_h^i} \| [\![\mathbf{\Pi e}_{\boldsymbol{\sigma}}^n]\!] \|_{L^2(e)}^2 \right)$$

$$\leq C(\Delta t)^2 \sum_{n=1}^{N} \|\partial_{\tau\tau} u\|_{L^2(J^n;L^2(\Omega))}^2 + C \sum_{n=1}^{N} \|\partial_t (\Pi u - u)\|_{L^2(J^n;L^2(\Omega))}^2$$

$$+ C\Delta t \sum_{n=1}^{N} | \Pi u^{n-1} - u^{n-1} |_{H^1(\Omega)}^2 + C_\epsilon h^{2k+1}.$$

By the triangle inequality, we obtain the desired result. □

5.4.3 *Numerical results*

We illustrate the numerical performance of the proposed schemes by two examples. We will use the piecewise linear basis functions to simulate the solution in triangular meshes. When we calculate the matrix involving the fractional integrals, both the Gaussian quadrature formulae on triangle [Hesthaven and Warburton (2008)] and on interval (see Appendix A) are used, and the implement processes are similar to [Qiu *et al.* (2015)]. The meshes used in this work and the affected elements of a specific quadrature point are presented in Figs. 5.1 and 5.2.

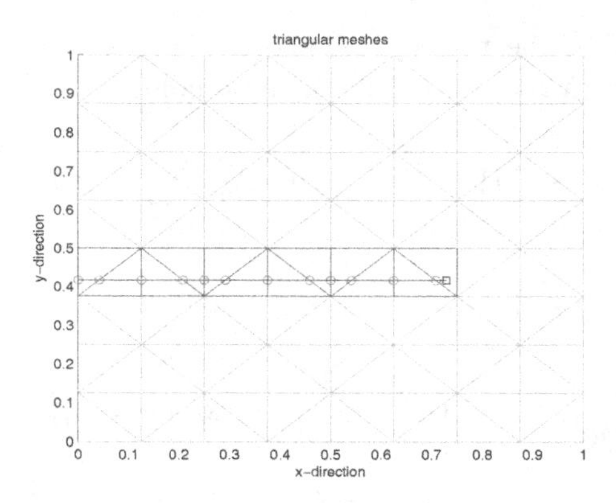

Fig. 5.1: All triangles in x-direction affected by the Gauss points (denoted by black square).

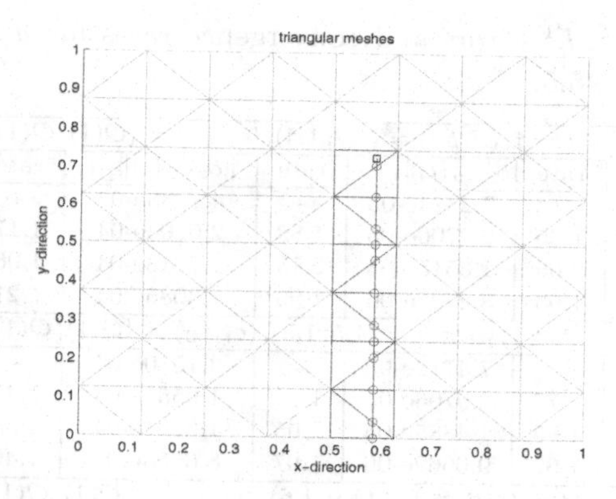

Fig. 5.2: All triangles in y-direction affected by the Gauss points (denoted by black square).

Example 5.5. Consider (5.195) in domain $\Omega = (0,1) \times [0,1]$. The initial condition and the exact solution in Ω are specified as

$$\begin{cases} u(x,y,t) = e^{-t}x^2(x-1)^2y^2(y-1)^2, & (x,y) \in \Omega, \\ u_0(x,y) = x^2(x-1)^2y^2(y-1)^2, & (x,y) \in \Omega, \\ b(x,y,t) = (0,0), \end{cases} \tag{5.233}$$

and the source term f is determined accordingly from (5.195).

Tables 5.9 and 5.10 illustrate that the schemes have a good convergence order with the different choices of the fluxes and $\Delta t = \mathcal{O}(h^{3/2})$, where the experimental L^2 convergence rate is given by

$$\text{rate} = \frac{\log\left(\| u(t) - u_{h_1}(t) \|_{L^2(\mathscr{E}_{h_1})} / \| u(t) - u_{h_2}(t) \|_{L^2(\mathscr{E}_{h_2})} \right)}{\log(h_1/h_2)}.$$

The rate for L^1 error is similar. Table 5.9 shows that the convergence rates at least have an order of $\mathcal{O}(h^{3/2})$; in Table 5.10 we take the same choice of ϵ_1, ϵ_2 as Ref. [Castilo *et al.* (2000)] and see that the convergence rates increase to $\mathcal{O}(h^2)$. Comparing the numerical results with the work [Qiu *et al.* (2015)], we can see that the DG method here has smaller numerical errors for the first order polynomial approximation.

Table 5.9: The L^2, L^1 errors and convergence rates for u and u_x, u_y (at $t=1$) for Example 5.5.

	$t=1, (\beta,\beta_1)=(1.2,1.4), (\epsilon_1,\epsilon_2)=(\mathcal{O}(1),\mathcal{O}(1))$						
h	$\Vert e_u(t)\Vert_{L^2}$	rate	$\Vert e_u(t)\Vert_{L^1}$	rate	$\Vert\partial_x e_u(t)\Vert_{L^2}$	rate	$\Vert\partial_y e_u(t)\Vert_{L^2}$
1/6	8.2881e-05	-	7.2374e-05	-	4.7733e-04	-	2.5556e-03
1/10	3.2222e-05	1.85	2.7700e-05	1.88	2.6204e-04	1.17	2.7560e-03
1/14	1.6162e-05	2.05	1.3515e-05	2.13	1.7748e-04	1.06	2.8077e-03
1/18	9.9448e-06	1.93	8.2787e-06	1.95	1.3085e-04	1.21	2.8291e-03
	$t=1, (\beta,\beta_1)=(1.5,1.5), (\epsilon_1,\epsilon_2)=(\mathcal{O}(1),\mathcal{O}(1))$						
1/6	8.7928e-05	-	7.4712e-05	-	4.1510e-04	-	2.5466e-03
1/10	3.5668e-05	1.77	2.9706e-05	1.81	1.9555e-04	1.47	2.7408e-03
1/14	1.8524e-05	1.95	1.4885e-05	2.05	1.2291e-04	1.38	2.8010e-03
1/18	1.1432e-05	1.92	9.0662e-06	1.97	8.6553e-05	1.40	2.8244e-03
	$t=1, (\beta,\beta_1)=(1.9,1.6), (\epsilon_1,\epsilon_2)=(\mathcal{O}(1),\mathcal{O}(1))$						
1/6	1.1250e-04	-	8.4393e-05	-	4.6409e-04	-	2.5983e-03
1/10	4.7998e-05	1.67	3.6510e-05	1.64	2.1614e-04	1.50	2.7410e-03
1/14	2.7916e-05	1.61	2.0971e-05	1.65	1.3512e-04	1.40	2.7985e-03
1/18	1.8767e-05	1.58	1.4104e-05	1.58	9.5762e-05	1.37	2.8216e-03

Table 5.10: The L^2, L^1 errors and convergence rates for u and u_x, u_y (at $t=1$) for Example 5.5.

	$t=1, (\beta,\beta_1)=(1.9,1.6), (\epsilon_1,\epsilon_2)=(\mathcal{O}(h^{-1}),\mathcal{O}(1))$						
h	$\Vert e_u(t)\Vert_{L^2}$	rate	$\Vert e_u(t)\Vert_{L^1}$	rate	$\Vert\partial_x e_u(t)\Vert_{L^2}$	rate	$\Vert\partial_y e_u(t)\Vert_{L^2}$
1/6	5.2142e-05	-	4.1592e-05	-	4.0695e-04	-	2.5777e-03
1/10	1.9772e-05	1.90	1.5784e-05	1.90	1.7761e-04	1.62	2.7383e-03
1/14	9.5805e-06	2.15	7.5267e-06	2.20	1.1392e-04	1.32	2.7976e-03
1/18	5.8213e-06	1.98	4.6596e-06	1.91	7.8388e-05	1.49	2.8210e-03
	$t=1, (\beta,\beta_1)=(1.9,1.6), (\epsilon_1,\epsilon_2)=(\mathcal{O}(h^{-1}),\mathcal{O}(h))$						
1/6	4.8702e-05	-	3.9358e-05	-	4.2857e-04	-	2.5666e-03
1/10	1.9169e-05	1.83	1.5766e-05	1.79	1.9937e-04	1.50	2.7356e-03
1/14	9.2520e-06	2.17	7.6271e-06	2.16	1.2817e-04	1.31	2.7962e-03
1/18	5.6525e-06	1.96	4.6650e-06	1.96	8.9725e-05	1.42	2.8202e-03
	$t=1, (\beta,\beta_1)=(1.9,1.6), (\epsilon_1,\epsilon_2)=(\mathcal{O}(1),\mathcal{O}(h))$						
1/6	8.6286e-05	-	6.6645e-05	-	4.5649e-04	-	2.5761e-03
1/10	3.4635e-05	1.79	2.8021e-05	1.70	2.0660e-04	1.55	2.7365e-03
1/14	1.7787e-05	1.98	1.4393e-05	1.98	1.2993e-04	1.38	2.7964e-03
1/18	1.0969e-05	1.92	8.9170e-06	1.91	8.9646e-05	1.48	2.8200e-03

Example 5.6. In this example, we let $\Omega=(0,1)\times(0,1)$. The exact solution u, initial value u_0 and the vector function $\boldsymbol{b}$ are given by

$$\begin{cases} u(x,y,t) = e^{-t}x^2(x-0.5)^2(x-1)^2y^2(y-0.5)^2(y-1)^2, & (x,y)\in\Omega; \\ u_0(x,y) = x^2(x-0.5)^2(x-1)^2y^2(y-0.5)^2(y-1)^2, & (x,y)\in\Omega; \\ \boldsymbol{b} = ((x-0.5),-(y-0.5)). \end{cases} \tag{5.234}$$

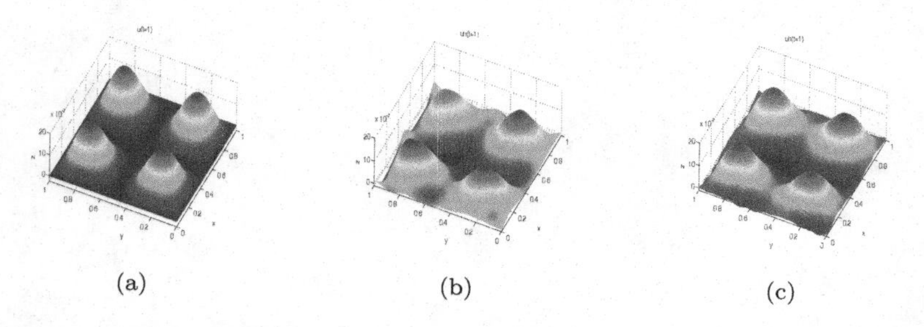

(a) (b) (c)

Fig. 5.3: Exact solution u and the numerical solutions u_h at $t = 1$ for Example 5.6.

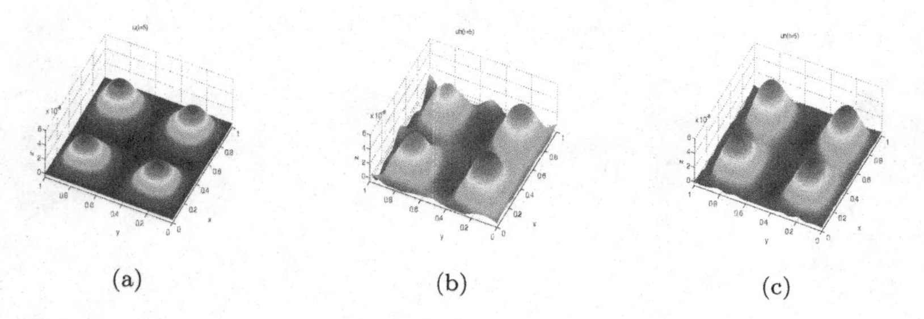

(a) (b) (c)

Fig. 5.4: Exact solution u and the numerical solutions u_h at $t = 5$ for Example 5.6.

Figure 5.3 displays the exact solution u and the numerical solutions u_h based on different space stepsizes $h = \frac{1}{8}, \frac{1}{16}$ at $t = 1$ with $\alpha = 1.2, \beta = 1.4$, $\epsilon_1 = \epsilon_2 = 1$. Figure 5.4 displays the exact solution u and the numerical solutions u_h based on different space stepsizes $h = \frac{1}{8}, \frac{1}{16}$ at $t = 5$ with $\alpha = 1.9, \beta = 1.6, \epsilon_1 = h^{-1}, \epsilon_2 = 1$. The numerical solutions recover the exact solution perfectly with all four hills in coarse meshes and the approximations are more and more accurate with the refining of the meshes.

Appendix A

Basic concepts for functional and numerical analysis

This appendix is devoted to introduce the knowledges of functional analysis and numerical analysis involved in this book, including the Lebesgue integration theory, the Sobolev spaces, the Fourier and Laplace transforms, the Grönwall lemmas, the Jocobi-Gauss-type quadratures, and so on.

A.1 Lebesgue integrals and continuous functions

A.1.1 *Lebesgue spaces*

We now review the basic concepts of Lebesgue integration theory, cf. [Halmos (1991); Rudin (1987)]. Let Ω be an open (finite or infinite) subset of $\mathbb{R}^d$ with non-empty interior, and $\mathbb{M}(\Omega)$ be the space of functions on Ω that are Lebesgue measurable. In this appendix, x may be a scalar or vector, depending on the circumstance. Letting $1 \leq p \leq \infty$, the Lebesgue space $L^p(\Omega)$ is given by

$$L^p(\Omega) := \{f \in \mathbb{M}(\Omega); \|f\|_{L^p(\Omega)} < \infty\}, \tag{A.1}$$

where for $1 \leq p < \infty$

$$\|f\|_{L^p(\Omega)} := \left(\int_a^b |f(x)|^p dx\right)^{1/p}, \tag{A.2}$$

and for $p = \infty$

$$\|f\|_{L^\infty(\Omega)} := \text{ess sup}\left\{|f(x)| : x \in \Omega\right\}. \tag{A.3}$$

Equipped with the norm $\|\cdot\|_{L^p(\Omega)}$, $L^p(\Omega)$ is a Banach space. Note that if f and g differ only on a set of measure zero, we view them as the same functions in $L^p(\Omega)$. There is the famous Hölder inequality: for $1 \leq p, q \leq \infty$

such that $1 = 1/p + 1/q$, if $f \in L^p(\Omega)$ and $g \in L^q(\Omega)$, then $fg \in L^1(\Omega)$ and

$$\|fg\|_{L^1(\Omega)} \leq \|f\|_{L^p(\Omega)} \|g\|_{L^q(\Omega)} \,. \tag{A.4}$$

Among all the Lebesgue spaces, $L^2(\Omega)$ is a Hilbert space equipped with the scalar product

$$(f, g) = \int_\Omega f(x)\overline{g(x)}\, dx, \tag{A.5}$$

and at this time the Hölder inequality becomes the Cauchy-Schwarz inequality, i.e.,

$$\forall f, g \in L^2(\Omega), \quad (f, g) \leq \|f\|_{L^2(\Omega)} \|g\|_{L^2(\Omega)} \,. \tag{A.6}$$

A.1.2 *Spaces of continuous functions*

There exist many different ways of measuring the smoothness of a function f. The most natural one is certainly the order of differentiability, i.e., the maximal index $n \in \mathbb{N}$ such that $\partial^\gamma f(|\gamma| = n)$ exist, where $\partial^\gamma f := \frac{\partial^{|\gamma|} f}{\partial x_1^{\gamma_1} \cdots \partial x_n^{\gamma_d}}$, $\gamma = (\gamma_1, \cdots, \gamma_d) \in \mathbb{N}^d$ and $|\gamma| := \gamma_1 + \cdots + \gamma_d$. To this particular measure of smoothness, we associate a class of functions space $C^n(\Omega)$, which is the space of functions that are n-times continuously differentiable on Ω. Note that $C^0(\Omega)$ is usually written as $C(\Omega)$. In addition, we also use $C^\infty(\Omega)$ to denote the space of functions that are infinity differentiable on Ω, and use $C_0^\infty(\Omega)$ to denote the set of $C^\infty(\Omega)$ functions with compact support in Ω.

In the case of $C^n(\Omega)$ spaces, we note that $\sup\limits_{x\in\Omega} |f(x+h) - f(x)| \leq \sup |f'|\,|h|$ if $f \in C^1(\Omega)$ for any $h \in \mathbb{R}^d$, whereas for an arbitrary $f \in C^0(\Omega)$, $\sup\limits_{x\in\Omega} |f(x+h) - f(x)|$ might go to zero arbitrarily slow as $|h| \to 0$. This motivates the definition of the Hölder space $C^{0,\mu}(\Omega), 0 < \mu \leq 1$, consisting of those $f \in C^0(\Omega)$ such that

$$\sup_{\substack{x,y\in\Omega \\ x\neq y}} \frac{|f(x) - f(y)|}{|x-y|^\mu} < \infty. \tag{A.7}$$

In particular, $C^{0,1}(\overline{\Omega})$ (resp., $C^{0,1}(\Omega)$) usually is called the space of globally (resp., locally) Lipschitz functions. If $n \in \mathbb{N}^+$, a natural definition of $C^{n,\mu}(\Omega)$ is given by $f \in C^n(\Omega)$ and $\partial^\gamma f \in C^{0,\mu}(\Omega)$ $(|\gamma| = n)$.

The smoothness could also be measured by the absolute continuity. Let $\Omega = [a, b]$ be a finite interval. We denote $AC[a, b]$ the space of functions f which are absolutely continuous on $[a, b]$. It is known that the space $AC[a, b]$ is equivalent to the Sobolev space $W^{1,1}(a, b)$ (see Sec. A.2), which can be characterized as: there exists $g(t) \in L^1(a, b)$, such that $f(x) = c + \int_a^x g(t)dt$.

A.2 Sobolev spaces

A.2.1 *Integer order Sobolev spaces*

In order to measure the smoothness properties of a function in an average sense, it is also natural to introduce the Sobolev spaces. Let $m \geq 0$ be an integer and $1 \leq p < \infty$. The so-called Sobolev space $W^{m,p}(\Omega)$ is given as

$$W^{m,p}(\Omega) = \{u \in L^p(\Omega) : \partial^\gamma u \in L^p(\Omega),\ |\gamma| \leq m\}. \tag{A.8}$$

Here the derivatives $\partial^\gamma u$ are understood in the weak (distributions) sense [Evans (2010)], i.e.,

$$\int_\Omega [\partial^\gamma u(x)]\, \phi(x)dx = (-1)^{|\gamma|} \int_\Omega u(x)\partial^\gamma \phi(x)dx \quad \text{for all } \phi \in C_0^\infty(\Omega),$$

and obviously, $W^{0,p}(\Omega) = L^p(\Omega)$. Equipped with norm

$$\|u\|_{W^{m,p}(\Omega)} := \left(\sum_{|\gamma| \leq m} \|\partial^\gamma u\|^p_{L^p(\Omega)} \right)^{\frac{1}{p}}, \tag{A.9}$$

$W^{m,p}(\Omega)$ is a Banach space, and $C^\infty(\Omega) \cap W^{m,p}(\Omega)$ is dense in $W^{m,p}(\Omega)$. When developing the variational formulation of differential equations involving the generalized homogeneous boundary conditions, we also need the space $W_0^{m,p}(\Omega)$ defined by the closure of $C_0^\infty(\Omega)$ with respect to norm $\|u\|_{W^{m,p}(\Omega)}$, i.e.,

$$W_0^{m,p}(\Omega) := \text{closure}\left\{u \in C_0^\infty(\Omega) : \|u\|_{W^{m,p}(\Omega)} < \infty\right\}, \tag{A.10}$$

and its dual space $W^{-m,q}(\Omega) = (W_0^{m,p}(\Omega))'$ (the space of continuous linear form on $W_0^{m,q}(\Omega)$) with the norm

$$\|u\|_{W^{-m,q}(\Omega)} := \sup_{\substack{v \in W_0^{m,p}(\Omega) \\ v \neq 0}} \frac{\langle u, v\rangle}{\|v\|_{W^{m,p}(\Omega)}},\ u \in W^{-m,q}(\Omega), \tag{A.11}$$

where p and q satisfy $1/p + 1/q = 1$.

Among all the Sobolev spaces, the case $p = 2$ is particularly interesting since the space $W^{m,2}(\Omega)$ is a Hilbert space when equipped with the scalar product

$$(u, v)_m = \sum_{|\gamma| \leq m} (\partial^\gamma u, \partial^\gamma v). \tag{A.12}$$

We usually use the notations $H^m(\Omega), H_0^m(\Omega)$ and $H^{-m}(\Omega)$ to represent the spaces $W^{m,2}(\Omega), W_0^{m,2}(\Omega)$ and $W^{-m,2}(\Omega)$, respectively. Note that if Ω is bounded, there exist the inclusion relations $L^q(\Omega) \subset L^p(\Omega)\,(1 \leq p <$

$q \leq \infty$) and $W^{m_1,p}(\Omega) \subset W^{m_2,p}(\Omega)\,(m_2 < m_1)$. Moreover, for any $u \in W_0^{1,p}(\Omega)$, $1 \leq p < \infty$, it follows that the Poincaré inequality

$$C_{\Omega,p} \|u\|_{L^p(\Omega)} \leq \|\nabla u\|_{L^p(\Omega)} \,. \tag{A.13}$$

Therefore, for $p = 2$ one can also equip $H_0^1(\Omega)$ with the scalar product $(u, v)_1 = (\nabla u, \nabla v)$, which defines an equivalent norm to $\|\cdot\|_{H^1(\Omega)}$. For details, refer to [Adams (1975); Brenner and Scott (1994); Evans (2010)].

A.2.2 *Fractional Sobolev spaces*

In many instances of theoretical and numerical analysis of fractional PDEs, one is interested in describing the regularity of a function through fractional order of smoothness. Let

$$|u|_{W^{\mu,p}(\Omega)} := \left(\int_\Omega \int_\Omega \frac{|u(x) - u(y)|^p}{|x - y|^{d+p\mu}}\right)^{\frac{1}{p}}, \quad 0 < \mu < 1 \tag{A.14}$$

be the Slobodeckiî seminorm. In the following, we give the definition of the fractional Sobolev space $W^{s,p}(\Omega)$ $(s \notin \mathbb{Z})$. If the case $s \in \mathbb{Z}$ is involved, it will be the integer order Sobolev space given above.

A.2.2.1 *Fractional Sobolve spaces on $\mathbb{R}^d$*

Let $\Omega = \mathbb{R}^d$. For $s > 0, s \notin \mathbb{N}$ and $1 \leq p < \infty$, one can define the fractional Sobolev space $W^{s,p}(\mathbb{R}^d)$ by

$$W^{s,p}(\mathbb{R}^d) := \left\{u \in W^{[s],p}(\mathbb{R}^d) : |\partial^\gamma u|_{W^{s-[s],p}(\mathbb{R}^d)} < \infty \text{ for } |\gamma| = [s]\right\}, \tag{A.15}$$

and equip this space with norm

$$\|u\|_{W^{s,p}(\mathbb{R}^d)} = \left(\|u\|^p_{W^{[s],p}(\mathbb{R}^d)} + \sum_{|\gamma|=[s]} |\partial^\gamma u|^p_{W^{s-[s],p}(\mathbb{R}^d)}\right)^{1/p}, \tag{A.16}$$

where $[s]$ is the integer part of s.

For the special case $p = 2$, $W^{s,2}(\mathbb{R}^d)$ coincides with the space $H^s(\mathbb{R}^d)$ defined by the Fourier transform [McLean (2000), Theorem 3.16]

$$H^s(\mathbb{R}^d) := \left\{u \in L_2(\mathbb{R}^d) : \|u\|_{H^s(\mathbb{R}^d)} < \infty\right\}, \tag{A.17}$$

where

$$\|u\|_{H^s(\mathbb{R}^d)} = \int_{\mathbb{R}^d} \left(1 + |\omega|^2\right)^s |\hat{u}(\omega)|^2 d\omega \tag{A.18}$$

with $\hat{u}(\omega) = \int_{\mathbb{R}^d} e^{-i\omega\cdot x}u(x)dx$, $\omega \in \mathbb{R}^d$ being the Fourier transform of $u(x)$. For any $s \geq 0$, we have $\|u\|_{W^{s,2}(\mathbb{R}^d)} \simeq \|u\|_{H^s(\mathbb{R}^d)}$, and for $0 < s < 1$, we further have

$$|u|_{W^{s,2}(\mathbb{R}^d)} \simeq |u|_{H^s(\mathbb{R}^d)}, \tag{A.19}$$

where

$$|u|^2_{H^s(\mathbb{R}^d)} = \int_{\mathbb{R}^d} |\omega|^{2s}|\hat{u}(\omega)|^2 d\omega. \tag{A.20}$$

Note that $H^s(\mathbb{R}^d)$ is a Hilbert space under the inner product

$$(u,v)_s := \int_{\mathbb{R}^d} \left(1+|\omega|^2\right)^s \hat{u}(\omega)\overline{\hat{v}(\omega)}d\omega,$$

or the equivalent one

$$(u,v)_s := (u,v)_{[s]} + \sum_{|\gamma|=[s]} \int_{\mathbb{R}^d}\int_{\mathbb{R}^d} \frac{\partial^\gamma\left(u(x)-u(y)\right)\overline{\partial^\gamma\left(v(x)-v(y)\right)}}{|x-y|^{d+2(s-[s])}}dx\,dy.$$

In fact, fractional Sobolev spaces could also be defined by the space interpolation. Interpolation spaces are useful technical tools. They allow one to bridge between known results, yielding results that could not be obtained directly. Let A_0, A_1 are two Banach spaces satisfying $A_1 \subset A_0$. Then the interpolation space between A_0, A_1 can be defined by

$$[A_0, A_1]_{\theta,p} = A_{\theta,p} := \left\{u \in A_0 : \|u\|_{[A_0,A_1]_{\theta,p}} < \infty\right\}, \tag{A.21}$$

where $1 \leq p < \infty$, $0 < \theta < 1$, and the norm

$$\|u\|_{[A_0,A_1]_{\theta,p}} := \left(\int_0^\infty t^{-\theta p}K(t,u)^p\frac{dt}{t}\right)^{1/p} \tag{A.22}$$

with the K-function $K(t,u) := \inf_{v\in A_1}\left(\|u-v\|_{A_0} + t\|v\|_{A_1}\right)$. With this definition, for the fractional Sobolev space $H^s(\mathbb{R}^d)$ given above, it holds that [McLean (2000), Appendix B]

$$H^s(\mathbb{R}^d) = \left[H^{s_0}(\mathbb{R}^d), H^{s_1}(\mathbb{R}^d)\right]_{\theta,2} \tag{A.23}$$

with equivalent norms, where $s_0, s_1, s \geq 0$ and $\theta \in (0,1)$ satisfying $s = (1-\theta)s_0 + \theta s_1$. However, if $p \neq 2$, the above statement may not hold for all s_0, s_1 [Brenner and Scott (1994)].

A.2.2.2 *Fractional Sobolev spaces in bounded domain*

If Ω is an open subset of $\mathbb{R}^d$, there are also several different ways to introduce the fractional space $W^{s,p}(\Omega)$ ($s > 0$ and $s \notin \mathbb{N}$):

(1) Reproduce (A.15) by restricting the domain of integration in Ω, i.e.,

$$W^{s,p}(\Omega) := \left\{ u \in W^{[s],p}(\Omega) : |\partial^\gamma u|_{W^{s-[s],p}(\Omega)} < \infty \text{ for } |\gamma| = [s] \right\}.$$

Then $W^{s,p}(\Omega)$ is a Banach space under the norm

$$\|u\|_{W^{s,p}(\Omega)} := \left(\|u\|^p_{W^{[s],p}(\Omega)} + \sum_{|\gamma|=[s]} |\partial^\gamma u|^p_{W^{s-[s],p}(\Omega)} \right)^{\frac{1}{p}},$$

and $C^\infty(\Omega) \cap W^{s,p}(\Omega)$ is dense in $W^{s,p}(\Omega)$.

(2) Define $W^{s,p}(\Omega)$ as the space of all distributions in Ω which are restrictions of elements of $W^{s,p}(\mathbb{R}^d)$, equipped with the norm

$$\|u\|_{W^{s,p}(\Omega)} := \inf_{U|_\Omega = u} \|U\|_{W^{s,p}(\mathbb{R}^d)}, \quad U \in W^{s,p}(\mathbb{R}^d). \tag{A.24}$$

(3) Define $W^{s,p}(\Omega)$ as real interpolation space, i.e.,

$$W^{s,p}(\Omega) := \left[W^{k,p}(\Omega), W^{k+1,p}(\Omega) \right]_{s-k,p}, \tag{A.25}$$

where $k < s < k+1$, $k \in \mathbb{N}$.

By Theorem 1.4.3.1 of [Grisvard (1985), p. 25], Theorem 3.30 of [McLean (2000), p. 92], and Theorems 14.2.3 and 14.2.7 of [Brenner and Scott (1994), p. 374–375], we have the following results:

Proposition A.2. *Let Ω be a bounded open subset of $\mathbb{R}^d$ with a Lipschitz boundary. Then the three definitions yield the same fractional space $W^{s,p}(\Omega)$, and the norms are equivalent. In particular, for the case $p = 2$, one has*

$$H^s(\Omega) = [H^{s_0}(\Omega), H^{s_1}(\Omega)]_{\theta,2}, \tag{A.26}$$

where $s_0, s_1, s \geq 0$, and $\theta \in (0,1)$ such that $s = (1-\theta)s_0 + \theta s_1$.

For $s > 0$, we can also define the Sobolev space $W_0^{s,p}(\Omega)$ and its dual space $W^{-s,q}(\Omega)$ as were done in the integer cases. By Corollary 1.4.4.5 of [Grisvard (1985), p. 31] and Theorem 3.33 of [McLean (2000), p. 95], it holds that

Proposition A.3. *Let Ω be a bounded open subset of $\mathbb{R}^d$ with a Lipschitz boundary, and $s - 1/p$ be not an integer. Then it holds that*

$$W_0^{s,p}(\Omega) = \left\{ u \in W^{s,p}(\Omega) : \tilde{u} \in W^{s,p}(\mathbb{R}^d) \right\}$$

and

$$\|u\|_{W^{s,p}(\Omega)} \simeq \|\tilde{u}\|_{W^{s,p}(\mathbb{R}^d)}, \quad u \in W_0^{s,p}(\Omega),$$

where $\tilde{u}$ is the zero-extension of u. Furthermore, when $0 < s < \frac{1}{p}$ we have

$$W^{s,p}(\Omega) = W_0^{s,p}(\Omega). \tag{A.27}$$

In the literature, we usually use the notation $H_0^s(\Omega)$ to represent the space $W_0^{s,2}(\Omega)$, and it is a very important space in solving fractional PDEs. Moreover, when $s - \frac{1}{2} \notin \mathbb{N}$, $H_0^s(\Omega)$ is also an interpolation space (see McLean, 2000, Theorems 3.33 and B.9) and it holds that

$$H_0^s(\Omega) = \left[L^2(\Omega), H_0^k(\Omega)\right]_{\theta,2}, \tag{A.28}$$

where $k \in \mathbb{N}^+$, and $\theta \in (0,1)$ such that $s = \theta k$. Note that when $s = n + \frac{1}{2}\,(n \in \mathbb{N})$, $\left[L^2(\Omega), H_0^k(\Omega)\right]_{\frac{2n+1}{2k},2}$ is the Lions-Magenes space $H_{00}^{n+\frac{1}{2}}(\Omega)$, which is strictly contained in $H_0^{n+\frac{1}{2}}(\Omega)$ [Lions and Magenes (1972), p. 66].

Lemma A.12 ([Nezza *et al.* (2004)]). *For $x \in \mathbb{R}^d$, let $1 \le p < \infty, s \in (0,1)$ and $E \subset \mathbb{R}^d$ be a measurable set with finite measure. Then*

$$\int_{\mathbb{R}^d \setminus E} \frac{dy}{|x-y|^{d+sp}} \ge C|E|^{-sp/d}, \tag{A.29}$$

for a suitable constant $C = C(d,p,s) > 0$.

Proposition A.4. *Let Ω be a bounded open subset of $\mathbb{R}^d$ with Lipschitz boundary and $1 \le p < \infty, s \in (0, \frac{1}{2}) \cup (\frac{1}{2}, 1)$. Then for $u \in W_0^{s,p}(\Omega)$, we have*

$$\|u\|_{L^p(\Omega)} \lesssim |\tilde{u}|_{W^{s,p}(\mathbb{R}^d)} \lesssim \|u\|_{W^{s,p}(\Omega)}. \tag{A.30}$$

Proof. By Proposition A.3, $\tilde{u} \in W^{s,p}(\mathbb{R}^d)$. Making use of Lemma A.12, there exists some constant $C(d,p,s) > 0$ such that for all $x \in \Omega$

$$C|\Omega|^{-sp/d} \le \int_{\mathbb{R}^d \setminus \Omega} \frac{1}{|x-y|^{d+sp}} dy.$$

Therefore

$$C|\Omega|^{-sp/d} \int_\Omega |u(x)|^p dx \le \int_\Omega \int_{\mathbb{R}^d \setminus \Omega} \frac{|u(x)-u(y)|^p}{|x-y|^{d+sp}} dy dx,$$

where $u = 0$ on $\mathbb{R}^d \setminus \Omega$ has been used. In addition, using Proposition A.3 again, we have

$$|\tilde{u}|_{W^{s,p}(\mathbb{R}^d)} \lesssim \|u\|_{W^{s,p}(\Omega)}.$$

Thus, we complete the proof. □

The following properties are useful; see [Bergh and Löfström (1976); Brenner and Scott (1994); Nezza *et al.* (2004)] for details.

Proposition A.5. *Let $A_{\theta,q} := [A_0, A_1]_{\theta,q}$ with $0 < \theta < 1$ and $1 \leq q < \infty$.*

(1) $[A_0, A_1]_{\theta,q} = [A_1, A_0]_{1-\theta,q}$. If A_0 and A_1 are complete then so is $A_{\theta,q}$.
(2) If $A_1 \subset A_0$, then $A_1 \subset A_{\theta,q} \subset A_0$, and A_1 is dense in $A_{\theta,q}$. In particular, $A_0 = A_1$ implies $A_{\theta,q} = A_0$.
(3) If $A_1 \subset A_0$ and $0 < \theta_0 < \theta_1 < 1$, one has $A_{\theta_1,q} \subset A_{\theta_0,q}$. Furthermore, if $B_1 \subset B_0$ satisfying $B_i \subset A_i \, (i = 0, 1)$, then $B_{\theta,q} \subset A_{\theta,q}$.

Proposition A.6 (Embeddedness). *Let $1 \leq p < \infty$ and Ω be a bounded open subset of $\mathbb{R}^d$ with Lipschitz boundary.*

(1) The embeddings

$$W^{s',p}(\Omega) \hookrightarrow W^{s,p}(\Omega), \quad 0 < s \leq s' < 1, \tag{A.31}$$

$$W^{1,p}(\Omega) \hookrightarrow W^{s,p}(\Omega), \quad 0 < s < 1, \tag{A.32}$$

$$W^{s',p}(\Omega) \hookrightarrow W^{s,p}(\Omega), \quad 1 < s \leq s' \tag{A.33}$$

are all continuous.

(2) If $s \in (0, 1)$ and $sp > d$, then the embedding

$$W^{s,p}(\Omega) \hookrightarrow C^{0,\beta}(\Omega) \tag{A.34}$$

is compact for every $\beta < \alpha$, with $\alpha := (sp - d)/p$.

Proposition A.7 (Interpolation Theorem). *Suppose that A_i and $B_i \, (i = 0, 1)$ are two pairs of Banach spaces with $A_1 \subset A_0$, $B_1 \subset B_0$, and that $\mathcal{T}$ is a linear operator that maps A_i to $B_i \, (i = 0, 1)$. Then $\mathcal{T}$ maps $A_{\theta,q}$ to $B_{\theta,q}$. Moreover,*

$$\|\mathcal{T}\|_{A_{\theta,q} \to B_{\theta,q}} \leq \|\mathcal{T}\|_{A_0 \to B_0}^{1-\theta} \|\mathcal{T}\|_{A_1 \to B_1}^{\theta}. \tag{A.35}$$

A.3 Integral transforms

A.3.1 *Fourier transform*

Here we only consider the case in $\mathbb{R}$. The Fourier transform of a function $f(x)$ of real variable is defined by

$$\mathscr{F}[f(x)](\omega) = \hat{f}(\omega) := \int_{-\infty}^{\infty} f(x) e^{-ix\omega} dx, \quad \omega \in \mathbb{R}. \tag{A.36}$$

The inverse Fourier transform is given by the formula

$$\mathscr{F}^{-1}[g(\omega)](x) := \frac{1}{2\pi}\int_{-\infty}^{\infty} g(\omega)e^{ix\omega}d\omega, \quad x \in \mathbb{R}. \tag{A.37}$$

The integrals in (A.36) and (A.37) converge absolutely for functions $f, g \in L^1(\mathbb{R})$ and in the norm of the space $L^2(\mathbb{R})$ for $f, g \in L^2(\mathbb{R})$ (see [Kilbas *et al.* (2006); Sneddon (1995)]). In particular, if $f(x), \hat{f}(\omega) \in L^1(\mathbb{R})$, then $\hat{f}$ and f are bounded continuous, and

$$\mathscr{F}^{-1}[\hat{f}(\omega)](x) = f(x) \quad \text{a.e. } x \in \mathbb{R}. \tag{A.38}$$

The following results are useful in the analysis.

Lemma A.13 (Riemann-Lebesgue Lemma [Zorich (2004)]). *If a locally integrable function $f : (x_1, x_2) \to \mathbb{R}$ is absolutely integrable on an open interval (x_1, x_2), then*

$$\int_{x_1}^{x_2} f(x)e^{i\lambda x}dx \to 0 \quad \text{as } \lambda \to \infty, \ \lambda \in \mathbb{R}. \tag{A.39}$$

Lemma A.14 (Young Theorem [McLean (2000)]). *Let $1 \le p, q, \gamma \le \infty$, and suppose that*

$$\frac{1}{p} + \frac{1}{q} = 1 + \frac{1}{\gamma}. \tag{A.40}$$

If $f \in L^p(\mathbb{R})$ and $g \in L^q(\mathbb{R})$, then the Fourier convolution $f * g := \int_{\mathbb{R}} f(x - \xi)g(\xi)d\xi$ exists almost everywhere in $\mathbb{R}$ belongs to $L^\gamma(\mathbb{R})$, with

$$\|f * g\|_{L^\gamma(\mathbb{R})} \le \|f\|_{L^p(\mathbb{R})} \|g\|_{L^q(\mathbb{R})}. \tag{A.41}$$

Lemma A.15 ([Zorich (2004)]). *(1) If $f \in C^k(\mathbb{R})(k \in \mathbb{N})$ and all the functions $f, f', \cdots, f^{(k)} \in L^1(\mathbb{R})$, then for every $n \in \{0, 1, \cdots, k\}$*

$$\widehat{f^{(n)}}(\omega) = (i\omega)^n \hat{f}(\omega) \tag{A.42}$$

and $\hat{f}(\omega) = o\left(\frac{1}{\omega^k}\right)$ as $\omega \to \infty$.

(2) If $x^k f \in L^1(\mathbb{R})$, $k \in \mathbb{N}$, then $\hat{f} \in C^k(\mathbb{R})$ and $\hat{f}^{(n)} = (-i)^n \widehat{x^n f}$ for $n = 0, 1, \cdots, k$.

Lemma A.16 ([Guo *et al.* (2015)]). *$\mathscr{F}$ is an isomorphism of $L^2(\mathbb{R})$ onto to $L^2(\mathbb{R})$; and if $f, g \in L^2(\mathbb{R})$, it holds that the Parseval identity*

$$\int_{\mathbb{R}} f(x)\overline{g(x)}dx = \frac{1}{2\pi}\int_{\mathbb{R}} \hat{f}(\omega)\overline{\hat{g}(\omega)}d\omega. \tag{A.43}$$

Especially, if $f(x) = g(x)$, one obtains the Plancherel identity

$$\|f(x)\|_{L^2(\mathbb{R})} = \frac{1}{\sqrt{2\pi}}\left\|\hat{f}(\omega)\right\|_{L^2(\mathbb{R})}. \tag{A.44}$$

A.3.2 *Laplace transform*

As we see, if $f \in L^1(\mathbb{R})$ then its Fourier transform exists in the classical sense. However, many simple functions cannot satisfy this strict requirement to be integrable. To remedy this drawback, the Laplace transform is proposed. The Laplace transform of a function $f(t), 0 < t < \infty$, is defined as follows

$$\mathscr{L}\left[f(t)\right](s) = \int_0^\infty e^{-st} f(t)dt. \tag{A.45}$$

It can be shown that if $f \in L^1(0, b)$ for any $b > 0$, and $f(t)$ grows at most exponentially (i.e., there exist constants $A > 0$ and $p_0 \geq 0$ such that $|f(t)| \leq Ae^{p_0 t}(t > b > 0)$), then the Laplace transform of $f(t)$ exists for all s with $\text{Re}(s) \geq p_0$. Moreover, the integral (A.45) converges absolutely and uniformly in $G = \{s : \text{Re}(s) \geq \sigma > p_0\}$ and $\mathscr{L}\left[f(t)\right](s)$ is an analytic function. In practical problems, we usually need to find $f(t)$ from $\mathscr{L}\left[f(t)\right](s)$, which is generally given by the Bromwich integral

$$f(t) = \frac{1}{2\pi i}\int_{\gamma - i\infty}^{\gamma + i\infty} \mathscr{L}\left[f(t)\right](s)e^{st}ds \quad \forall t > 0, \tag{A.46}$$

where γ is chosen such that all singularities of $\mathscr{L}\left[f(t)\right](s)$ fall to the left of the vertical integration path. Note that when all the singularities are in the left half plane, we can take $\gamma = 0$, and the integral reduces to the inverse Fourier transform of $\mathscr{L}\left[f(t)\right](s)$. In practice, when $\mathscr{L}\left[f(t)\right](s)$ satisfies certain conditions, by the Cauchy theorem in the complex integration theory, for effectively performing numerical inverse Laplace transform, the integration path is usually deformed to an appropriate Hankel contour.

It is helpful to know some conclusions and properties of the Laplace transform, such as

$$\mathscr{L}\left[e^{at}f(t)\right](s) = \mathscr{L}\left[f(t)\right](s - a),$$

$$\mathscr{L}\left[t^k\right](s) = \frac{\Gamma(k+1)}{s^{k+1}}\,(k > -1, \text{Re}(s) > 0),$$

$$\mathscr{L}\left[(f * g)(t)\right] = \mathscr{L}\left[f(t)\right](s)\mathscr{L}\left[g(t)\right](s),$$

where $(f * g)(t) := \int_0^t f(t - \xi)g(\xi)d\xi$ denotes the Laplace convolution. In addition, the Mittag-Leffler function with two parameters is defined by

$$E_{\alpha,\beta}(z) = \sum_{k=1}^{\infty} \frac{z^k}{\Gamma(k\alpha + \beta)}\;(z \in \mathbb{C};\; \alpha, \beta > 0) \tag{A.47}$$

plays a very important role in the fractional calculus. It is an entire function, and there hold the result [Podlubny (2012)]

$$ {}_0D_t^{\gamma}\left(t^{an+\beta-1}E_{\alpha,\beta}^{(n)}(zt^{\alpha})\right)=t^{an+\beta-\gamma-1}E_{\alpha,\beta-\gamma}^{(n)}(zt^{\alpha}) \quad \text{(A.48)} $$

and the Laplace transform formula

$$ \mathscr{L}\left[t^{n\alpha+\beta-1}\left(\frac{\partial}{\partial z}\right)^n E_{\alpha,\beta}(zt^{\alpha})\right](s)=\frac{n!s^{\alpha-\beta}}{(s^{\alpha}-z)^{n+1}}, \quad \text{(A.49)} $$

where $z \in \mathbb{C}, n \in \mathbb{N}, \gamma \in \mathbb{R}$, and $\mathrm{Re}(s) > |z|^{\frac{1}{\alpha}}$. In particular, for $n = 0$ and $t = 1$, one has the inverse transform formula

$$ E_{\alpha,\beta}(z)=\frac{1}{2\pi i}\int_{\Gamma}\frac{s^{\alpha-\beta}e^{s}}{s^{\alpha}-z}ds, \quad \text{(A.50)} $$

which can be used to design algorithm to compute the Mittag-Leffler function. Here the path of integration Γ is a loop which starts and ends at $-\infty$ and encircles the circular disk $|s| \leq |z|^{\frac{1}{\alpha}}$ in the positive sense: $|\arg(s)| \leq \pi$ on Γ. For more details on the Mittag-Leffler function we refer to [Kilbas *et al.* (2006); Podlubny (2012)].

A.4 The Young and Grönwall inequalities

Lemma A.17 (Young's inequality). *For $1 < p, q < \infty$ with $1/p+1/q = 1$ and all $a, b \geq 0$, we have*

$$ ab \leq \frac{a^p}{p}+\frac{b^q}{q}. \quad \text{(A.51)} $$

The Grönwall type inequalities are very useful in the stability and convergence analysis of initial-boundary value problem. The following is a typical Grönwall inequality [Quarteroni and Valli (2008)]:

Lemma A.18 (Continuous Grönwall's Lemma). *Let $f(t)$ be a non-negative integrable function over $(t_0, T]$, and let $g(t)$ and $\phi(t)$ be continuous functions on $[t_0, T]$. If $\phi(t)$ satisfies*

$$ \phi(t) \leq g(t)+\int_{t_0}^{t} f(\tau)\phi(\tau)d\tau \quad \forall t \in [t_0, T], $$

then we have

$$ \phi(t) \leq g(t)+\int_{t_0}^{t} f(s)g(s)e^{\int_s^t f(\tau)d\tau}ds \quad \forall t \in [t_0, T]. $$

If, in addition, g is non-decreasing, then

$$ \phi(t) \leq g(t)e^{\int_{t_0}^t f(\tau)d\tau} \quad \forall t \in [t_0, T]. $$

On the other hand, the discrete Grönwall inequalities are often used in the stability and convergence analysis of time discretization schemes.

Lemma A.19 (Discrete Grönwall's Lemma). *Assume that* $\{k_n\}$ *and* $\{p_n\}$ *are nonnegative sequences, and the sequence* $\{\phi_n\}$ *satisfies*

$$\phi_0 \leq g_0, \qquad \phi_n \leq g_0 + \sum_{l=0}^{n-1} p_l + \sum_{l=0}^{n-1} k_l \phi_l, \quad n \geq 1,$$

where $g_0 \geq 0$. *Then there exists*

$$\phi_n \leq \Big(g_0 + \sum_{l=0}^{n-1} p_l\Big) \exp\Big(\sum_{l=0}^{n-1} k_l\Big), \quad n \geq 1.$$

A.5 Jacobi polynomials and Gauss-type quadratures

Let $w(x)$ be a non-negative, continuous and integrable real-valued function in the interval $\Omega = (-1, 1)$. For $1 \leq p < \infty$, the associated weighted Banach space of real-valued function [Guo (1998); Canuto *et al.* (2006)] is

$$L^p_w(\Omega) := \left\{ v : \|v\|_{L^p_w(\Omega)} < \infty \right\},$$

equipped with the norm

$$\|v\|_{L^p_w(\Omega)} = \left(\int_\Omega |v(x)|^p w(x) dx \right)^{\frac{1}{p}}.$$

If $p = 2$, then $L^2_w(\Omega)$ is a Hilbert space for the weighted inner product

$$(u, v)_w = \int_{-1}^{1} u(x) v(x) w(x) dx.$$

A.5.1 *Jacobi polynomials*

From the theory of orthogonal polynomials, we know for any given $w(x)$, there exists an orthogonal polynomials system with respect to the weighted inner product $(\cdot, \cdot)_w$ [Guo (1998); Shen *et al.* (2011)]. They are complete in $L^2_w(\Omega)$ and can be constructed by a three-term recurrence formula. In particular, if we let $w(x) = (1-x)^\alpha (1+x)^\beta$, $\alpha, \beta > -1$, then the Jocobi orthogonal polynomials system $J^{\alpha,\beta}_j(x)$ is obtained. When normalized by

$$J^{\alpha,\beta}_j(1) = \frac{\Gamma(j+\alpha+1)}{\Gamma(j+1)\Gamma(\alpha+1)},$$

the three-term recurrence relation for the Jacobi polynomials is

$$J_0^{\alpha,\beta}(x) = 1, \quad J_1^{\alpha,\beta}(x) = \frac{1}{2}(\alpha+\beta+2)\,x + \frac{1}{2}(\alpha-\beta),$$
$$J_{j+1}^{\alpha,\beta}(x) = \left(a_j^{\alpha,\beta}x - b_j^{\alpha,\beta}\right) J_j^{\alpha,\beta}(x) - c_j^{\alpha,\beta} J_{j-1}^{\alpha,\beta}(x), \quad j \geq 1,$$

where a_j, b_j, and c_j are given by

$$a_j^{\alpha,\beta} = \frac{(2j+\alpha+\beta+1)(2j+\alpha+\beta+2)}{2(j+1)(j+\alpha+\beta+1)},$$
$$b_j^{\alpha,\beta} = \frac{(\beta^2-\alpha^2)(2j+\alpha+\beta+1)}{2(j+1)(j+\alpha+\beta+1)(2j+\alpha+\beta)},$$
$$c_j^{\alpha,\beta} = \frac{(j+\alpha)(j+\beta)(2j+\alpha+\beta+2)}{(j+1)(j+\alpha+\beta+1)(2j+\alpha+\beta)}.$$

In fact, if $\alpha = \beta = 0$, then the Jacobi polynomials $J_{j+1}^{\alpha,\beta}(x)$ reduce to the Legendre polynomials $L_j(x)$.

The Jacobi polynomials satisfy

$$\int_{-1}^{1} J_j^{\alpha,\beta}(x) J_m^{\alpha,\beta}(x)(1-x)^{\alpha}(1+x)^{\beta}dx = r_j^{\alpha,\beta}\delta_{j,m},$$
$$\frac{d}{dx}J_j^{\alpha,\beta}(x) = \frac{1}{2}(j+\alpha+\beta+1)J_{j-1}^{\alpha+1,\beta+1}(x), \quad j \geq 1,$$

where

$$r_j^{\alpha,\beta} = \frac{2^{\alpha+\beta+1}\Gamma(j+\alpha+1)\Gamma(j+\beta+1)}{(2j+\alpha+\beta+1)j!\Gamma(j+\alpha+\beta+1)}.$$

A.5.2 *Jacobi-Gauss-type quadratures*

Define

$$A_{N+1}^{\alpha,\beta} := \begin{pmatrix} a_0 & b_1 & & & \\ b_1 & a_1 & b_2 & & \\ & \ddots & \ddots & \ddots & \\ & & b_{N-1} & a_{N-1} & b_N \\ & & & b_N & a_N \end{pmatrix},$$

where

$$a_j = \frac{b_j^{\alpha,\beta}}{a_j^{\alpha,\beta}}, \quad j \geq 0; \quad b_j = \sqrt{\frac{c_j^{\alpha,\beta}}{a_{j-1}^{\alpha,\beta}a_j^{\alpha,\beta}}}, \quad j \geq 1$$

with $a_0^{\alpha,\beta} = \frac{1}{2}(\alpha+\beta+2)$ and $b_0^{\alpha,\beta} = \frac{1}{2}(\beta-\alpha)$.

Let $\{x_j\}_{j=0}^N$ be the eigenvalues of $A_{N+1}^{\alpha,\beta}$, and

$$\omega_j = r_0^{\alpha,\beta} \left[Q_0(x_j)\right]^2 = \frac{2^{\alpha+\beta+1}\Gamma(\alpha+1)\Gamma(\beta+1)}{\Gamma(\alpha+\beta+2)} \left[Q_0(x_j)\right]^2$$

with $Q_0(x_j)$ being the first component of orthonormal eigenvector corresponding to the eigenvalue x_j. Then

$$\int_{-1}^{1} p(x)\,(1-x)^\alpha(1+x)^\beta dx = \sum_{j=0}^{N} p(x_j)\omega_j + R.$$

Rounding off the remainder R, we obtain the so-called Jacobi-Gauss quadrature formula, and this moment $\{x_j\}_{j=0}^N$ and $\{w_j\}_{j=0}^N$ are called the Jacobi-Gauss quadrature nodes and weights, respectively. Note that if $p(x)$ is an algebraic polynomials with the degree at most $2N+1$, then the Jacobi-Gauss quadrature formula is exact.

Let $\left\{x_{G,j-1}^{\alpha+1,\beta+1}, \omega_{G,j-1}^{\alpha+1,\beta+1}\right\}_{j=1}^{N-1}$ be the Jacobi-Gauss quadrature nodes and weights corresponding to matrix $A_{N-1}^{\alpha+1,\beta+1}$, and define

$$x_0 = -1,\; x_N = 1,\; x_j = x_{G,j-1}^{\alpha+1,\beta+1}\; (j = 1,\cdots,N-1)$$

and

$$\omega_0 = \frac{2^{\alpha+\beta+1}(\beta+1)\Gamma^2(\beta+1)\Gamma(N)\Gamma(N+\alpha+1)}{\Gamma(N+\beta+1)\Gamma(N+\alpha+\beta+2)},$$

$$\omega_j = \frac{\omega_{G,j-1}^{\alpha+1,\beta+1}}{1-\left(x_{G,j-1}^{\alpha+1,\beta+1}\right)^2}, \quad 1 \le j \le N-1,$$

$$\omega_N = \frac{2^{\alpha+\beta+1}(\alpha+1)\Gamma^2(\alpha+1)\Gamma(N)\Gamma(N+\beta+1)}{\Gamma(N+\alpha+1)\Gamma(N+\alpha+\beta+2)}.$$

Then

$$\int_{-1}^{1} p(x)\,(1-x)^\alpha(1+x)^\beta dx = \sum_{j=0}^{N} p(x_j)\omega_j + R_1.$$

Rounding off the remainder R_1, we obtain the so-called Jacobi-Gauss-Lobatto quadrature, which is exact for any $p \in P_{2N-1}$.

The weights of the Jacobi-Gauss-type quadrature are all positive, which means that the quadrature formulae have good numerical stability. In addition, if N is very big, the eigenvalue method introduced above may suffer from round-off errors, this moment the root-finding iterative approaches [Glaser *et al.* (2007); Hale and Townsend (2013); Press *et al.* (2007)] should be used, but here it is enough for our purpose.

Appendix B

Some properties of the fractional calculus

There are several different definitions of fractional derivatives, which result in different discretization schemes and different stability and convergence analyses. In this appendix, we first introduce some basic facts on the fractional calculus, including the R-L fractional integrals and derivatives, and the Caputo derivatives, then the results are extended to the tempered fractional calculus. For the proof of these results, the readers may refer to specialized [Baeumera and Meerschaert (2010); Cartea and del Castillo-Negrete (2007); Diethelm (2010); Kilbas *et al.* (2006); Podlubny (2012); Samko *et al.* (1993); Sabzikar *et al.* (2015)].

B.1 Riemann-Liouville fractional integrals

For $\beta > 0$, the left and right R-L fractional integrals on $\mathbb{R}$ are defined by

$$
{}_{-\infty}D_x^{-\beta}u(x) := \frac{1}{\Gamma(\beta)}\int_{-\infty}^{x}(x-\xi)^{\beta-1}u(\xi)d\xi \tag{B.1}
$$

and

$$
{}_{x}D_{\infty}^{-\beta}u(x) := \frac{1}{\Gamma(\beta)}\int_{x}^{\infty}(\xi-x)^{\beta-1}u(\xi)d\xi, \tag{B.2}
$$

respectively. Here $\Gamma(z)$ is Euler's Gamma function, it is defined in the whole complex plane except zero and negative integers, being its poles [Podlubny (1999)].

Proposition B.8 ([Samko *et al.* (1993)]). *If $\beta \in (0,1)$ and $u(x) \in L^1(\mathbb{R})$, then the R-L fractional integrals ${}_{-\infty}D_x^{-\beta}u$ and ${}_{x}D_{\infty}^{-\beta}u$ exist almost everywhere, and*

$$
\mathscr{F}\left[{}_{-\infty}D_x^{-\beta}u(x)\right](\omega) = (i\omega)^{-\beta}\hat{u}(\omega), \tag{B.3}
$$

$$
\mathscr{F}\left[{}_{x}D_{\infty}^{-\beta}u(x)\right](\omega) = (-i\omega)^{-\beta}\hat{u}(\omega). \tag{B.4}
$$

Remark B.11. The results in Proposition B.8 can not be extended directly to values $\beta \geq 1$, even for $u(x) \in C_0^\infty(\mathbb{R})$. For example, if $\beta = 1$, we have ${}_{-\infty}D_x^{-\beta}u(x) = \int_{-\infty}^x u(\xi)d\xi$, so ${}_{-\infty}D_x^{-\beta}u(x) \to$ const as $x \to +\infty$ and then the Fourier transform $\mathscr{F}\left[{}_{-\infty}D_x^{-\beta}u\right](\omega)$ does not exist in the usual sense.

Obviously, if $u(x) = 0$ for $x \in \mathbb{R}\backslash[a,b]$, when $x \in (a,b)$, we have

$$ {}_{-\infty}D_x^{-\beta}u(x) = {}_aD_x^{-\beta}u(x) := \frac{1}{\Gamma(\beta)}\int_a^x \frac{u(\xi)d\xi}{(x-\xi)^{1-\beta}} \tag{B.5} $$

and

$$ {}_xD_\infty^{-\beta}u(x) = {}_xD_b^{-\beta}u(x) := \frac{1}{\Gamma(\beta)}\int_x^b \frac{u(\xi)d\xi}{(\xi-x)^{1-\beta}}. \tag{B.6} $$

Proposition B.9. *Suppose that $-\infty < a < c < b$ and $u(x) \in C[a,c]$. Then*

$$ \lim_{x\to a^+} {}_aD_x^{-\beta}u(x) = 0. \tag{B.7} $$

Proof. For $a < x \leq c$, we have

$$ \begin{aligned} \left|{}_aD_x^{-\beta}u(x)\right| &\leq \frac{1}{\Gamma(\beta)}\int_a^x \left|\frac{u(\xi)}{(x-\xi)^{1-\beta}}\right| d\xi \\ &\leq \frac{\sup_{x\in[a,c]}|u(x)|}{\Gamma(\beta+1)}(x-a)^\beta. \end{aligned} $$

Taking $x \to a^+$ completes the proof. □

Remark B.12. If $u(x)$ is not continuous near a^+, Proposition B.9 may not hold (such as $u(x) = (x-a)^\mu$ with $-1 < \mu < 0$ and $\mu + \beta < 0$). One can see [Li and Xu (2010), Lemma 2.2] for the more general results. For convenience, we will denote $\lim_{x\to a^+} {}_aD_x^{-\beta}u(x)$ as ${}_aD_x^{-\beta}u(a)$.

Proposition B.10. *Suppose that (a,b) is a bounded interval and $u(x) \in L^2(a,b)$. Then*

(1) ${}_aD_x^{-\beta}u$ and ${}_xD_b^{-\beta}u$ are bounded in $L^2(a,b)$, i.e.,

$$ \left\|{}_aD_x^{-\beta}u\right\|_{L^2(a,b)} \leq \frac{(b-a)^\beta}{\Gamma(\beta+1)}\|u\|_{L^2(a,b)}, \tag{B.8} $$

$$ \left\|{}_xD_b^{-\beta}u\right\|_{L^2(a,b)} \leq \frac{(b-a)^\beta}{\Gamma(\beta+1)}\|u\|_{L^2(a,b)}. \tag{B.9} $$

(2) If $\beta, \beta_1 \in \mathbb{R}^+$, then for $u \in L^2(a,b)$,

$$ {}_aD_x^{-\beta}\left({}_aD_x^{-\beta_1}u(x)\right) = {}_aD_x^{-(\beta+\beta_1)}u(x), \tag{B.10} $$

$$ {}_xD_b^{-\beta}\left({}_xD_b^{-\beta_1}u(x)\right) = {}_xD_b^{-(\beta+\beta_1)}u(x) \tag{B.11} $$

are satisfied at almost every point $x \in (a,b)$.

(3) The operators ${}_aD_x^{-\beta}$ and ${}_xD_b^{-\beta}$ are adjoint in $L^2(a,b)$, i.e.,

$$\left({}_aD_x^{-\beta}u(x), v(x)\right)_{L^2(a,b)} = \left(u(x), {}_xD_b^{-\beta}v\right)_{L^2(a,b)}. \tag{B.12}$$

Proof. See Lemmas 2.1, 2.3 and 2.7 in [Kilbas *et al.* (2006)]. □

B.2 Riemann-Liouville and Caputo fractional derivatives

For any $\beta \geq 0$, the left and right R-L fractional derivatives ${}_{-\infty}D_x^\beta u$ and ${}_xD_\infty^\beta u$ of order β are given by

$$\begin{aligned} {}_{-\infty}D_x^\beta u(x) &:= \frac{d^n}{dx^n}\left({}_{-\infty}D_x^{-(n-\beta)}u\right) \\ &= \frac{1}{\Gamma(n-\beta)}\frac{d^n}{dx^n}\left(\int_{-\infty}^x (x-\xi)^{n-\beta-1}u(\xi)d\xi\right) \end{aligned} \tag{B.13}$$

and

$$\begin{aligned} {}_xD_\infty^\beta u(x) &:= (-1)^n\frac{d^n}{dx^n}\left({}_xD_\infty^{-(n-\beta)}u\right) \\ &= \frac{(-1)^n}{\Gamma(n-\beta)}\frac{d^n}{dx^n}\left(\int_x^\infty (\xi-x)^{n-\beta-1}u(\xi)d\xi\right), \end{aligned} \tag{B.14}$$

respectively, where $n = [\beta]+1$ and $[\beta]$ is the integer part of β. In particular, when $\beta \in \mathbb{N}$, it holds that

$${}_{-\infty}D_x^n u(x) = u^{(n)}(x), \tag{B.15}$$

$${}_xD_\infty^n u(x) = (-1)^n u^{(n)}(x), \tag{B.16}$$

where $u^{(n)}(x)$ is the usual derivative of $u(x)$ of order n. Obviously, if $u(x) = 0$ for $x \in \mathbb{R}\backslash[a,b]$, when $x \in (a,b)$, we have

$$\begin{aligned} {}_{-\infty}D_x^\beta u(x) = {}_aD_x^\beta u(x) &:= \frac{d^n}{dx^n}\left({}_aD_x^{-(n-\beta)}u\right) \\ &= \frac{1}{\Gamma(n-\beta)}\frac{d^n}{dx^n}\left(\int_a^x (x-\xi)^{n-\beta-1}u(\xi)d\xi\right) \end{aligned} \tag{B.17}$$

and

$$\begin{aligned} {}_xD_\infty^\beta u(x) = {}_xD_b^\beta u(x) &:= (-1)^n\frac{d^n}{dx^n}\left({}_xD_b^{-(n-\beta)}u\right) \\ &= \frac{(-1)^n}{\Gamma(n-\beta)}\frac{d^n}{dx^n}\left(\int_x^b (\xi-x)^{n-\beta-1}u(\xi)d\xi\right). \end{aligned} \tag{B.18}$$

For $\alpha > 0$ and $\alpha \notin \mathbb{N}$, the Caputo fractional derivative ${}_0^C D_t^\alpha v(t)$ of order α is defined by

$$\begin{aligned} {}_0^C D_t^\alpha v(t) &:= {}_0 D_t^{-(n-\alpha)} \left(\frac{d^n v(t)}{dt} \right) \\ &= \frac{1}{\Gamma(n-\alpha)} \int_0^t (t-s)^{n-\alpha-1} \frac{d^n v(s)}{ds^n} ds, \quad n = [\alpha] + 1, \end{aligned} \tag{B.19}$$

and for $\alpha \in \mathbb{N}$, it is natural to define ${}_0 D_t^\alpha v(t) := v^{(\alpha)}(t)$.

Proposition B.11 ([Samko *et al.* (1993)]). *If $\beta \in \mathbb{R}^+$ and $u(x)$ is 'sufficiently good', then*

$$\mathscr{F} \left[{}_{-\infty} D_x^\beta u(x) \right] (\omega) = (i\omega)^\beta \hat{u}(\omega), \tag{B.20}$$

$$\mathscr{F} \left[{}_x D_\infty^\beta u(x) \right] (\omega) = (-i\omega)^\beta \hat{u}(\omega). \tag{B.21}$$

Remark B.13. The 'sufficiently good' functions maybe those which are differentiable up to the order $[\beta] + 1$, and vanish sufficiently rapidly at infinity together with their derivatives. At this moment, for $\beta \notin \mathbb{N}$, one may verify e.g. (B.20) by rewriting ${}_{-\infty} D_x^\beta u$ as ${}_{-\infty} D_x^{\beta - [\beta] - 1} \left(u^{([\beta]+1)} \right)$ and applying (A.42) and (B.3).

Proposition B.12 ([Kilbas *et al.* (2006)]). *Let $\alpha \in \mathbb{R}^+, n - 1 < \alpha \leq n \, (n \in \mathbb{N})$ be such that the $\mathscr{L}[u(t)](s)$ and $\mathscr{L}\left[u^{(n)}(t)\right](s)$ exist, and $\lim_{t \to +\infty} (u^{(n)}(t)) = 0$ for $k = 0, 1, \cdots, n-1$. Then*

$$\mathscr{L} \left[{}_0^C D_t^\alpha u(t) \right] (s) = s^\alpha \mathscr{L} [u(t)] (s) - \sum_{k=0}^{n-1} s^{\alpha - k - 1} u^{(k)}(0). \tag{B.22}$$

In particular, if $0 < \alpha \leq 1$, then

$$\mathscr{L} \left[{}_0^C D_t^\alpha u(t) \right] (s) = s^\alpha \mathscr{L} [u(t)] (s) - s^{\alpha - 1} u(0). \tag{B.23}$$

Proposition B.13. *Let $0 < \beta, \beta_1 < 1$ and $u(a) = 0$. One has*

$${}_a D_x^\beta \left({}_a D_x^{\beta_1} u(x) \right) = {}_a D_x^{\beta_1} \left({}_a D_x^\beta u(x) \right) = {}_a D_x^{\beta + \beta_1} u(x). \tag{B.24}$$

Proof. See the proof in [Podlubny (1999), p. 74–75]. □

Proposition B.14. *Let $\beta > 0$.*

(1) If $u(x) \in L^2[a, b]$, then

$${}_a D_x^\beta \left({}_a D_x^{-\beta} u(x) \right) = u(x). \tag{B.25}$$

(2) If $0 < \beta < 1$ and $u(x) \in C^1[a,b]$, then

$$ {}_aD_x^{-\beta}\left({}_aD_x^{\beta}u(x)\right) = u(x). \tag{B.26}$$

Proof. For (B.25),

$$\begin{aligned} {}_aD_x^{\beta}\,{}_aD_x^{-\beta}u(x) &= \frac{d^n}{dx^n}\left({}_aD_x^{-(n-\alpha)}\,{}_aD_x^{-\alpha}u(x)\right) \\ &= \frac{d^n}{dx^n}\left({}_aD_x^{-n}u(x)\right) = u(x), \end{aligned} \tag{B.27}$$

where $n = [\beta] + 1$. For (B.26),

$$\begin{aligned} {}_aD_x^{-\beta}\left({}_aD_x^{\beta}u(x)\right) &= \frac{1}{\Gamma(\beta)}\int_a^x (x-\xi)^{\beta-1}\,{}_aD_\xi^{\beta}u(\xi)d\xi \\ &= \frac{d}{dx}\left\{\frac{1}{\Gamma(\beta+1)}\int_a^x (x-\xi)^{\beta}\,{}_aD_\xi^{\beta}u(\xi)d\xi\right\} \\ &= \frac{d}{dx}\left\{{}_aD_x^{-1}u(x) - {}_aD_x^{-(1-\beta)}u(a)\,\frac{(x-a)^{\beta}}{\Gamma(\beta+1)}\right\} = u(x), \end{aligned}$$

where ${}_aD_x^{-(1-\beta)}u(a) = 0$ (see Proposition B.9) has been used in the last step. □

Proposition B.15. *If $0 < \alpha < 1$, then*

$$ {}_0^CD_t^{\alpha}v(t) = {}_0D_t^{\alpha}\left[v(t) - v(0)\right]. \tag{B.28}$$

Proof. Denote $g(t) = v(t) - v(0)$. Then

$$\begin{aligned} {}_0D_t^{\alpha}g(t) &= \frac{1}{\Gamma(1-\alpha)}\frac{d}{dt}\int_0^t (t-s)^{-\alpha}g(s)ds \\ &= \frac{1}{\Gamma(2-\alpha)}\frac{d}{dt}\left((t-s)^{1-\alpha}g(s)\,\Big|_{s=0}^{t} + \int_0^t (t-s)^{1-\alpha}\frac{dg(s)}{ds}ds\right) \\ &= \frac{1}{\Gamma(1-\alpha)}\int_0^t (t-s)^{-\alpha}\frac{dv(s)}{ds}ds. \end{aligned}$$

□

Proposition B.16. *If $u(x) \in AC[a,b]$ and $0 < \alpha < 1$, then*

$$ {}_0D_x^{-\alpha}\left({}_0^CD_x^{\alpha}u(x)\right) = u(x) - u(0). \tag{B.29}$$

Proof.

$$ {}_0D_x^{-\alpha}\left({}_0^CD_x^{\alpha}u(x)\right) = {}_0D_x^{-1}u^{(1)}(x) = u(x) - u(0).$$

□

B.3 Tempered fractional calculus

In the following, we give/collect the definitions and properties of the tempered fractional integrals and derivatives. For the details, see, e.g., [Baeumera and Meerschaert (2010); Cartea and del Castillo-Negrete (2007); Li and Deng (2016); Sabzikar *et al.* (2015)].

B.3.1 *Tempered fractional integrals*

Given $\lambda > 0$, the left and right tempered R-L fractional integrals are defined by

$$_{-\infty}D_x^{-\beta,\lambda}u(x) = \frac{1}{\Gamma(\beta)}\int_{-\infty}^{x}(x-\xi)^{\beta-1}e^{-\lambda(x-\xi)}u(\xi)d\xi \tag{B.30}$$

and

$$_{x}D_{\infty}^{-\beta,\lambda}u(x) = \frac{1}{\Gamma(\beta)}\int_{x}^{\infty}(\xi-x)^{\beta-1}e^{-\lambda(\xi-x)}u(\xi)d\xi, \tag{B.31}$$

respectively. It is easy to check that

$$_{-\infty}D_x^{-\beta,\lambda}u(x) = e^{-\lambda x}{}_{-\infty}D_x^{-\beta}\left(e^{\lambda x}u(x)\right), \tag{B.32}$$

$$_{x}D_{\infty}^{-\beta,\lambda}u(x) = e^{\lambda x}{}_{x}D_{\infty}^{-\beta}\left(e^{-\lambda x}u(x)\right), \tag{B.33}$$

and when $\lambda = 0$, they all reduce to the R-L fractional integrals.

Proposition B.17. *Let $u(x) \in L^1(\mathbb{R})$ and $_{-\infty}D_x^{-\beta,\lambda}u(x), {}_{x}D_{\infty}^{-\beta,\lambda}u(x)$ exist. Then*

$$\mathscr{F}\left[_{-\infty}D_x^{-\beta,\lambda}u(x)\right](x) = (\lambda + i\omega)^{-\beta}\hat{u}(\omega), \tag{B.34}$$

$$\mathscr{F}\left[_{x}D_{\infty}^{-\beta,\lambda}u(x)\right](x) = (\lambda - i\omega)^{-\beta}\hat{u}(\omega). \tag{B.35}$$

Proof. Let $h_+(x) = \begin{cases} \frac{x^{\beta-1}}{\Gamma(\beta)}e^{-\lambda x} & x > 0 \\ 0 & x \le 0 \end{cases}$. Then

$$\begin{aligned}
&\mathscr{F}\left[_{-\infty}D_x^{-\beta,\lambda}u(x)\right](\omega) = \mathscr{F}\left[h_+(x) * u(x)\right](\omega) \\
&= \mathscr{F}\left[h_+(x)\right](\omega)\mathscr{F}\left[u(x)\right](\omega) = \mathscr{L}\left[h_+(x)\right](i\omega)\mathscr{F}[u(x)](\omega) \\
&= (\lambda + i\omega)^{-\beta}\hat{u}(\omega).
\end{aligned}$$

The proof for (B.35) is similar. $\square$

Remark B.14. Unlike the Fourier transforms of the R-L fractional integrals, we don't require the condition $0 < \beta < 1$.

Obviously, if $u(x) = 0$ for $x \in \mathbb{R}\backslash[a,b]$, when $x \in (a,b)$, it holds that

$$ {}_{-\infty}D_x^{-\beta,\lambda}u(x) = {}_aD_x^{-\beta,\lambda}u(x) := e^{-\lambda x}{}_aD_x^{-\beta}\left(e^{\lambda x}u(x)\right) \tag{B.36} $$

and

$$ {}_xD_\infty^{-\beta,\lambda}u(x) = {}_xD_b^{-\beta,\lambda}u(x) := e^{\lambda x}{}_xD_b^{-\beta}\left(e^{-\lambda x}u(x)\right). \tag{B.37} $$

Proposition B.18. *Suppose that (a,b) is a bounded interval. Then*

(1) ${}_aD_x^{-\beta,\lambda}u$ and ${}_xD_b^{-\beta,\lambda}u$ are bounded in $L^2(a,b)$, i.e.,

$$ \left\|{}_aD_x^{-\beta,\lambda}u\right\|_{L^2(a,b)} \le \frac{(b-a)^\beta}{\Gamma(\beta+1)}\|u\|_{L^2(a,b)}, \tag{B.38} $$

$$ \left\|{}_xD_b^{-\beta,\lambda}u\right\|_{L^2(a,b)} \le \frac{(b-a)^\beta}{\Gamma(\beta+1)}\|u\|_{L^2(a,b)}. \tag{B.39} $$

(2) If $\beta, \beta_1 \in \mathbb{R}^+$, then for $u \in L^2(a,b)$,

$$ {}_aD_x^{-\beta,\lambda}\left({}_aD_x^{-\beta_1,\lambda}u(x)\right) = {}_aD_x^{-(\beta+\beta_1),\lambda}u(x), \tag{B.40} $$

$$ {}_xD_b^{-\beta,\lambda}\left({}_xD_b^{-\beta_1,\lambda}u(x)\right) = {}_xD_b^{-(\beta+\beta_1),\lambda}u(x) \tag{B.41} $$

are satisfied at almost every point $x \in (a,b)$.

(3) The operators ${}_aD_x^{-\beta,\lambda}$ and ${}_xD_b^{-\beta,\lambda}$ are adjoint in $L^2(a,b)$, i.e.,

$$ \left({}_aD_x^{-\beta,\lambda}u(x), v(x)\right)_{L^2(a,b)} = \left(u(x), {}_xD_b^{-\beta,\lambda}v\right)_{L^2(a,b)}. \tag{B.42} $$

Proof. By (B.10)-(B.12), (B.32), and (B.33), (B.40)-(B.42) can be proved easily. Here we only give the proofs of (B.38) and (B.39). Define

$$ \tilde{u}(x) = \begin{cases} u(x), & x \in (a,b), \\ 0, & \text{else}; \end{cases} \quad \tilde{v} = \begin{cases} \frac{e^{-\lambda x}x^{\beta-1}}{\Gamma(\beta)}, & x \in (0, b-a], \\ 0, & \text{else}. \end{cases} \tag{B.43} $$

Then for $x \in (a,b)$, it holds that ${}_aD_x^{-\beta,\lambda}u(x) = \tilde{u} * \tilde{v}$. Using the Young theorem (A.14), we have

$$ \begin{aligned} \left\|{}_aD_x^{-\beta,\lambda}u(x)\right\|_{L^2(a,b)} &\le \|\tilde{u} * \tilde{v}\|_{L^2(\mathbb{R})} \\ &\le \|\tilde{v}\|_{L^1(\mathbb{R})}\|\tilde{u}\|_{L^2(\mathbb{R})} \le \frac{(b-a)^\beta}{\Gamma(\beta+1)}\|u\|_{L^2(a,b)}. \end{aligned} \tag{B.44} $$

For the right tempered R-L fractional integral we can prove

$$ \left\|{}_xD_b^{-\beta,\lambda}u(x)\right\|^2_{L^2(a,b)} = \int_{-b}^{-a}\left(\int_{-b}^{x}\frac{(x-\xi)^{\beta-1}}{\Gamma(\beta)}e^{-\lambda(x-\xi)}u(-\xi)d\xi\right)^2 dx. $$

Letting $g(x) = u(-x)$, then

$$ \left\|{}_xD_b^{-\beta,\lambda}u(x)\right\|^2_{L^2(a,b)} = \left\|{}_{-b}D_x^{-\beta,\lambda}g(x)\right\|^2_{L^2(-b,-a)} \le \frac{(b-a)^\beta}{\Gamma(\beta+1)}\|u\|_{L^2(a,b)}. $$

□

B.3.2 *Tempered fractional derivatives*

Let $\beta \in (0,1) \cup (1,2)$ and $\lambda > 0$. Define

$$\tilde{H}^{\beta}(\mathbb{R}) := \left\{ u \in L^2(\mathbb{R}) : \int_{\mathbb{R}} \left(\lambda^2 + |\omega|^2\right)^{\beta} |\hat{u}(\omega)|^2 \, d\omega < \infty \right\}.$$

It is easy to check that

$$\min\{1, \lambda^{2\beta}\} \left(1 + |\omega|^2\right)^{\beta} \leq \left(\lambda^2 + |\omega|^2\right)^{\beta} \leq \max\{1, \lambda^{2\beta}\} \left(1 + |\omega|^2\right)^{\beta}$$

and

$$2^{-\beta} \left(1 + |\omega|^2\right)^{\beta} \leq \left(1 + |w|^{2\beta}\right) \leq \max\{1, 2^{1-\beta}\} \left(1 + |\omega|^2\right)^{\beta}.$$

Therefore, $\tilde{H}^{\beta}(\mathbb{R}) = H^{\beta}(\mathbb{R})$. The tempered fractional derivatives are defined by the inverse Fourier transform. More specifically, for $0 < \beta < 1$, the left and right tempered fractional derivatives are given by

$$_{-\infty}D_x^{\beta,\lambda} u(x) = \mathscr{F}^{-1}[\left((\lambda + i\omega)^{\beta} - \lambda^{\beta}\right) \hat{u}(\omega)](x), \tag{B.45}$$

$$_{x}D_{\infty}^{\beta,\lambda} u(x) = \mathscr{F}^{-1}[\left((\lambda - i\omega)^{\beta} - \lambda^{\beta}\right) \hat{u}(\omega)](x), \tag{B.46}$$

and for $1 < \beta < 2$, the left and right tempered fractional derivatives are given by

$$_{-\infty}D_x^{\beta,\lambda} u(x) = \mathscr{F}^{-1}[\left((\lambda + i\omega)^{\beta} - i\omega\beta\lambda^{\beta-1} - \lambda^{\beta}\right) \hat{u}(\omega)](x), \tag{B.47}$$

$$_{x}D_{\infty}^{\beta,\lambda} u(x) = \mathscr{F}^{-1}[\left((\lambda - i\omega)^{\beta} + i\omega\beta\lambda^{\beta-1} - \lambda^{\beta}\right) \hat{u}(\omega)](x). \tag{B.48}$$

Since

$$\int_{\mathbb{R}} e^{-i\omega x} \left(e^{\pm\lambda x} u(x)\right) dx = \hat{u}(\omega \pm i\lambda),$$
$$\mathscr{F}\left[{}_{-\infty}D_x^{\beta} u(x)\right](\omega) = (i\omega)^{\beta} \hat{u}(\omega),$$

we have

$$\begin{aligned}
&\mathscr{F}[e^{-\lambda x}{}_{-\infty}D_x^{\beta}(e^{\lambda x}u(x))](\omega) \\
&= (i(\omega - i\lambda))^{\beta} \hat{u}(\omega - i\lambda + i\lambda) \\
&= (\lambda + i\omega)^{\beta} \hat{u}(\omega).
\end{aligned} \tag{B.49}$$

Similarly,

$$\mathscr{F}[e^{\lambda x}{}_{x}D_{\infty}^{\beta}(e^{-\lambda x}u(x))](\omega) = (\lambda - i\omega)^{\beta} \hat{u}(\omega). \tag{B.50}$$

Therefore, for $\beta \in (0, 2]$ and $\beta \neq 1$, the tempered fractional derivatives could also be given as

- for $0 < \beta < 1$

$$
{}_{-\infty}D_x^{\beta,\lambda}u(x) = {}_{-\infty}\mathbb{D}_x^{\beta,\lambda}u(x) - \lambda^\beta u(x), \tag{B.51}
$$

$$
{}_xD_\infty^{\beta,\lambda}u(x) = {}_x\mathbb{D}_\infty^{\beta,\lambda}u(x) - \lambda^\beta u(x); \tag{B.52}
$$

- for $1 < \beta < 2$

$$
{}_{-\infty}D_x^{\beta,\lambda}u(x) = {}_{-\infty}\mathbb{D}_x^{\beta,\lambda}u(x) - \beta\lambda^{\beta-1}\frac{du(x)}{dx} - \lambda^\beta u(x), \tag{B.53}
$$

$$
{}_xD_\infty^{\beta,\lambda}u(x) = {}_x\mathbb{D}_\infty^{\beta,\lambda}u(x) + \beta\lambda^{\beta-1}\frac{du(x)}{dx} - \lambda^\beta u(x), \tag{B.54}
$$

where

$$
{}_{-\infty}\mathbb{D}_x^{\beta,\lambda}u(x) := e^{-\lambda x}{}_{-\infty}D_x^\beta\left(e^{\lambda x}u(x)\right), \tag{B.55}
$$

$$
{}_x\mathbb{D}_\infty^{\beta,\lambda}u(x) := e^{\lambda x}{}_xD_\infty^\beta\left(e^{-\lambda x}u(x)\right). \tag{B.56}
$$

Letting $\beta = 2$ in (B.53) and (B.54), it is easy to check that

$$
{}_{-\infty}\mathbb{D}_x^{\beta,\lambda}u(x) = \left(\frac{d}{dx} + \lambda\right)^2 u(x), \quad {}_x\mathbb{D}_\infty^{\beta,\lambda}u(x) = \left(\frac{d}{dx} - \lambda\right)^2 u(x). \tag{B.57}
$$

Then

$$
{}_{-\infty}D_x^{\beta,\lambda}u(x) = \frac{d^2u(x)}{dx^2} = {}_xD_\infty^{\beta,\lambda}u(x). \tag{B.58}
$$

So the tempered parameter has no impact on the classic second order derivative.

As before, if $u(x) = 0$ for $x \in \mathbb{R}\backslash[a,b]$, when $x \in (a,b)$, it holds that

$$
{}_{-\infty}\mathbb{D}_x^{\beta,\lambda}u(x) = {}_a\mathbb{D}_x^{\beta,\lambda}u(x) := e^{-\lambda x}{}_aD_x^\beta\left(e^{\lambda x}u(x)\right) \tag{B.59}
$$

and

$$
{}_x\mathbb{D}_\infty^{\beta,\lambda}u(x) = {}_x\mathbb{D}_b^{\beta,\lambda}u(x) := e^{\lambda x}{}_xD_b^\beta\left(e^{-\lambda x}u(x)\right); \tag{B.60}
$$

and for ${}_a\mathbb{D}_x^{\beta,\lambda}u(x)$ and ${}_x\mathbb{D}_b^{\beta,\lambda}u(x)$, we also have the results similar to Propositions B.13-B.15, which are omitted here.

Proposition B.19. *If $\beta > 0$ and $\gamma > 0$, then*

$$
{}_aD_x^{-\beta,\lambda}\left(e^{-\lambda x}(x-a)^{\gamma-1}\right) = \frac{\Gamma(\gamma)}{\Gamma(\gamma+\beta)}e^{-\lambda x}(x-a)^{\gamma+\beta-1}, \tag{B.61}
$$

$$
{}_xD_b^{-\beta,\lambda}\left(e^{\lambda x}(b-x)^{\gamma-1}\right) = \frac{\Gamma(\gamma)}{\Gamma(\gamma+\beta)}e^{\lambda x}(b-x)^{\gamma+\beta-1}. \tag{B.62}
$$

Proposition B.20. *If $\beta \geq 0$ and $\gamma > 0$, then*

$$ {}_a\mathbb{D}_x^{\beta,\lambda}\left(e^{-\lambda x}(x-a)^{\gamma-1}\right) = \frac{\Gamma(\gamma)}{\Gamma(\gamma-\beta)} e^{-\lambda x}(x-a)^{\gamma-\beta-1}, \tag{B.63} $$

$$ {}_x\mathbb{D}_b^{\beta,\lambda}\left(e^{\lambda x}(b-x)^{\gamma-1}\right) = \frac{\Gamma(\gamma)}{\Gamma(\gamma-\beta)} e^{\lambda x}(b-x)^{\gamma-\beta-1}. \tag{B.64} $$

Since for $\gamma = \beta - [\beta], \beta - [\beta] + 1, \cdots, \beta$, the term $\Gamma(\gamma-\beta)$ in the denominators of these expressions are infinite, for these γ, we actually have

$$ {}_a\mathbb{D}_x^{\beta,\lambda}\left(e^{-\lambda x}(x-a)^{\gamma-1}\right) = 0, \tag{B.65} $$

$$ {}_x\mathbb{D}_b^{\beta,\lambda}\left(e^{\lambda x}(b-x)^{\gamma-1}\right) = 0. \tag{B.66} $$

Appendix C

Iteration methods for linear systems generated from fractional PDEs

In this appendix, we introduce some of the fast algorithms that can be used to solve the algebraic system

$$AU_* = F \tag{C.1}$$

with $A = A_1 + A_2$ appeared in this book effectively. Here A_1 is a (block) Toeplitz matrix multiplied by a diagonal one, and A_2 is a sparse Teoplitz matrix (maybe an identity matrix, a tridiagonal matrix, and so on) or a sparse block Toeplitz matrix with Toeplitz blocks. Due to the nonlocal nature of fractional operators, A is a dense or even full matrix. If the direct methods are used, the computational cost of $\mathcal{O}(N^3)$ (or $\mathcal{O}(N \log^2 N)$ for the special case that A is a positive definite Toeplitz matrix [Ammar and Gragg (1988)]) per time step and the memory of $\mathcal{O}(N^2)$ are needed. Here N denotes the number of unknowns. The significant computational cost and memory requirement impose a serious challenge for the numerical simulation of the fractional PDEs, especially for the high-dimensional fractional models. Therefore, by careful analysis of the structure of matrix A, we introduce the Krylov subspace methods and the multigrid methods, which reduce the computational cost to $\mathcal{O}(N \log N)$ per iteration and the memory requirement to $\mathcal{O}(N)$. These discussions stay at a tutorial level. For an extensive presentation and a thorough analysis, the reader may refer to [Barrett *et al.* (1994); Briggs *et al.* (2007); Chan and Jin (2007); Chen and Deng (2014b); Pang and Sun (2012); Pan *et al.* (2016); Saad (2003); Wang and Basu (2012)], and to the ample literatures cited therein.

C.1 Basic concepts of the Krylov subspace methods

The Krylov subspace methods are considered currently to be the most important iteration techniques available for solving linear systems. In general, the Krylov subspace approximation is defined via a projection property, which has the general form of Galerkin or Petrov-Galerkin orthogonality requirement:

$$\text{find } U_m \in U_0 + \mathcal{K}_m \text{ such that } (F - AU_m, V) = 0 \ \ \forall V \in \mathcal{L}_m. \quad (C.2)$$

Here, $(U, V) := V^{\mathrm{T}}U$, and the ansatz space $\mathcal{K}_m$ is the Krylov subspace

$$\mathcal{K}_m(A, r_0) := \mathrm{span}\left\{r_0, Ar_0, A^2r_0, \cdots, A^{m-1}r_0\right\} \quad (C.3)$$

with $r_0 = F - AU_0$. Different Krylov methods differ in the choice of the test space $\mathcal{L}_m$. Three broad choices for $\mathcal{L}_m$ give rise to the best known techniques.

(1) The first is simply $\mathcal{L}_m = \mathcal{K}_m$. If A is symmetric positive definite, letting $\|U\|_A := \sqrt{(AU, U)}$, then (C.2) is equivalent to find U_m satisfying

$$\|U_m - U_*\|_A = \min_{U \in U_0 + \mathcal{K}_m} \|U - U_*\|_A, \quad (C.4)$$

i.e., the mth iteration solution minimize the error (under energy norm) over $U_0 + \mathcal{K}_m$. In addition, U_m also satisfies

$$\phi(U_m) = \min_{U = U_0 + \mathcal{K}_m} \phi(U), \quad (C.5)$$

where $\phi(U) = \frac{1}{2}(AU, U) - (F, U)$. This choice leads to the well known CG method.

(2) The second is $\mathcal{L}_m = AK_m$. At this time, (C.2) is equivalent to find U_m satisfying

$$\|F - AU_m\|_2 = \min_{U \in U_0 + \mathcal{K}_m} \|F - AU\|_2, \quad (C.6)$$

i.e., the mth iteration solution minimizes the residual over $U_0 + \mathcal{K}_m$. Here $\|U\|_2 := \sqrt{(U, U)}$. This choice leads to the minimal residual methods, such as the MINRES (for the nonsingular symmetric matrix A) and the GMRES method (for nonsingular A).

(3) The third is $\mathcal{L}_m = \mathcal{K}_m(A^{\mathrm{T}}, \tilde{r_0})$ for some $\tilde{r}_0$. This choice will lead to the BICG method and its variants, such as the CGS method and the Bi-CGSTAB method. They also can be used to solve the linear system with the nonsingular matrix.

The speed of convergence of the Krylov subspace methods generally depends on spectral properties and condition number of the given matrix. In practice, they are always applied in collaboration with some preconditioners (especially for the algebraic system obtained from the discretization of PDEs), i.e., transforming the original system $AU_* = F$ into an equivalent one which is (hopefully) more effective to solve by an iteration technique. There are three general types of preconditioning:

(1) left preconditioning by an M_L, i.e.,

$$M_L^{-1}AU_* = M_L^{-1}F. \quad (C.7)$$

(2) right preconditioning by an M_R, i.e.,

$$AM_R^{-1}V_* = F, \quad U_* = M_R^{-1}V_*. \quad (C.8)$$

This involves a substitution V_* for the original variable U_*.

(3) split (two-sided) preconditioning, i.e.,

$$M_L^{-1}AM_R^{-1}V_* = M_L^{-1}F, \quad U_* = M_R^{-1}V_*. \quad (C.9)$$

Split preconditioning encompasses both the left and the right methods by setting $M_R = I$ or $M_L = I$, respectively.

We should note that in many cases, the convergence behavior of the three type of preconditioners is not significantly different. This is not completely unexpected in view of the fact that the spectra of AM^{-1} (corresponding to right preconditioning) and $M^{-1}A$ (corresponding to left preconditioning) coincide. In practice, the preconditioner is chosen in dependent on properties of the matrix A. A few general comments concerning the choice/design of a preconditioner are:

- Matrices $M_L^{-1}A$, AM_R^{-1}, or $M_L^{-1}AM_R^{-1}$ should be close to the identity matrix I in some sense or at least has a clustered spectrum or well conditioned number.
- The operations $U \to M_L^{-1}U$ and/or $U \to M_R^{-1}U$ should be easy to perform, where U denotes some vector appeared in the computation process.

In the following, we list the preconditioned CG (PCG) algorithm and the preconditioned GMRES (PGMRES) algorithm. They can be used to solve the linear systems with the symmetric positive definite matrix and the more general nonsingular matrix, respectively. Compared to the BiCG type algorithms, they are robust (this means that, theoretically, the algorithms will not break down unless the exact solution is obtained).

Firstly, letting A and $M = LL^{\mathrm{T}}$ be symmetric positive definite matrices (M 'near' to A in some sense), the CG algorithm for solving the preconditioned system

$$L^{-1}AL^{-T}V_* = L^{-1}F, \quad U_* = L^{-T}V_* \quad (C.10)$$

is given in Algorithm C.1, the mth iteration solution satisfies

$$\|U_m - U_*\|_A = \inf_{U \in U_0 + \hat{\mathcal{K}}_m} \|U - U_*\|_A, \quad (C.11)$$

where

$$\hat{\mathcal{K}}_m = \{M^{-1}r_0, (M^{-1}A)M^{-1}r_0, \cdots, (M^{-1}A)^{m-1}M^{-1}r_0\} \quad \text{(C.12)}$$

with $r_0 = F - AU_0$. Noting that in Algorithm C.1, L and L^T are not appeared explicitly, and in each iteration, we need store only the five vectors U, w, p, r and z and compute two matrix-vector products. In addition, the matrices A and M^{-1} themselves need not be formed or stored, only the routines for matrix-vector products are required. Obviously, when $M = I$, Algorithm C.1 reduces to the CG algorithm without preconditioning, at this time only a single matrix-vector product is required. The convergence result of the PCG algorithm is

$$\|U_m - U_*\|_A \leq 2\left(\frac{\sqrt{\kappa}-1}{\sqrt{\kappa}+1}\right)^k \|U_0 - U_*\|_A, \quad \text{(C.13)}$$

where $\kappa = \text{cond}_2(M^{-1}A)$, and a more precise characterization of the convergence rate with the spectral distribution of $M^{-1}A$ could also be found in [Chan and Jin (2007)].

Algorithm C.1 $[U_m, m] = \text{PCG}(U_0, F, A, M, \epsilon, K_{\max})$

1: $r = F - AU_0, U = U_0, m = 0$
2: **while** $(\|r\|_2 > \epsilon \|F\|_2)$ and $(m < K_{\max})$ **do**
3: $\quad m = m + 1, z = M^{-1}r$
4: $\quad$ **if** $m = 1$ **then**
5: $\quad\quad \tau = (r, z), p = z$
6: $\quad$ **else**
7: $\quad\quad \tilde{\tau} = \tau, \tau = (r, z), \beta = \tau/\tilde{\tau}, p = z + \beta p$
8: $\quad$ **end if**
9: $\quad w = Ap, \alpha = \tau/(p, w), U = U + \alpha p$
10: $\quad r = r - \alpha w$
11: **end while**
12: Return $U_m = U$ and m

Note: U_0 denotes the initial guess.

In addition, if A is nonsingular and nonsymmetric, one might consider solving (C.1) by applying the CG algorithm to its normal equation

$$A^{\mathrm{T}}AU_* = A^{\mathrm{T}}b. \quad \text{(C.14)}$$

This approach is called CGNR, which produces the mth approximate solution in the Krylov subspace

$$U_0 + \mathcal{K}_m := U_0 + \text{span}\left\{A^{\mathrm{T}}r_0, A^{\mathrm{T}}A(A^{\mathrm{T}}r_0), \cdots, (A^{\mathrm{T}}A)^{m-1}A^{\mathrm{T}}r_0\right\}$$

satisfying

$$\|F - AU_m\|_2 = \min_{U \in U_0 + \mathcal{K}_m} \|F - AU\|_2, \tag{C.15}$$

where $r_0 = b - AU_0$. The advantages of the CGNR are that all the theory for CG can be carried over and the simple implementation for both CG and PCG can be used. However, the condition number of the coefficient matrix $A^{\mathrm{T}}A$ is the square of that of A and two matrix-vector products AV and $A^{\mathrm{T}}V$ are needed for each CG iteration [Saad (2003)].

Secondly, for any nonsingular matrix A, the GMRES algorithms for solving the preconditioned system

$$M_L^{-1}AM_R^{-1}V_* = M_L^{-1}F, \ \ U_* = M_R^{-1}V_* \tag{C.16}$$

is given in Algorithm C.2, and the mth iteration solution satisfies

$$\left\|M_L^{-1}F - M_L^{-1}AU_m\right\|_2 = \inf_{U \in U_0 + \hat{\mathcal{K}}_m} \left\|M_L^{-1}F - M_L^{-1}AU\right\|_2, \tag{C.17}$$

where

$$\hat{\mathcal{K}}_m = \left\{M^{-1}r_0, (M^{-1}A)M^{-1}r_0, \cdots, (M^{-1}A)^{m-1}M^{-1}r_0\right\}, \tag{C.18}$$

with $r_0 = F - AU_0$ and $M = M_L M_R$. The steps 3-7 is the Arnoldi loop, and it produces a sequence of matrix $\{V_k\}$ with orthonormal columns such that

$$AV_m = V_{m+1}H_m, \tag{C.19}$$

where H_m is a $(m+1) \times m$ upper Hessenberg matrix defined by

$$H_m = \begin{bmatrix} h_{1,1} & h_{1,2} & h_{1,3} & \cdots & h_{1,m} \\ h_{2,1} & h_{2,2} & h_{2,3} & \cdots & h_{2,m} \\ & h_{3,2} & h_{3,3} & \cdots & h_{3,m} \\ & & \ddots & \ddots & \vdots \\ & & & h_{m,m-1} & h_{m,m} \\ & & & & h_{m+1,m} \end{bmatrix}; \tag{C.20}$$

and the step 8 incrementally performs the QR factorization as the GMRES iteration process by the Givens rotations, which translate the H_m in (C.20) to a upper triangular matrix (i.e., $Q_m H_m$ is upper triangular, in practice, we still store its entries in H_m) in a clever way so that the cost for a single GMRES iteration is $\mathcal{O}(Nm)$ floating-point operations. Meanwhile, $\rho = |(g)_{m+1}|$ is the residual $\left\|M_L^{-1}(AU_m - F)\right\|_2$. This allows us to perform the $\mathcal{O}(m^2)$ cost of the triangular solver and the $\mathcal{O}(mN)$ cost of the construction of U_k only after the algorithm termination; see [Kelley (1987), p. 43–46] and

[Saad (2003), p. 160–163] for details. However, unlike the CG algorithm, the basis for the Krylov space must be stored as the iteration process. This means that in order to perform m GMRES iterations one must stored m vector of length N. Therefore, for very large systems this becomes prohibitive and the iteration usually is restarted when the available room for basis vectors is exhausted [Saad and Schultz (1986)], but there is no general convergence theorem for the restarted algorithm and restarting will slow the convergence down.

C.2 Basic concepts of the multigrid methods

Multigrid methods are nowadays among the fastest and most efficient numerical solvers for linear systems of equations. The main ingredient for multigrid methods is the observations that classical iteration schemes like Jacobi and Gauss-Seidel schemes have a smoothing property. This means that a few iterations of such schemes reduce the high-frequency part of the error but the low-frequency part remains. However, when passing from the fine grid to the coarse grid, the smooth (low-frequency) part usually becomes more oscillatory. Hence, the idea is to perform few iterations (called pre-smoothing) and then switch to the next coarser level and perform few iterations there, and so on. After the so-called coarse grid correction one switches back to the fine grid and performs a few smoothing steps, also called post-smoothing.

In order to formulate the multigrid algorithm, we denote Ω^h as the partition of Ω associated with a stepsize h, and let A^h and F^h be the corresponding A and F, respectively, etc. We use the weighted Jacobi iteration as the smoothing operator, which is the best choice for dense Toeplitz matrix. Therefore, the smoother is defined by

$$\begin{aligned}
&\text{smooth}\,(A^h, U^h, F^h, \omega) \\
&= (1-\omega)U^h + \omega\left((D^h)^{-1}(D^h - A^h)U^h + (D^h)^{-1}F^h\right) \\
&= \left(1 - \omega(D^h)^{-1}A^h\right)U^h + \omega(D^h)^{-1}F^h, \qquad \text{(C.21)}
\end{aligned}$$

where D^h denotes the diagonal of matrix A^h. The transition operators between fine and coarse grids, so-called prolongation and restriction operators, are denoted by I_{2h}^h and I_h^{2h}, respectively. For one-dimensional system, letting $\Omega = (a, b)$, h_0 be the coarsest grid size, $N = \frac{b-a}{h}$ with h satisfying $\log_2(h_0/h) \in \mathbb{N}^+$, and

$$\Omega^h := \left\{x_i\,\big|_{i=0}^{N} : x_i = a + ih\right\}, \qquad \text{(C.22)}$$

$$U^h := \left(U_1^h, \cdots, U_{N-1}^h\right)^{\mathrm{T}}, \qquad \text{(C.23)}$$

Algorithm C.2 $[U_m, m] = \text{PGMRES}(U_0, F, A, M_L, M_R, \epsilon, K_{\max})$

1: $r'_0 = M_L^{-1}(F - AU_0), \rho = \|r'_0\|_2, v_1 = r'_0/\rho, g = \rho(1, 0, \cdots, 0)^{\mathrm{T}} \in \mathbb{R}^{K_{\max}+1}, m = 0$
2: **while** $(\rho > \epsilon \|F\|_2)$ and $(m < K_{\max})$ **do**
3: $m = m + 1, v_{m+1} = M_L^{-1} A M_R^{-1} v_m$
4: **for** $j = 1, \cdots, m$ **do**
5: $h_{j,m} = (v_j, v_{m+1}), v_{m+1} = v_{m+1} - h_{j,m} v_j$
6: **end for**
7: $h_{m+1,m} = \|v_{m+1}\|_2, v_{m+1} = v_{m+1} / \|v_{m+1}\|_2$
8:
- If $m > 1$, apply Q_{m-1} on $(h_{1,m}, \cdots, h_{m+1,m})^{\mathrm{T}}$ and denote the results still as $(h_{1,m}, \cdots, h_{m+1,m})^{\mathrm{T}}$
- $\nu = \sqrt{h_{m,m}^2 + h_{m+1,m}^2}$
- $c_m = h_{m,m}/\nu, \quad s_m = -h_{m+1,m}/\nu, \quad h_{m,m} = c_k h_{m,m} - s_k h_{m+1,m}, \ h_{m+1,m} = 0$
- $g = G_m(c_m, s_m) g$

9: $\rho = |(g)_{m+1}|$
10: **end while**
11: compute y as the solution of $\tilde{H} y = \tilde{g}$, in which the $m \times m$ upper triangular matrix $\tilde{H}$ has $h_{i,j}$ as its elements and $\tilde{g} = g(1 : m)$.
12: $U_m = U_0 + M_R^{-1} V_k y$

Note: For each m,

$$Q_{m-1} = G_{m-1}(c_{m-1}, s_{m-1}) G_{m-2}(c_{m-2}, s_{m-2}) \cdots G_1(c_1, s_1),$$

and

$$V_m = (v_1, v_2, \cdots, v_{m+1}).$$

Here, $G_i(c_i, s_i)$ denotes the Givens rotation [Kelley (1987), p. 44–45], which translates the $(i+1, i)$ entry of the H_m in (C.20) to 0, and the column vectors $v_1, v_2, \cdots, v_{m+1}$ (unless the $h_{m+1,m}$ in step 5 is 0, at this time the exact solution is obtained) form an orthonormal basis for the Krylov subspace $\left\{r'_0, (M_L^{-1} A M_R^{-1}) r'_0, \cdots, (M_L^{-1} A M_R^{-1})^m r'_0\right\}$.

---

we usually set

$$I_{2h}^h U^{2h} = \frac{1}{2} \begin{bmatrix} 1 & 2 & 1 & & & & \\ & & 1 & 2 & 1 & & \\ & & \cdots & \cdots & \cdots & & \\ & & & & 1 & 2 & 1 \end{bmatrix}^{\mathrm{T}} U^{2h} = U^h, \qquad \text{(C.24)}$$

and

$$I_h^{2h}U^h = \frac{1}{4}\begin{bmatrix} 1\ 2\ 1 & & & \\ & 1 \quad 2 \quad 1 & & \\ & \cdots\ \cdots\ \cdots & & \\ & & & 1\ 2\ 1 \end{bmatrix} U^h = U^{2h}. \tag{C.25}$$

The pseudocode of the multigrid method with V-cycle for solving linear system

$$A^h U^h = F^h \tag{C.26}$$

is given in algorithm C.3, where the damp parameters w_1, w_2 maybe problem dependent (see [Chen and Deng (2014b); Pang and Sun (2012)] for the discussion).

Algorithm C.3 $[U^h, m] = $ MGM $(A^h, F^h, U_0, h, h_0, \epsilon, \nu_1, \omega_1, \nu_2, K_{\max}, \omega_2)$

1: $U^h = U_0$, $m = 0$
2: $r_1 = \left\|A^h U_0 - F^h\right\|_2$
3: $r_2 = r_1$
4: **while** $\frac{r_2}{r_1} > \epsilon$ and $m < K_{\max}$ **do**
5: $\quad U^h = $ V-cycle$(A^h, U^h, F^h, h, h_0, \nu_1, \omega_1, \nu_2, \omega_2)$
6: $\quad m = m + 1$
7: $\quad r_2 = \left\|A_h U^h - F^h\right\|_2$
8: **end while**
9: Return U^h and m

Note: In this algorithm, U_0 denotes the initial guess, h_0 and h are the coarsest and finest grid sizes, respectively, and ν_1 and ν_2 denote the numbers of the pre-smoothing and post-smoothing steps, respectively.

C.3 Effective implementation of the Krylov subspace methods and multigrid methods

The efficiency of any of the iteration methods considered in previous sections is determined primarily by the performance of the matrix-vector product and the preconditioner solver (if there is one), and the storage scheme used for the matrix and the preconditioner. Regarding the algebraic equations appeared in Chap. 4, we will consider matrix A_1 with the forms

$$A_1 = \text{diag}(\mathbf{d}_1)B_1 + \text{diag}(\mathbf{d}_2)B_2 \tag{C.27}$$

Algorithm C.4 U^h =V-cycle($A^h, U^h, F^h, h, h_0, \nu_1, \omega_1, \nu_2, \omega_2$)

1: Pre-smooth: $U^h := \texttt{smooth}^{\nu_1}(A^h, U^h, F^h, \omega_1)$
2: Get residual: $r_h := F^h - A^h U^h$
3: Coarsen: $r_{2h} = I_h^{2h} r_h$
4: **if** $2 \times h = h_0$ **then**
5: Solve: $A^{2h}\xi = r^{2h}$
6: **else**
7: Recursion: $\xi = \text{V-cycle}(A^{2h}, 0, r^{2h}, 2 * h, h_0, \nu_1, \omega_1, \nu_2, \omega_2)$
8: **end if**
9: Correct: $U^h := U^h + I_{2h}^h \xi$
10: Post-smooth: $U^h := \texttt{smooth}^{\nu_2}(A^h, U^h, F^h, \omega_2)$
11: Return U^h

Note: A^{2h} denotes the matrix associated with the grid size $2\times h$. In practice, they can be obtained by a geometry way, i.e., A^{2h} are constructed from the fractional PDE itself. Therefore, all the matrices A^h, A^{2h} have the same structure.

or

$$A_1 = \text{diag}(\mathbf{d}_3)\,(C_2 \otimes B_3) + \text{diag}(\mathbf{d}_4)\,(B_4 \otimes C_1)\,, \tag{C.28}$$

where $\mathbf{d}_1, \mathbf{d}_2, \mathbf{d}_3, \mathbf{d}_4$ are column vectors, B_1, B_2, B_4, C_2 are N-by-N Toeplitz matrices, and B_3, C_1 are N_1-by-N_1 Toeplitz matrices. In particular, B_1, B_2, B_3, B_4 represent the differential matrices from the discretization of the R-L fractional order derivatives, and C_1 and C_2 are the identity matrices. The symbol '$\otimes$' denotes Kronecker product. We recall that the Kronecker product of matrices $C = (c_{i,j}) \in \mathbb{C}^{N\times N}$ and $B \in \mathbb{C}^{N_1\times N_1}$ is defined as follows:

$$C \otimes B = \begin{bmatrix} c_{1,1}B & c_{1,2}B & \cdots & c_{1,N}B \\ c_{2,1}B & c_{2,2}B & \cdots & c_{2,N}B \\ \vdots & \vdots & & \vdots \\ c_{N,1}B & c_{N,2}B & \cdots & c_{N,N}B \end{bmatrix}. \tag{C.29}$$

Remark C.15. Besides the linear system generated from the discretization of the one-dimensional fractional PDE, (C.27) also includes the linear systems generated from the ADI schemes for the high-dimensional space fractional models; see [Chen and Deng (2014b); Jin *et al.* (2015b)] for details.

Definition C.1. A matrix T_N is called a Toeplitz matrix if its entries are constant along each diagonal, i.e.,

$$T_N = \begin{bmatrix} t_0 & t_{-1} & \cdots & t_{2-N} & t_{1-N} \\ t_1 & t_0 & t_{-1} & \ddots & t_{2-N} \\ \vdots & t_1 & \ddots & \ddots & \vdots \\ t_{N-2} & \ddots & \ddots & t_0 & t_{-1} \\ t_{N-1} & t_{N-2} & \cdots & t_1 & t_0 \end{bmatrix}. \tag{C.30}$$

The Toeplitz matrix T_N is circulant if $t_{-k} = t_{N-k}$ for all $1 \leq k \leq N-1$, and in this case we will denote it as C_N.

It should be noticed that a Toeplitz matrix is determined by its first column and first row while a circulant matrix is determined only by its first column. Therefore, they can be stored with cost $\mathcal{O}(N)$.

Definition C.2. The N-by-N block-Toeplitz-Toeplitz-block (BTTB) matrix with N_1-by-N_1 Toeplitz block is defined as follows:

$$T_{NN_1} = \begin{bmatrix} T_{(0)} & T_{(-1)} & \cdots & T_{(2-N)} & T_{(1-N)} \\ T_{(1)} & T_{(0)} & T_{(-1)} & \ddots & T_{(2-N)} \\ \vdots & T_{(1)} & \ddots & \ddots & \vdots \\ T_{(N-2)} & \cdots & \ddots & T_{(0)} & T_{(-1)} \\ T_{(N-1)} & T_{(N-2)} & \cdots & T_{(1)} & T_{(0)} \end{bmatrix}, \tag{C.31}$$

where the blocks $T_{(l)}$, for $|l| \leq N-1$, are themselves Toeplitz matrices of order N_1. Similarly, we can define the block-Toeplitz-circulant-block (BTCB) matrix, the block-circulant-Toeplitz-block (BCTB) matrix, and the block-circulant-circulant-block (BCCB) matrix. We will denote the BCCB matrix T_{NN_1} as C_{NN_1}.

Obviously, the BTTB matrix can be stored with the cost $\mathcal{O}(NN_1)$. In the following, for clarity, we let $\mathbf{u}$ be some column vector that appears in the computational process, and the matrix vector products like $A\mathbf{u}$ will be involved.

C.3.1 *Fast Toeplitz matrix-vector products*

It is well known that circulant matrices can be diagonalized by the discrete Fourier transform (DFT) matrix F_N, i.e.,

$$C_N = F_N^{-1} \text{diag}(F_N \mathbf{c}) F_N, \tag{C.32}$$

where the entries of F_N are given by

$$(F_N)_{j,k} = e^{\frac{2\pi J(j-1)(k-1)}{N}}, \quad J = \sqrt{-1}, \tag{C.33}$$

with $1 \le j, k \le N$, and $\mathbf{c}$ denotes the first column of C_N. Therefore, the matrix-vector products

$$C_N \mathbf{u} = F_N^{-1} \text{diag}(F_N \mathbf{c}) F_N \mathbf{u}, \tag{C.34}$$

$$C_N^{-1} \mathbf{u} = F_N^{-1} \text{diag}(F_N \mathbf{c})^{-1} F_N \mathbf{u} \tag{C.35}$$

can be computed easily by the fast Fourier transform (FFT) algorithms in $\mathcal{O}(N \log N)$ operations. For using FFTs, we construct the $2N$-by-$2N$ circulant matrix

$$\begin{bmatrix} T_N & T_N' \\ T_N' & T_N \end{bmatrix} \begin{pmatrix} \mathbf{u} \\ \mathbf{0} \end{pmatrix} = \begin{pmatrix} T_N \mathbf{u} \\ T_N' \mathbf{u} \end{pmatrix}, \tag{C.36}$$

where the Toeplitz matrix T_N' can be given by

$$T_N' = \begin{bmatrix} 0 & t_{N-1} & \cdots & t_2 & t_1 \\ t_{1-N} & 0 & t_{N-1} & \ddots & t_2 \\ \vdots & t_{1-N} & \ddots & \ddots & \vdots \\ t_{-2} & \ddots & \ddots & 0 & t_{N-1} \\ t_{-1} & t_{-2} & \cdots & t_{1-N} & 0 \end{bmatrix}. \tag{C.37}$$

Therefore, the multiplication can be carried out by using the decomposition as in (C.32), and the cost for calculating $T_N \mathbf{u}$ with the DFTs is $\mathcal{O}(2N \log(2N))$.

Let C_{NN_1} be a N-by-N BCCB matrix with N_1-by-N_1 circulant blocks and let $\mathbf{c}^N$ be its first column vector. The corresponding two dimensional DFT matrix is $F_N \otimes F_{N_1}$, and we have [Davis (1979); Gray (2006)]

$$C_{NN_1} = (F_N \otimes F_{N_1})^{-1} \text{diag}\left((F_N \otimes F_{N_1}) \mathbf{c}^N \right) (F_N \otimes F_{N_1}). \tag{C.38}$$

Therefore, for any vector $\mathbf{u}$ of length NN_1, the matrix vector multiplications $C_{NN_1}\mathbf{u}$ and $C_{NN_1}^{-1}\mathbf{u}$ can also be carried out in $\mathcal{O}(NN_1 \log(NN_1))$ operations via the two dimensional DFTs. If T_{NN_1} is a N-by-N BTCB matrix with N_1-by-N_1 circulant blocks, and T_{NN_1}' is a N-by-N BTCB matrix with N_1-by-N_1 circulant blocks, i.e.,

$$T_{NN_1}' = \begin{bmatrix} 0 & T_{(N-1)} & \cdots & T_{(2)} & T_{(1)} \\ T_{(1-N)} & 0 & T_{(N-1)} & \ddots & T_{(2)} \\ \vdots & T_{(1-N)} & \ddots & \ddots & \vdots \\ T_{(-2)} & \ddots & \ddots & 0 & T_{(N-1)} \\ T_{(-1)} & T_{(-2)} & \cdots & T_{(1-N)} & 0 \end{bmatrix}, \tag{C.39}$$

then we can embed T_{NN_1} into a $2N$-by-$2N$ BCCT matrix with N_1-by-N_1 circulant blocks and $\mathbf{u}$ into a vector of length $2NN_1$ as follows:

$$\begin{bmatrix} T_{NN_1} & T'_{NN_1} \\ T'_{NN_1} & T_{NN_1} \end{bmatrix} \begin{pmatrix} \mathbf{u} \\ \mathbf{0} \end{pmatrix} = \begin{pmatrix} T_{NN_1}\mathbf{u} \\ T'_{NN_1}\mathbf{u} \end{pmatrix}. \tag{C.40}$$

Thus (C.40) can be carried out by using the decomposition as in (C.38) with the two dimensional DFT matrix $F_{2N} \otimes F_{N_1}$, and the first half of the matrix-vector products yields $T_{NN_1}\mathbf{u}$ with the cost $\mathcal{O}(NN_1 \log(NN_1))$. For the general BTTB matrix T_{NN_1}, by first embedding every block $T_{(k)}, k = 1-N, 2-N, \cdots, N-1$ into a $2N_1$-by-$2N_1$ circulant matrix (see (C.36)) and then embedding the obtained BTCB matrix into a $2N$-by-$2N$ block circulant matrix with $2N_1$-by-$2N_1$ circulant blocks (see (C.40)), and further extending $\mathbf{u}$ to a $4NN_1$-vector by putting zeros in the appropriate places and using the DFT matrix $F_{2N} \otimes F_{2N_1}$ to diagonalize the just obtained $4NN_1$-by-$4NN_1$ BCCB matrix, we can also obtain $T_{NN_1}\mathbf{u}$ in $\mathcal{O}(NN_1 \log(NN_1))$ operations. More precisely, we construct the $4NN_1$-vector by

$$\Bigg(\underbrace{\mathbf{u_1}^{\mathrm{T}}, \mathbf{0}^{\mathrm{T}}}_{2N_1}, \underbrace{\mathbf{u_2}^{\mathrm{T}}, \mathbf{0}}_{2N_1}, \cdots, \underbrace{\mathbf{u_N}, \mathbf{0}^{\mathrm{T}}}_{2N_1}, \underbrace{\mathbf{0}^{\mathrm{T}}, \mathbf{0}^{\mathrm{T}}}_{2N_1}, \underbrace{\mathbf{0}^{\mathrm{T}}, \mathbf{0}^{\mathrm{T}}}_{2N_1}, \cdots, \underbrace{\mathbf{0}^{\mathrm{T}}, \mathbf{0}^{\mathrm{T}}}_{2N_1}\Bigg)^{\mathrm{T}}, \tag{C.41}$$

where the column vectors $\mathbf{u_1}, \mathbf{u_2}, \cdots, \mathbf{u_N}$ of length N_1 satisfy

$$\mathbf{u} = \left(\mathbf{u_1}^{\mathrm{T}}, \mathbf{u_2}^{\mathrm{T}}, \cdots, \mathbf{u_N}^{\mathrm{T}}\right)^{\mathrm{T}}. \tag{C.42}$$

Thus, for $A_1 = \mathrm{diag}(\mathbf{d}_1)B_1 + \mathrm{diag}(\mathbf{d}_2)B_2$, we can calculate $A\mathbf{u}$ by

$$A\mathbf{u} = \mathbf{d}_1.(B_1\mathbf{u}) + \mathbf{d}_2.(B_2\mathbf{u}) + A_2\mathbf{u} \tag{C.43}$$

with the cost $\mathcal{O}(N \log N)$, where '.' denotes the dot product and the sparsity of A_2 has been used. For $A_1 = \mathrm{diag}(\mathbf{d}_3)(C_2 \otimes B_3) + \mathrm{diag}(\mathbf{d}_4)(B_4 \otimes C_1)$, since $(C_2 \otimes B_3)$ and $(B_4 \otimes C_1)$ are N-by-N BTTB matrix with N_1-by-N_1 Toeplitz blocks, we can also compute $A\mathbf{u}$ by

$$A\mathbf{u} = \mathbf{d}_3.((C_2 \otimes B_3)\mathbf{u}) + \mathbf{d}_4.((B_4 \otimes C_1)\mathbf{u}) + A_2\mathbf{u} \tag{C.44}$$

with the cost $\mathcal{O}(NN_1 \log(NN_1))$.

Note that the main cost on each grid level in Algorithm C.4 depends on the matrix-vector multiplication of $A^{2^k h}\mathbf{u}$ for $k = 0, 1, \cdots, \log_2(h_0/h)$. Recall that $A^{2^k h}$ has the same structures with A^h, for one dimensional system the multiplication $A^{2^k h}\mathbf{u}$ can be calculated with $\mathcal{O}(N/2^k \log(N/2^k))$ complexity. Therefore, the total computational cost over per V-cycle is roughly $\mathcal{O}(N \log N)$. Thus, if the V-cycle multigrid method convergent uniformly [Chen and Deng (2014b); Pang and Sun (2012)], the computational cost for solving (C.26) is of $\mathcal{O}(N \log N)$ in total.

Remark C.16. The techniques of the fast Toeplitz matrix-vector product may also be used to calculate the right term F.

C.3.2 *Some remarks on preconditioning*

In order to accelerate the convergence of the Krylov subspace method, the preconditioning is needed. Many circulant preconditioners have been proposed for solving Toeplitz systems, such as Strang's circulant preconditioner and T. Chan's (optimal) circulant preconditioner [Chan (1988); Chan and Jin (2007); Strang (1986)]. For the Toeplitz matrices T_N defined by (C.30), the entries in the first column of the Strang preconditioner $S(T_N)$ are given by

$$s_k = \begin{cases} t_k, & 0 \le k \le N/2 - 1, \\ 0, & k = N/2, \\ t_{k-N}, & N/2 < k \le N-1, \end{cases} \qquad \text{if } N \text{ is even,} \tag{C.45}$$

or

$$s_k = \begin{cases} t_k, & 0 \le k \le (N+1)/2, \\ t_{k-N}, & (N+1)/2 < k \le N-1, \end{cases} \qquad \text{if } N \text{ is odd;} \tag{C.46}$$

and the entries in the first column of the T. Chan circulant preconditioner $C_F(T_N)$ are given by

$$c_0 = t_0 \quad \text{and} \quad c_k = \frac{(N-k)t_k + k t_{k-N}}{N}, \quad 1 \le k \le N-1. \tag{C.47}$$

It is easy to check that $S(T_N^{\mathrm{T}}) = S(T_N)^{\mathrm{T}}$ and $C_F(T_N^{\mathrm{T}}) = C_F(T_N)^{\mathrm{T}}$. In particular, if T_N is Hermitian, then $S(T_N)$ and $C_F(T_N)$ are also Hermitian (at least for large N). In addition, both of the preconditioners can be easily adapted to solve the equation (C.1):

For the case $A_1 = \mathrm{diag}(\mathbf{d}_1)B_1 + \mathrm{diag}(\mathbf{d}_2)B_2$, let

$$\overline{d}_1 = \frac{\mathrm{sum}(\mathbf{d}_1)}{N}, \quad \overline{d}_2 = \frac{\mathrm{sum}(\mathbf{d}_2)}{N}, \tag{C.48}$$

and

$$\tilde{A} = \overline{d}_1 B_1 + \overline{d}_2 B_2 + A_2. \tag{C.49}$$

Then $\tilde{A}$ is a Toeplitz matrix, and a circulant preconditioner of A may be given as

$$M = C(\tilde{A}) = \overline{d}_1\, C(B_1) + \overline{d}_2\, C(B_2) + C(A_2), \tag{C.50}$$

where $C(B_1)$ denotes $S(B_1)$ or $C_F(B_1)$, etc. If M is non-singular (see [Lei and Sun (2013)] for the discussion), by (C.35), the matrix-vector product $M^{-1}\mathbf{u}$ can be calculated with $\mathcal{O}(N \log N)$ complexity. Therefore, in mth iteration, the cost of the preconditioned CG method is $\mathcal{O}(N \log N)$, and the cost of the preconditioned GMRES method is $\mathcal{O}(N \log N + Nm)$.

For the case $A_1 = \text{diag}(\mathbf{d}_3)\,(C_2 \otimes B_3) + \text{diag}(\mathbf{d}_4)\,(B_4 \otimes C_1)$, we can let

$$\overline{d}_3 = \frac{\text{sum}(\mathbf{d_3})}{NN_1}, \quad \overline{d}_4 = \frac{\text{sum}(\mathbf{d_4})}{NN_1}, \tag{C.51}$$

and

$$\tilde{A} = \overline{d}_3\,(C_2 \otimes B_3) + \overline{d}_4\,(B_4 \otimes C_1) + A_2. \tag{C.52}$$

Then $\tilde{A}$ is a block Toeplitz matrix, i.e.,

$$\tilde{A} = \begin{bmatrix} \tilde{A}_{(0)} & \tilde{A}_{(-1)} & \cdots & \tilde{A}_{(2-N)} & \tilde{A}_{(1-N)} \\ \tilde{A}_{(1)} & \tilde{A}_{(0)} & \tilde{A}_{(-1)} & \ddots & \tilde{A}_{(2-N)} \\ \vdots & \tilde{A}_{(1)} & \ddots & \ddots & \vdots \\ \tilde{A}_{(N-2)} & \cdots & \ddots & \tilde{A}_{(0)} & \tilde{A}_{(-1)} \\ \tilde{A}_{(N-1)} & \tilde{A}_{(N-2)} & \cdots & \tilde{A}_{(1)} & \tilde{A}_{(0)} \end{bmatrix}, \tag{C.53}$$

and a N-by-N $BTTB$ preconditioner of A may be given as

$$M = \begin{bmatrix} C_F(\tilde{A}_{(0)}) & C_F(\tilde{A}_{(-1)}) & \cdots & C_F(\tilde{A}_{(2-N)}) & C_F(\tilde{A}_{(1-N)}) \\ C_F(\tilde{A}_{(1)}) & C_F(\tilde{A}_{(0)}) & C_F(\tilde{A}_{(-1)}) & \ddots & C_F(\tilde{A}_{(2-N)}) \\ \vdots & C_F(\tilde{A}_{(1)}) & \ddots & \ddots & \vdots \\ C_F(\tilde{A}_{(N-2)}) & \cdots & \ddots & C_F(\tilde{A}_{(0)}) & C_F(\tilde{A}_{(-1)}) \\ C_F(\tilde{A}_{(N-1)}) & C_F(\tilde{A}_{(N-2)}) & \cdots & C_F(\tilde{A}_{(1)}) & C_F(\tilde{A}_{(0)}) \end{bmatrix},$$

where the blocks $\tilde{A}_{(l)}$, for $|l| \leq N-1$, are themselves Toeplitz matrices of order N_1. If M is inverse, then $M^{-1}\mathbf{u}$ makes sense and it can be calculated with the cost $\mathcal{O}(NN_1 \log(NN_1))$ [Chan and Jin (2007)]. In addition, based on (C.52), a N-by-N T. Chan's BCCB preconditioner with N_1-by-N_1 T. Chan's circulant blocks are also proposed in [Jia and Wang (2016)], but the related theoretical analysis is absent. In the special cases $C_1 = I, C_2 = I$ and $A_2 = I$, (C.52) reduces to

$$\tilde{A} = \overline{d}_3\,(I \otimes B_3) + \overline{d}_4\,(B_4 \otimes I) + I; \tag{C.54}$$

then a BCCB preconditioner of A may be given as [Lei *et al.* (2016)]

$$M = \overline{d}_3\,(I \otimes C(B_3)) + \overline{d}_4\,(C(B_4) \otimes I) + I, \tag{C.55}$$

where $C(B_3)$ denotes $S(B_3)$ or $C_F(B_3)$, and so on. Making use of (C.38), $M^{-1}\mathbf{u}$ could also be calculated with the cost $\mathcal{O}(NN_1 \log(NN_1))$.

Finally, we point out that there are some other types of preconditioners, such as the banded preconditioners proposed in [Lin *et al.* (2014); Jin *et al.* (2015b)]. Combining with the LU factorization or the incomplete LU factorization with no fills in (i.e., ILU(0)), for large N and N_1, the cost per iteration of the preconditioned CG method and the GMRES method also is $\mathcal{O}(N \log N)$ or $\mathcal{O}(NN_1 \log(NN_1))$.

Bibliography

Adams, R. A. (1975). *Sobolev Spaces* (Academic Press, New York).

Adjerid, S., Devine, K. D., Flaherty, J. E., and Krivodonova, L. (2002). A posteriori error estimation for discontinuous Galerkin solution of hyperbolic problems, *Comput. Methods. Appl. Mech. Engrg.* **191**, pp. 1097–1112.

Alikhanov, A. A. (2015). A new difference scheme for the fractional diffusion equation, *J. Comput. Phys.* **280**, pp. 424–438.

Ammar, G. S. and Gragg, W. B. (1988). Superfast solution of real positve definition Toeplitz systems, *SIAM J. Matrix Anal. Appl.* **9**, pp. 61–76.

Applebaum, D. (2009). *Lévy Processes and Stochastic Calculus* (Cambridge University Press, UK).

Baeumera, B. and Meerschaert, M. M. (2010). Tempered stable Lévy motion and transient super-diffusion, *J. Comput. Appl. Math.* **233**, 10, pp. 2438–2448.

Barkai, E., Metzler, R., and Klafter, J. (2000). From continuous time random walks to the fractional Fokker-Planck equation, *Phys. Rev. E* **61**, pp. 132–138.

Barrett, R., Berry, M., Chan, T. F., Demmel, J., Donato, J., Dongarra, J., Eijkhout, V., Pozo, R., Romine, C., and Vorst, H. (1994). *Templates for the Solution of Linear Systems: Building Blocks for Iteratative methods* (SIAM, Philadelphia).

Bassi, F. and Rebay, S. (1997). A high-order accurate discontinuous finite element method for the numerical solution of the compressible Navier-Stokes equations, *J. Comput. Phys.* **131**, pp. 267–279.

Bergh, J. and Löfström, J. (1976). *Interpolation Spaces* (Springer-Verlag, Berlin).

Brenner, S. C. and Scott, L. R. (1994). *The Mathematical Theory of Finite Element Methods* (Springer-Verlag, New York).

Briggs, W. L., Henson, V. E., and McCormick, S. F. (2007). *A Multigrid Tutorial*, 2nd edn. (SIAM, Philadelphia).

Brunner, H. and Schötzau, D. (2006). *hp*-discontinuous Galerkin time-stepping for Volterra integrodifferential equations, *SIAM J. Numer. Anal.* **44**, pp. 224–245.

Canuto, C., Hussaini, M. Y., Quarteroni, A., and Zhang, T. A. (2006). *Spectral Methods: Fundamentals in Single Domains* (Springer-Verlag, Berlin).

Carmi, S. and Barkai, E. (2011). Fractional Feynman-Kac equation for weak ergodicity breaking, *Phys. Rev. E* **84**, p. 061104.

Carmi, S., Turgeman, L., and Barkai, E. (2010). On distributions of functionals of anomalous diffusion paths, *J. Stat. Phys.* **141**, 6, pp. 1071–1092.

Cartea, Á. and del Castillo-Negrete, D. (2007). Fluid limit of the continuous-time random walk with general Lévy jump distribution functions, *Phys. Rev. E* **76**, p. 041105.

Castillo, P., Cockburn, B., Schötzau, D., and Schwab, C. (2001). Optimal a priori error esmates for the hp-version of the local discontinuous Galerkin method for convection-diffusion problem, *Math. Comp.* **71**, pp. 455–478.

Castilo, P., Cockburn, B., Perugia, I., and Schötzau, D. (2000). An a priori error analysis of the local discontinuous Galerkin method for elliptic problems, *SIAM J. Numer. Anal.* **38**, 5, pp. 1676–1706.

Çelik, C. and Duman, M. (2012). Crank-Nicolson method for the fractional diffusion equation with the Riesz fractional derivative, *J. Comput. Phys.* **231**, pp. 1743–1750.

Chan, R. H. and Jin, X. Q. (2007). *An Introduction to Iterative Toeplitz Solvers* (SIAM).

Chan, T. F. (1988). An optimal circulant preconditioner for Toeplitz systems, *SIAM. J. Sci. Stat. Comput.* **9**, pp. 766–771.

Chen, M. H. and Deng, W. H. (2014a). Fourth order difference approximations for space Riemann-Liouville derivatives based on weighted and shifted Lubich difference operators, *Commun. Comput. Phys.* **16**, pp. 516–540.

Chen, M. H. and Deng, W. H. (2014b). A second-order numerical method for two-dimensional two-sided space fractional convection diffusion equation, *Appl. Math. Model.* **38**, 13, pp. 3244–3259.

Chen, M. H. and Deng, W. H. (2015). High order algorithms for the fractional substantial diffusion equation with truncated Lévy flights, *SAIM J. Sci. Comput.* **37**, pp. A890–A917.

Chen, M. H., Wang, Y. T., Cheng, X., and Deng, W. H. (2014). Second-order LOD multigrid method for multidimensional Riesz fractional diffusion equation, *BIT Numer. Math.* **54**, pp. 623–647.

Chen, S., Shen, J., and Wang, L. L. (2016). Generalized Jacobi functions and their application to fractional differential equations, *Math. Comp.* **85**, pp. 1603–1638.

Chen, Z. X. (2002). Characteristic mixed discontinuous finite element methods for advection-dominated diffusion problems, *Comput. Meth. Appl. Mech. Engrg.* **191**, pp. 2509–2538.

Chen, Z. X., Ewing, R. E., Jiang, Q. Y., and Spagnuolo, M. (2003). Error analysis for characteristic-based methods for degenerate parabolic problems, *SIAM J. Numer. Anal.* **40**, pp. 1491–1515.

Ciarlet, P. (1975). *The Finite Element Method for Ellipic Problems* (North-Holland, Amsterdam).

Ciarlet, P. (2013). Analysis of the Scott-Zhang interpolation in the fractional order Sobolev spaces, *J. Numer. Math.* **21**, pp. 173–180.

Cockburn, B. and Shu, C.-W. (1998). The local discontinuous Galerkin method for time-dependent convection diffusion systems, *SIAM J. Numer. Anal.* **35**, pp. 2440–2463.

Cuesta, E., Lubich, C., and Palencia, C. (2006). Convolution quadrature time discretization of fractional diffusion-wave equations, *Math. Comp.* **75**, pp. 673–696.

Daftardar-Gejji, V. and Babakhani, A. (2004). Analysis of a system of fractional differential equations, *J. Math. Anal. Appl.* **293**, pp. 511–522.

Davis, P. J. (1979). *Circulant Matrices* (John Wiley & Sons, New York).

del Castillo-Negrete, D. (2009). Truncation effects in superdiffusive front propagation with Lévy flights, *Phys. Rev. E* **79**, p. 031120.

Delfour, M., Hager, W., and Trochu, F. (1981). Discontinuous Galerkin methods for ordinary differential equations, *Math. Comp.* **36**, pp. 455–473.

Deng, W. H. (2007a). Numerical algorithm for the time fractional Fokker-Planck equation, *J. Comput. Phys.* **227**, pp. 1510–1522.

Deng, W. H. (2007b). Short memory principle and a predictor-corrector aproach for fractional differential equations, *J. Comput. Appl. Math.* **206**, pp. 174–188.

Deng, W. H. (2008). Finite element method for the space and time fractional Fokker-Planck equation, *SIAM J. Numer. Anal.* **47**, pp. 204–226.

Deng, W. H. (2010). Smoothness and stability of the solutions for nonlinear fractional differential equations, *Nonl. Anal.* **72**, pp. 1768–1777.

Deng, W. H. and Barkai, E. (2009). Ergodic properties of fractional Brownian-Langevin motion, *Phys. Rev. E* **79**, p. 011112.

Deng, W. H. and Chen, M. H. (2014). Efficient numerical algorithms for three-dimensional fractional partial differential equations, *J. Comput. Math.* **32**, pp. 371–391.

Deng, W. H. and Hesthaven, J. S. (2013). Local discontious Galerkin methods for fractional diffusion equations, *ESAIM Math. Model. Numer Anal.* **47**, 6, pp. 1845–1864.

Deng, W. H., Li, B. Y., Tian, W. Y., and Zhang, P. W. (2018). Boundary problems for the fractional and tempered fractional operators, *Multiscale Model. Simul.* **16**(1), pp. 125–149.

Deng, W. H., Wu, X. C., and Wang, W. L. (2017). Mean exit time and escape probability for the anomalous processes with the tempered power-law waiting times, *EPL* **117**, p. 10009.

Deng, W. H. and Zhang, Z. J. (2018). Variational formulation and efficient implementation for solving the tempered fractional problems, *Numer. Methods Partial Differential Equations*, **34**, pp. 1224–1257.

Diethelm, K. (2010). *The Analysis of Fractional Differential Equations* (Springer-Verlag, Berlin).

Diethelm, K. and Ford, N. J. (2002). Analysis of fractional differential equations, *J. Math. Anal. Appl.* **265**, pp. 229–248.

Diethelm, K., Ford, N. J., and Freed, A. D. (2002). A predictor-corrector approach for the numerical solution of fractional differential eqations, *Nonlinear Dynam.* **29**, pp. 3–22.

Diethelm, K., Ford, N. J., and Freed, A. D. (2004). Detailed error analysis for a fractional Adams method, *Nonlinear Dynam.* **36**, pp. 31–52.

Douglas, J. (1955). On the numerical integration of $u_{xx} + u_{yy} = u_{tt}$ by implicit methods, *J. Soc. Indust. Appl. Math.* **3**, pp. 42–65.

Douglas, J. and Kimy, S. (2001). Improved accuracy for locally one-dimensional methods for parabolic equations, *Math. Models Methods Appl. Sci.* **11**, pp. 1563–1579.

Douglas, J. and Russell, T. F. (1982). Numerical method for convection-dominated diffusion problems based on combining the method of characteristics with finite element or finite difference procedures, *SIAM J. Numer. Anal.* **19**, 5, pp. 871–885.

Einstein, A. (1905). On the motion of small particles suspended in liquids at rest required by the molecular-kinetic theory of heat, *Ann. Phys.* **17**, pp. 891–921.

Ern, A. and Guermond, J.-L. (2004). *Theory and Practice of Finite Elements, Applied Mathematical Sciences*, Vol. 159 (Springer-Verlag, New York).

Ervin, V. and Roop, J. (2006). Variational formulation for the stationary fractional advection dispersion equation, *Numer. Methods Partial Differential Equations* **22**, 3, pp. 558–576.

Ervin, V. J., Heuer, N., and Roop, J. P. (2007). Numerical approximation of a time dependent nonlinear, space-fractional diffusion equation. *SIAM. J. Numer. Anal.* **45**, 2, pp. 572–591.

Evans, L. C. (2010). *Partial Differential Equations, Graduate Studies in Mathematics*, Vol. 19, 2nd edn. (AMS).

Fogedby, H. C. (1994). Langevin equations for continuous time Lévy flights, *Phys. Rev. E* **50**, p. 1657.

Ford, N. J. and Simpson, A. C. (2001). The numerical solution of fractional differential equations: speed versus accuracy, *Numer. Algorithms* **26**, pp. 333–346.

Friedman, A. (1975). *Stochastic Differential Equations and Applications* (Academic Press, New York).

Friedrich, R., Jenko, F., and Eule, A. B. S. (2006). Anomalous dffusion of inertial, weakly damped particles, *Phys. Rev. Lett.* **96**, p. 230601.

Garrappa, R. (2015). Numerical evaluation of two and three parameter Mittag-Leffler functions, *SIAM. J. Numer. Anal.* **53**, pp. 26–37.

Gavrilyuk, I. P. and Makarov, V. L. (2005). Exponentially convergent algorithms for the operator exponential with applications to inhomogeneous problems in banach spaces, *SIAM. Numer. Anal.* **43**, pp. 2144–2171.

Glaser, A., Liu, X., and Rokhlin, V. (2007). A fast algorithm for the calculation of the roots of special functions, *SIAM J. Sci. Comput.* **29**, pp. 1420–1438.

Gray, R. M. (2006). Toeplitz and circulant matrices: A review, *Foundations and Trends in Communications and Information Theory* **2**, pp. 155–239.

Grisvard, P. (1985). *Elliptic Problems in Nonsmooth Domains* (Pitman, Boston).

Guo, B., Pu, X. K., and Huang, F. H. (2015). *Fractional Partial Differential Equations and Their Numerical Solutions* (World Scientific, Singapore).

Guo, B. Q. and Heuer, N. (2006). The optimal convergence of the *h-p* version of the boundary element method with quasiuniform meshes for elliptic problems on polygonal domains, *Adv. Comput. Math.* **24**, pp. 353–374.

Guo, B. Y. (1998). *Spectral Methods and Their Applications* (World Scientific, Singapore).

Gustafsson, B. (2008). *High Order Difference Methods for Time Dependent PDE* (Springer-Verlag, Berlin).

Hale, N. and Townsend, A. (2013). Fast and accurate computation of Gauss-Legendre and Gauss-Jacobi quadrature nodes and weights, *SIAM J. Sci. Comput.* **35**, pp. A652–A674.

Halmos, P. R. (1991). *Measure Theory* (Springer-Verlag, New York).

Hanert, E. and Piret, C. (1995). Analytic approach to the problem of converence of truncatd Lévy flights towards the Gaussian stochastic process, *Phys. Rev. E* **52**, pp. 1197–1195.

Hanert, E. and Piret, C. (2014). A Chebyshev pseudo-spectral method to solve the space-time tempered fractional diffusion equation, *SIAM J. Sci. Comput.* **36**, pp. A1797–A1812.

Henrici, P. (1962). *Discrete Variable Methods in Ordinary Differential Equations* (John Wiley, New York).

Hesthaven, J. S. and Warburton, T. (2004). High-order nodal discontinuous Galerkin methods for Maxwell eigenvalue problem, *Philos. Trans. Roy. Soc. Lond. Ser. A: Math. Phys. Eng. Sci.* **362**, pp. 493–524.

Hesthaven, J. S. and Warburton, T. (2008). *Nodal Discontinuous Galerkin Methods: Algorithms, Analysis, and Applications* (Springer, Berlin).

Isaacson, E. and Keller, H. B. (1996). *Analysis of Numerical Methods* (Wiley, New York).

Ji, C. C. and Sun, Z. Z. (2015). A high-order compact finite difference scheme for the fractional sub-diffusion equation, *J. Sci. Comput.* **64**, pp. 959–985.

Jia, J. H. and Wang, H. (2016). A fast finite volume method for conservative space-fractinal diffusion equations in convex domains, *J. Comput. Phys.* **310**, pp. 63–84.

Jin, B., Lazarov, R., and Pasciak, J. (2015a). Variational formulation of problem invovlving fractional order differential operators, *Math. Comp.* **84**, pp. 2665–2700.

Jin, X. Q., Lin, F. R., and Zhao, Z. (2015b). Preconditioned iterative methods for two-dimensional space-fractional diffusion equations, *Commun. Comput. Phys.* **18**, pp. 469–488.

Kelley, C. (1987). *Iterative Methods for Linear and Nonlinear Equations* (SIAM).

Kilbas, A. A., Srivastava, H. M., and Trujillo, J. J. (2006). *Thoeory and Applications of Fractional Differential Equations* (Elsevier, Amsterdam).

Klafter, J. and Sokolov, I. M. (2011). *First Steps in Random Walks: From Tools to Applications* (Oxford University Press, Oxford).

Laub, A. J. (2005). *Matrix Analysis for Scientists and Engineers* (SIAM).

Le, K. N., Mclean, W., and Mustapha, K. (2016). Numerical solution of the time-fractional Fokker-Planck equation with general forcing, *SIAM J. Numer. Anal.* **54**, pp. 1763–1784.

Lei, S. L., Chen, X., and Zhang, X. H. (2016). Multilevel circulant preconditioner for high-dimensional fractional diffusion equations, *E. Asian J. Appl. Math.* **6**, pp. 109–130.

Lei, S. L. and Sun, H. W. (2013). A circulant preconditioner for fractional diffusion equations, *J. Comput. Phys.* **242**, pp. 715–725.

Li, C. and Deng, W. H. (2016). High order schemes for the tempered fractional diffusion equaitons, *Adv. Comput. Math.* **42**, pp. 543–572.

Li, X. J. and Xu, C. J. (2009). A space-time spectral method for the time fractional diffusion equation, *SIAM J. Numer. Anal.* **47**, 3, pp. 2108–2131.

Li, X. J. and Xu, C. J. (2010). Existence and uniquness of the weak solution of the space-time fractional diffusion equation and a spectral method approximation, *Commun. Comput. Phys.* **8**, 5, pp. 1016–1051.

Lin, F. R., Yang, S. W., and Jin, X. Q. (2014). Preconditioned iterative methods for fractional diffusion equation, *J. Comput. Phys.* **256**, pp. 109–117.

Lin, Y., Li, X. J., and Xu, C. J. (2011). Finite difference/spectral approximations for the fractional cable equation, *Math. Comp.* **80**, pp. 1369–1396.

Lions, J. L. and Magenes, E. (1972). *Non-Homogenous Boundary Value Problems and Applications*, Vol. 1 (Springer-Verlag).

Lubich, C. (1983). Runge-Kutta theory for Volterra and Abel integral equations of the second kind, *Math. Comp.* **41**, pp. 87–102.

Lubich, C. (1985). Fractional linear multistep methods for Abel-Volterra integral equations of the second kind, *Math. Comp.* **45**, pp. 463–469.

Lubich, C. (1986). Discretized fractional calculus, *SIAM J. Math. Anal.* **17**, pp. 704–719.

Magdziarz, M. and Weron, A. (2007). Competition between subdiffusion and L évy flights: a MonteCarlo approach, *Phys. Rev. E* **75**, p. 056702.

Mainardi, F. (2013). On some properties of the Mittag-Leffler function $e_\alpha(-t^\alpha)$, completely monotone for $t > 0$ with $0 < \alpha < 1$. *Discrete Contin Dyn Syst Ser B 2013* `https://arxiv.org/abs/1305.0161v3`.

Mandelbrot, E. B. and Van-Ness, J. W. (1968). Fractional Brownian motions, fractional noises and applications, *SIAM Rev.* **10**, pp. 422–437.

Marchuk, G. I. and Shaidurov, V. V. (1983). *Difference Methods and Their Extrapolations* (Springer-Verlag, New York).

Marcus, M. and Minc, H. (1964). *A Survey of Matrix Theory and Matrix Inequalities* (Allyn and Bacon, Boston).

McLean, W. (2000). *Strongly Ellipic Systems and Boundary Integral Equations* (Combridge university press, Cambrige).

Meerschaert, M. M. and Sabzikar, F. (2013). Tempered fractional Brownian motion, *Statist. Probab. Lett.* **83**, pp. 2269–2275.

Meerschaert, M. M., Scheffler, H. P., and Tadjeran, C. (2006). Finite difference methods for two-dimensional fractional dispersion equation, *J. Comput. Phys.* **211**, pp. 249–261.

Meerschaert, M. M. and Tadjeran, C. (2004). Finite difference approximations for fractional advection-dispersion flow equations, *J. Comput. Appl. Math.* **172**, 1, pp. 65–77.

Meerschaert, M. M. and Tadjeran, C. (2006). Finite difference approximations for two-sided space-fractional partial differential equations, *Appl. Numer. Math.* **56**, pp. 80–90.

Metzler, R. and Klafter, J. (2000). The random walk's guide to anomalous diffusion: a fractional dynamics approach, *Phys. Rep.* **339**, 1, pp. 1–77.

Metzler, R. and Nonnenmacher, T. F. (2002). Space- and time-fractional diffusion and wave equations, fractional Fokker-Planck equations, and physical motivation, *Chem. Phys.* **284**, pp. 67–90.

Montroll, E. W. and Weiss, G. H. (1965). Random walks on lattices. II, *J. Math. Phys.* **6**, p. 167.

Mustapha, K., Brunner, H., Mustapha, H., and Schötzau, D. (2011). *hp*-version discontinuous Galerkin method for integro-differential equations of parabolic type, *SIAM J. Numer. Anal.* **49**, pp. 1369–1396.

Mustapha, K. A. (2013). A superconvergent discontinuous Galerkin method for Volterra integro-differential equations, *Math. Comp.* **82**, pp. 1987–2005.

Nezza, E. D., Palatuci, G., and Valdinoci, E. (2004). Hitchhiker's guide to the fractional Sobolev spaces, *Bull. Sci. Math.* **136**, pp. 225–36.

Ortigueira, M. D. (2006). Riesz potential operators and inverses via fractional centred derivatives, *Int. J. Math. Math. Sci.* **2006**, pp. 1–12.

Pan, J. U., NG, M. K., and Wang, H. (2016). Fast iterative solvers for linear systems arising from time-dependent space-fractional diffusion equations, *SIAM. J. Sci. Comput.* **38**, pp. A2806–A2826.

Pang, H. K. and Sun, H. W. (2012). Multigrid method for fractional diffusion equations, *J. Comput. Phys.* **231**, 2, pp. 693–703.

Peaceman, D. W. and Rachford, H. H. (1955). The numerical solution of parabolic and elliptic differential equations, *J. Soc. Ind. Appl. Math.* **3**, pp. 28–41.

Pearson, K. (1905). The problem of the random walk, *Nature* **72(1865)**, p. 294.

Podlubny, I. (1999). *Fractional Differential Equations* (Academic Press, San Diego).

Podlubny, I. (2012). Mittag-leffler function, `http://www.mathworks.com/matlabcentral/fileexchange/8738`.

Press, W. H., Teukolsky, S. A., Vetterling, W. T., and Flannery, B. P. (2007). *Numerical Recipes in C: The Art of Scientific Computing* (Cambridge University Press, Cambridge, UK).

Qiu, L. L., Deng, W. H., and Hesthaven, J. S. (2015). Nodal discontinuous Galerkin methods for fractional diffusion equations on 2D domain with triangualr meshes, *J. Comput. Phys.* **298**, pp. 678–694.

Quarteroni, A., Sacco, R., and Saleri, F. (2007). *Numerical Mathematics*, 2nd edn. (Springer).

Quarteroni, A. and Valli, A. (2008). *Numerical Approximation of Partial Differential Equations* (Springer).

Risken, H. and Frank, T. (1996). *The Fokker-Planck equation: methods of solution and applications* (Springer-Verlag, New York).

Rudin, W. (1987). *Real and Complex analysis*, 3rd edn. (McGraw-Hill, New York).

Saad, Y. (2003). *Iterative Methods for Sparse Linear Systems*, 2nd edn. (SIAM).

Saad, Y. and Schultz, M. (1986). GMRES: A generalized minimal residual algorithm for solving nonsymmetric linear systems, *SIAM. J. Sci. Satist. Comput.* **7**, pp. 856–869.

Sabzikar, F., Meerschaert, M. M., and Chen, J. H. (2015). Tempered fractional calculus, *J. Comput. Phys.* **293**, pp. 14–28.

Samko, S. G., Kilbas, A. A., and l. Marichev, O. (1993). *Fractional Integrals and Derivatives, Theory and Applications* (Gordon and Breach Sci. Publ., Singapore).

Schmelzer, T. and Trefethen, L. N. (2007). Computing the Gamma function using contour integrals and rational approximations, *SIAM. J. Numer. Anal.* **45**, pp. 558–571.

Schötzau, D. and Schwab, C. (2000). An *hp* a priori error analysis of the DG time-stepping method for initial value problems, *Calcolo* **37**, pp. 207–232.

Scott, L. R. and Zhang, S. Y. (1990). Finite element interpolation of nonsmooth functions satisfying boundary conditions, *Math. Comp.* **54**, pp. 483–493.

Sheen, D., Sloan, I. H., and Thomée, V. (2000). A parallel method for time discretization of parabolic problems based on contour integral representation and quadrature, *Math. Comp.* **69**, pp. 177–195.

Sheen, D., Sloan, I. H., and Thomée, V. (2003). A parallel method for time discretization of parabolic problems based on Laplace transformation and quadrature, *IMA. J. Numer. Anal.* **23**, pp. 269–299.

Shen, J. and Tang, T. (2006). *Spectral and High-Order Methods with Applications* (Science Press, Beijing).

Shen, J., Tang, T., and Wang, L. L. (2011). *Spectral Methods: Algorithms, Analysis and Applications* (Springer, Heidelberg).

Silvestre, L. (2007). Regularity of the obstacle problem for a fractional power of the Laplace operator, *Commun. Pur. Appl. Math.* **60**, p. 67.

Sneddon, I. N. (1995). *Fourier Transforms* (Dover Publications, New York).

Sousa, E. and Li, C. (2015). A weighted finite difference method for the fractional diffusion equation based on the Riemann-Liouville derivative, *Appl. Numer. Math.* **90**, pp. 22–37.

Strang, G. (1986). A proposal of Toeplitz matrix calculations, *Stud. Appl. Math.* **74**, pp. 171–176.

Sun, Z. Z. (2005). *Numerical Methods of Partial Differential Equations (in Chinese)* (Science Press, Beijing).

Tadjeran, C., Meerschaert, M. M., and Scheffler, H. P. (2006). A second-order accurate numerical approximation for the fractional diffusion equation, *J. Comput. Phys.* **213**, pp. 205–213.

Tian, W. Y., Deng, W. H., and Wu, Y. J. (2014). Polynomial spectral collocation method for space fractional advection-diffusion equation, *Numer. Methods Partial Differential Equations* **30**, 2, pp. 514–535.

Tian, W. Y., Zhou, H., and Deng, W. H. (2015). A class of second order difference approximations for solving space fractional diffusion equations, *Math. Comp.* **84**, pp. 1703–1727.

Trefethen, L. N. and Weideman, J. (2014). The exponentially convergent trapezoidal rule, *SIAM. Rev.* **56**, pp. 385–458.

Trefethen, L. N., Weideman, J., and Schemelzer, T. (2006). Talbot quadratures and rational approximations, *BIT* **46**, pp. 653–670.

Turgeman, L., Carmi, S., and Barkai, E. (2009). Fractional Feynman-Kac equation for non-brownian functionals, *Phys. Rev. Lett.* **103**, p. 190201.

Wang, H. and Basu, T. S. (2012). A fast finite diffence method for two-dimensional space-fractional diffusion equations, *SIAM. J. Sci. Comput.* **34**, pp. A2444–A2458.

Wang, H., Yang, D. P., and Zhu, S. F. (2014). Inhomogeneous Dirichlet bounary-value problems of space-fractinal diffusion equations and their finite element approximation, *SIAM. J. Numer. Anal.* **52**, 3, pp. 1292–1310.

Wang, H. and Zhang, X. H. (2015). A high-accuracy preserving spectral Galerkin method for the Dirichlet boundary-value problem of variable-coefficient conservative fractional diffusion equations, *J. Comput. Phys.* **281**, pp. 67–81.

Wang, S., Yuan, J. Y., Deng, W. H., and Wu, Y. J. (2016). A hybridized discontinuous Galerkin method for 2d fractional convection-diffusion equations, *J. Sci. Comput.* **68**, pp. 826–847.

Weilbee, M. (2005). Efficient Numerical Methods for Fractional Differential Equations and their Analytical Background (Technichal University of Braunschweig).

Wu, X. C., Deng, W. H., and Barkai, E. (2016). Tempered fractional Feynman-Kac equation: Theory and examples, *Phys. Rev. E* **93**, p. 032151.

Xu, Q. and Hesthaven, J. S. (2014). Discontinuous Galerkin method for fractional convection-diffusion equations, *SIAM J. Numer. Anal.* **52**, 1, pp. 405–423.

Yan, J. and Shu., C.-W. (2002). A local discontinuous Galerkin method for KdV type equations, *SIAM J. Numer. Anal.* **40**, pp. 769–791.

Yang, Q., Liu, F., and Turner, I. (2010). Numerical methods for fractional partial differential equations with Riesz space fractional derivatives, *Appl. Math. Model.* **34**, pp. 200–218.

Yuste, S. B. and Acedo, L. (2005). An explicit finite difference method and a new Von Neumann-type stablity analysis for fractional diffusion equations, *SIAM. J. Numer. Anal.* **42**, pp. 1862–1874.

Zayernouri, M. and Karniadakis, G. E. (2013). Fractional Sturm-Liouville eigen-problems: theory and numrical approximation, *J. Comput. Phys.* **252**, pp. 495–517.

Zayernouri, M. and Karniadakis, G. E. (2014a). Exponentially accuarate spectral and spectral element methods for fractional ODEs, *J. Comput. Phys.* **257**, pp. 460–480.

Zayernouri, M. and Karniadakis, G. E. (2014b). Fractional spectral collocation method, *SIAM. J. Sci. Comput.* **2014**, pp. A40–A62.

Zhang, Y. N., Sun, Z. Z., and Wu, H. W. (2011). Error estimates of Crank-Nicolson-type difference schemes for the subdiffusion equation, *SIAM J. Numer. Anal.* **49**, 6, pp. 2302–2322.

Zhang, Z. J. and Deng, W. H. (2017). Numerical approaches to the functional distribution of anomalous diffusion with both traps and flights, *Adv. Comput. Math.* **43**, 4, pp. 699–732.

Zhao, L. J. and Deng, W. H. (2014). Jacobian-predictor-corrector approach for fractional differential equations, *Adv. Comput. Math.* **40**, pp. 137–165.

Zhao, L. J. and Deng, W. H. (2016). High order finite difference methods on non-uniform meshes for space fractional operators, *Adv. Comput. Math.* **42**, 2, pp. 425–468.

Zhou, H., Tian, W. Y., and Deng, W. H. (2013). Quasi-compact finite difference schemes for space fractional diffusion equations, *J. Sci. Comput.* **56**, pp. 45–46.

Zhuang, P., Liu, F., Anh, V., and Turner, I. (2008). New solution and analytical techniques of the implicit numerical method for the anomalous subdiffusion equation, *SIAM J. Numer. Anal.* **46**, 2, pp. 1079–1095.

Zorich, V. (2004). *Mathematical Analysis II* (Springer, Berlin).

Index